Reviews in Modern Astronomy

15

Further Titles in Astronomy

Holliday, K.
Introductory Astronomy
1998. XII, 314 pages. Softcover.
ISBN 0-471-98332-2

Plait, PC
Bad Astronomy – Misconceptions & Misuese Revealed from Astrology to the Moon Landing "Hoax"
2002. 288 pages. Softcover.
ISBN 0-471-40976-6

Scientific American / Carlson, S. (eds.)
The Amateur Astronomer
2001. XIV, 272 pages. Softcover.
ISBN 0-471-38282-5

Moché, D. L.
Astronomy
A Self-Teaching Guide
5th edition
2000. XII, 352 pages. Softcover.
ISBN 0-471-38353-8

Coles, P. / Lucchin, F.
Cosmology
The Origin and Evolution of Cosmic Structure
2nd edition
2002. XX, 492 pages. Hardcover.
ISBN 0-471-48909-3

Maran
Astronomy For Dummies
1999. 350 pages. Softcover.
ISBN 0-7645-5155-8

Journal:
Astronomische Nachrichten/Astronomical Notes
ISSN 0004-6337

Reinhard E. Schielicke (Ed.)

Reviews in Modern Astronomy 15

Astronomy with Large Telescopes from Ground and Space

 WILEY-VCH

Edited on behalf of the *Astronomische Gesellschaft* by
Dr. *Reinhard E. Schielicke*
Universitäts-Sternwarte Jena
Schillergäßchen 2, D-07745 Jena
Germany

Cover picture:
Colour rendering of Tarantula Nebula in LMC
Composite of three narrow-band filter exposures of the famous Nebula (30 Doradus) in our closest neighbouring galaxy, the Large Magellanic Cloud. The filters are centred on the 2.166 micron Br-gamma line of atomic hydrogen (blue); the 1.644 micron (FeII) line (green) and the 2.12 micron 1-0 S(1) line of molecular hydrogen (red). The colour coding reflects the ionization state of the gas.
The scale is 0.26 arcsec/pixel and the field is 4.5 x 4.5 arcmin with North at the top and East to the left.
(Colour composite by F. Selman)

Library of Congress Card No. applied for.

British Library Cataloguing-in-Publication Data:
A catalogue record for this book is available from the British Library.

Die Deutsche Bibliothek – CIP Cataloguing-in-Publication Data:
A catalogue record for this publication is available from Die Deutsche Bibliothek.

ISBN 3-527-40404-X

© 2002 WILEY-VCH GmbH & Co. KGaA, Weinheim

Printed on acid-free paper.

Printing: Druckhaus Darmstadt GmbH, Darmstadt
Bookbinding: Litges & Dopf Buchbinderei GmbH, Heppenheim

Printed in the Federal Republic of Germany.

Preface

The annual series *Reviews in Modern Astronomy* of the Astronomische Gesellschaft was established in 1988 in order to bring the scientific events of the meetings of the Society to the attention of the worldwide astronomical community. *Reviews in Modern Astronomy* is devoted exclusively to the Karl Schwarzschild Lectures, the Ludwig Bierman Award Lectures, the invited reviews, and to the Highlight Contributions from leading scientists reporting on recent progress and scientific achievements at their respective research institutes.

The Karl Schwarzschild Lectures constitute a special series of invited reviews delivered by outstanding scientists who have been awarded the Karl Schwarzschild Medal of the Astronomische Gesellschaft, whereas excellent young astronomers are honoured by the Ludwig Biermann Prize.

Volume 15 continues the series with fourteen invited reviews and Highlight Contributions which were presented during the Joint European and National Astronomical Meeting JENAM 2001 of the ASTRONOMISCHE GESELLSCHAFT and the EUROPEAN ASTRONOMICAL SOCIETY on "Five Days of Creation: Astronomy with Large Telescopes from Ground and Space" held at Munich, September 10 to 15, 2001.

The Karl Schwarzschild medal 2001 was awarded to Professor Keiichi Kodaira, Japan. His lecture with the title "Macro- and Microscopic Views of Nearby Galaxies" opened the meeting.

The talk presented by the Ludwig Biermann Prize winner 2001, Dr Stefanie Komossa, Garching, dealt with the topic "X-ray Evidence for Supermassive Black Holes at the Centers of Nearby, Non-active Galaxies".

Other contributions to the meeting published in this volume discuss, among other subjects, the search for extrasolar planets, formation of stars and galaxies, physics of active galactic nuclei, and, last but not least, new telescope and sensor technologies for various wavelengths. To the regret of the members of the Board of the Astronomische Gesellschaft, two of our invited review speakers, Michel Mayor (extrasolar planets) and Massimo Tarenghi (new telescope technology) were not in a position to prepare the manuscripts of their lectures, as was agreed before.

Jena, May 2002   *Reinhard E. Schielicke*

The Astronomische Gesellschaft awards the **Karl Schwarzschild Medal**. Awarding of the medal is accompanied by the Karl Schwarzschild lecture held at the scientific annual meeting and the publication.

Recipients of the Karl Schwarzschild Medal are

1959   Martin Schwarzschild:
Die Theorien des inneren Aufbaus der Sterne.
Mitteilungen der AG 12, 15

1963   Charles Fehrenbach:
Die Bestimmung der Radialgeschwindigkeiten
mit dem Objektivprisma.
Mitteilungen der AG 17, 59

1968   Maarten Schmidt:
Quasi-stellar sources.
Mitteilungen der AG 25, 13

1969   Bengt Strömgren:
Quantitative Spektralklassifikation und ihre Anwendung
auf Probleme der Entwicklung der Sterne und der Milchstraße.
Mitteilungen der AG 27, 15

1971   Antony Hewish:
Tree years with pulsars.
Mitteilungen der AG 31, 15

1972   Jan H. Oort:
On the problem of the origin of spiral structure.
Mitteilungen der AG 32, 15

1974   Cornelis de Jager:
Dynamik von Sternatmosphären.
Mitteilungen der AG 36, 15

1975   Lyman Spitzer, jr.:
Interstellar matter research with the Copernicus satellite.
Mitteilungen der AG 38, 27

1977   Wilhelm Becker:
Die galaktische Struktur aus optischen Beobachtungen.
Mitteilungen der AG 43, 21

1978   George B. Field:
Intergalactic matter and the evolution of galaxies.
Mitteilungen der AG 47, 7

1980   Ludwig Biermann:
Dreißig Jahre Kometenforschung.
Mitteilungen der AG 51, 37

1981   Bohdan Paczynski:
Thick accretion disks around black holes.
Mitteilungen der AG 57, 27

1982 Jean Delhaye:
Die Bewegungen der Sterne
und ihre Bedeutung in der galaktischen Astronomie.
Mitteilungen der AG 57, 123

1983 Donald Lynden-Bell:
Mysterious mass in local group galaxies.
Mitteilungen der AG 60, 23

1984 Daniel M. Popper:
Some problems in the determination
of fundamental stellar parameters from binary stars.
Mitteilungen der AG 62, 19

1985 Edwin E. Salpeter:
Galactic fountains, planetary nebulae, and warm H I.
Mitteilungen der AG 63, 11

1986 Subrahmanyan Chandrasekhar:
The aesthetic base of the general theory of relativity.
Mitteilungen der AG 67, 19

1987 Lodewijk Woltjer:
The future of European astronomy.
Mitteilungen der AG 70, 21

1989 Sir Martin J. Rees:
Is there a massive black hole in every galaxy.
Reviews in Modern Astronomy 2, 1

1990 Eugene N. Parker:
Convection, spontaneous discontinuities,
and stellar winds and X-ray emission.
Reviews in Modern Astronomy 4, 1

1992 Sir Fred Hoyle:
The synthesis of the light elements.
Reviews in Modern Astronomy 6, 1

1993 Raymond Wilson:
Karl Schwarzschild and telescope optics.
Reviews in Modern Astronomy 7, 1

1994 Joachim Trümper:
X-rays from Neutron stars.
Reviews in Modern Astronomy 8, 1

1995 Henk van de Hulst:
Scaling laws in multiple light scattering under very small angles.
Reviews in Modern Astronomy 9, 1

1996 Kip Thorne:
Gravitational Radiation – A New Window Onto the Universe.
Reviews in Modern Astronomy 10, 1

1997   Joseph H. Taylor:
       Binary Pulsars and Relativistic Gravity.
       not published

1998   Peter A. Strittmatter:
       Steps to the LBT – and Beyond.
       Reviews in Modern Astronomy 12, 1

1999   Jeremiah P. Ostriker:
       Historical Reflections
       on the Role of Numerical Modeling in Astrophysics.
       Reviews in Modern Astronomy 13, 1

2000   Sir Roger Penrose:
       The Schwarzschild Singularity:
       One Clue to Resolving the Quantum Measurement Paradox.
       Reviews in Modern Astronomy 14, 1

2001   Keiichi Kodaira:
       Macro- and Microscopic Views of Nearby Galaxies.
       Reviews in Modern Astronomy 15, 1

The **Ludwig Biermann Award** was established in 1988 by the Astronomische Gesellschaft to be awarded in recognition of an outstanding young astronomer. The award consists of financing a scientific stay at an institution of the recipient's choice.

Recipients of the Ludwig Biermann Award are

1989   Dr. Norbert Langer (Göttingen),

1990   Dr. Reinhard W. Hanuschik (Bochum),

1992   Dr. Joachim Puls (München),

1993   Dr. Andreas Burkert (Garching),

1994   Dr. Christoph W. Keller (Tucson, Arizona, USA),

1995   Dr. Karl Mannheim (Göttingen),

1996   Dr. Eva K. Grebel (Würzburg) and
       Dr. Matthias L. Bartelmann (Garching),

1997   Dr. Ralf Napiwotzki (Bamberg),

1998   Dr. Ralph Neuhäuser (Garching),

1999   Dr. Markus Kissler-Patig (Garching),

2000   Dr. Heino Falcke (Bonn),

2001   Dr. Stefanie Komossa (Garching).

Contents

ASTRONOMISCHE GESELLSCHAFT: Reviews in Modern Astronomy **15**, 1–26 (2002)

Karl Schwarzschild Lecture

Macro- and Microscopic Views
of Nearby Galaxies

Keiichi Kodaira

The Graduate University for Advanced Studies (SOKEN-DAI)
Hayama, Miuragun, Kanagawaken, Japan 240-0193

It is really a great honor for me to be nominated for the Karl-Schwarzschild Prize and to be invited to give this lecture at JENAM 2001 in Munich. On this special occasion granted to me, I would like to talk about the issues which have been deeply moving me, rather than to present a round paper on a fixed subject. Since I wrote a thesis on a high-velocity population-II star HD161817 at the University of Kiel under the supervision of late Prof. A. Unsöld, the dynamic structure and evolution of galaxies as stellar ensembles have been stimulating my imagination all through my research career, while I was largely engaged in optical observational astronomy and instrumentation. I will start my talk with a review of my statistical works on the global structures of nearby galaxies, move on to the discussion of recent progresses in the same topic, and at the end, I wish to introduce some of the high-resolution images of the Andromeda galaxy obtained by the Subaru Telescope, to the realization of which I devoted myself for the last 20 years.

1 Empirical Relations
about Galaxies in Macroscopic View

In contrast to stars, galaxies show varying morphology, which has been fascinating astronomers and astrophysicists. With increasing capability in observation technology, more and more detailed features were detected among galaxies of various types. That puzzled me to ask what the essential of a galaxy was. If astronomers could have observed details on the surface of stars, like in the case of the sun, the spectacular surface phenomena might have caught too much attention of observers, to refrain them from comprehending the basic nature of the stars. The basic physical nature of stars was

inferred by interpreting the empirical correlations among observational parameters exhibited in the Hertzsprung-Russell Diagram and the Mass-Luminosity
Relation, recurring to the theories of plasma and nuclear physics: The stars
are self-gravitating gas spheres of varying mass in the thermal equilibrium
with central energy generation. The basic nature of stars in a slow evolution
phase is primarily characterized by their mass, which determines the physical state of the main sequence in the thermal equilibrium. The observational
parameters of stars along the main sequence, therefore, show strong inter-
correlation, thus the distribution of stars forms a quasi-linear domain in the
multi-dimensional space of the observational parameters.

In order to identify significant independent variables, which control the
observed properties of galaxies, Watanabe, Kodaira, & Okamura (1985) applied the "Principal Factor Analysis", following Brosche (1973), to the global
surface-photometric parameters of nearby galaxies which had been accumulated by their group (Watanabe, Kodaira, & Okamura 1982; Watanabe 1983;
Okamura, Kodaira, & Watanabe 1984; Takase, Kodaira, & Okamura 1984; Kodaira, Watanabe, & Okamura 1986; also see, Kodaira, Okamura, & Ichikawa
1990). In doing so, they ignored detailed textures and patterns of galactic
morphology, adopting only the backbone structures such as spheroid- and
disk-components, their sizes, mean surface-brightness, light-concentration,
and photometric magnitudes. The majority of the bright nearby galaxies
may be regarded to be in a slow evolution phase so that they are in a quasi-
equilibrium stage. They found out two principal factors for S galaxies, and one
principal and one subordinate factor for E galaxies. The two-dimensional surface, that is spun by the two corresponding eigen-vectors of the correlation ma-
trix, forms a plane in the multi-dimensional space of the surface-photometric
parameters of galaxies. This plane was approximated as the "Diameter versus
Surface-Brightness Diagram (DSBD)" collectively for E, S0, and S galaxies by
Kodaira, Okamura, & Watanabe (1983).

As for E galaxies, more comprehensive analyses including the spectro-
scopic parameters by Dressler et al. (1987), Djorgovski & Davis (1987), and
Faber et al. (1987) led to the concept of the "Fundamental Plane (FP)" of E
galaxies. The correlations between luminosity and internal-velocity parameters had been recognized as the "Faber-Jackson Relation" for E galaxies (Faber
& Jackson 1976) and, similarly, as the "Tully-Fisher Relation" for S galaxies
(Tully & Fisher 1977) which was widely utilized to yield distance estimates.
Dressler et al. (1987) empirically devised a more precise "$D_n - \sigma_0$ Relation"
for E galaxies for the same application purpose, by recruiting the third parameter, surface brightness, into the correlation study, referring to the FP
concept. This relation as well as the FP itself was applied to derive cluster
distances (cf. Kelson et al. 2000a, and others), and also to study luminosity
evolution of remote galaxies (cf. Illingworth et al. 1999; Kelson et al. 2000b).

The FP concept was later worked out by Bender, Burstein, & Faber (1992)
to introduce the concept of the "κ-Space" for E and S0 galaxies, a three
dimensional Cartesian space that had coordinates along mass parameter κ_1,
mass-to-luminosity ratio parameter κ_3, and $\kappa_2 = I_e^3 \, m/L$, where I_e was the

average surface-brightness within the equivalent radius, r_e. The plane (κ_1, κ_2) in this κ-Space was empirically defined by the distribution of dynamically hot galaxies (Es and a part of S0s) in the Virgo Cluster, and is close to being a face-on view of FP. Its edge-on projection onto the (κ_1, κ_3) plane was to be $\kappa_3 = 0.15\kappa_1 + 0.36$, corresponding to the $D_n - \sigma_0$ Relation.

These relations for E (and a part of S0) galaxies were interpreted in a simple context of $L_B = m \times (L_B/m)$ with constant L_B/m within a galaxy and m being a kind of virial mass, $r_e\sigma_0^2/G$, under the assumption of the structural homology, leading to an empirical finding $m/L_B \propto m^\alpha$ with $\alpha \sim 0.15$. They also suggested that this small value of α indicated that the dynamically hot galaxies such as E and S0 might have been formed by almost dissipation-less merging, keeping proportional increase in m and L. This framework of the κ-Space was later applied by Burstein et al. (1997) also to S galaxies, and further to other stellar systems such as groups and clusters of galaxies, and globular clusters (see also Djorgovski 1995). The distribution of S galaxies in the κ-Space significantly deviated from the plane for E galaxies, so that its projection onto the (κ_1, κ_3) plane led to a $\kappa_1 - \kappa_3$ relation different from that of E galaxies with substantial dispersion.

As for S galaxies, Chiba & Yoshii (1995) proposed a relation for disk galaxies to yield distance estimates, similar to the $D_n - \sigma_0$ Relation for E galaxies, by assuming the homology of galaxy structure, constant m/L ratio within a galaxy, and the dynamics model of a rotating disk in the centrifugal equilibrium. An empirical correlation between the radial scale length of the disk, $r_{\rm d}$, and the specific combination of the central surface-brightness, I_0, and the rotation velocity, v, was derived as $r_{\rm d} \propto (v^2/I_0)^\alpha$ with $\alpha \sim 0.5$, suggesting $m/L \propto (L/I_0)^{1/2}$ for disks. The same context of argument was applied by Zou & Han (2000), to interprete the empirical relation for S galaxies, that luminosity was correlated with a specific-angular-momentum parameter vR as $L \propto (vR)^\alpha$ with $\alpha \sim 1.3$, which was found by Koda, Sofue, & Wada (2000a) as corresponding one to the $D_n - \sigma_0$ Relation of E galaxies. The chinese group recently proposed another empirical relation $L \propto v^2R$ (Han et al. 2001). Courteau & Rix (1999) had already pointed out in their analyses of the rotation curves of S galaxies using maximal disk models that disk size or surface-brightness should be a significant additional parameter to the rotational velocity in the TF Relation. The $D_n - \sigma_0$ Relation for E galaxies itself could be derived in the same manner as adopted for the disk galaxies by Chiba & Yoshii (1995), by assuming the structural homology and the virial equilibrium, and by empirically determining the dependence of m/L on m from observational data.

All of these statistical relations for E and S galaxies are a kind of respective edge-on projections of the basic two-dimensional distributions of galaxies in the multi-dimensional observational-parameter space onto appropriate planes in it. The tight correlations between the radial scale, the surface brightness, and the velocity parameter found for nearby, high-surface-brightness (HSB), E or S galaxies were conventionally interpreted in terms of the dynamic equilibrium of structurally homologous systems under the assumption of the constant m/L ratio within a galaxy, the value of which was assumed to be function of other global parameters of each galaxy, typically of mass (E), or of mass and surface brightness (S). This empirical dependence of the m/L ratio on other global parameter(s) of galaxies was attributed to the systematic variation of the way of formation and evolution of individual galaxies.

2 Luminosity versus Phase-Space-Density Relation

The TF Relation and the FP concept have been widely applied to the distance estimates of S and E galaxies respectively, while the underpinning physics of these relations are not yet fully understood. In the course of the search for the underpinning physics, Kodaira (1989) found another empirical relation between the luminosity and the phase-space-density (PSD) parameter, collectively for E and S galaxies of the Virgo Cluster. The PSD parameter was defined as $w \equiv 1/(vD^2)$, where D was the photometric diameter and v was the central velocity dispersion, σ_0, for E galaxies or the rotation velocity, $0.5 \, W_{20}(\mathrm{H\,I})/\sin i$, for S galaxies. Since quantity $1/(GvD^2)$ has a dimension of PSD of a self-gravitating single-particle ensemble in the virial equilibrium, w may represent a kind of an average PSD of a galaxy as a stellar system (cf. Carlberg 1986). The correlations between the luminosity and the mass parameter, Dv^2, or the specific-angular-momentum parameter, Dv, were also examined for the same galaxy sample, but they were judged to be less tight than the $L - w$ relation.

Kodaira, Kashikawa, & Misawa (2000; hereafter KKM) reexamined the $B_{\mathrm{T}} - \log w$ relation of galaxies for the Virgo, the Coma, the Fornax, and the Perseus Cluster. They adopted the internal velocity parameter, v_c, to be $0.5 \, W_{20}(\mathrm{H\,I})/\sin i$ for S galaxies, and $\sqrt{3} \, \sigma_0/1.1$ for E and S0 galaxies. These velocities are to correspond to the flat rotation velocity in the gravitational potential of an isothermal spherical dark halo of a galaxy (see Shimasaku 1993). The factor $\sqrt{3}/1.1 = 1.57$ corresponds to the factor of $\sqrt{2} = 1.41$, which was empirically found by Whitemore & Kirshner (1981), and was derived theoretically by Binney & Tremaine (1987). Kodaira (1989) also derived the same factor to convert σ_0 into $0.5 \, W_{20}(\mathrm{H\,I})/\sin i$ for S galaxies of the earliest morphological types. A tight correlation, $B_{\mathrm{T}} = a \, \log w + b$, common to different morphological types of galaxies (E, S0, S) was found by KKM for the Virgo and the Coma Cluster, with $a \simeq 1.87$ and 1.33, respectively. An investigation using only E galaxies was made for four clusters, and its result indicated that the empirical linear relation might be common among the Coma, the Fornax, and the Perseus Cluster, with the Virgo Cluster showing slight deviation.

I have tested the results of KKM by adding dwarf E galaxies, 3 in the Virgo Cluster and 4 in the Local Group (Appendix A, Table 1), in order to enlarge the parameter space. These additional dwarfs roughly matched the general trend, extending the sequence of KKM for normal galaxies almost by two orders of magnitude in luminosity, to cover in total almost 4 and 5 orders of magnitude in the luminosity and the PSD, respectively (Figure 1).

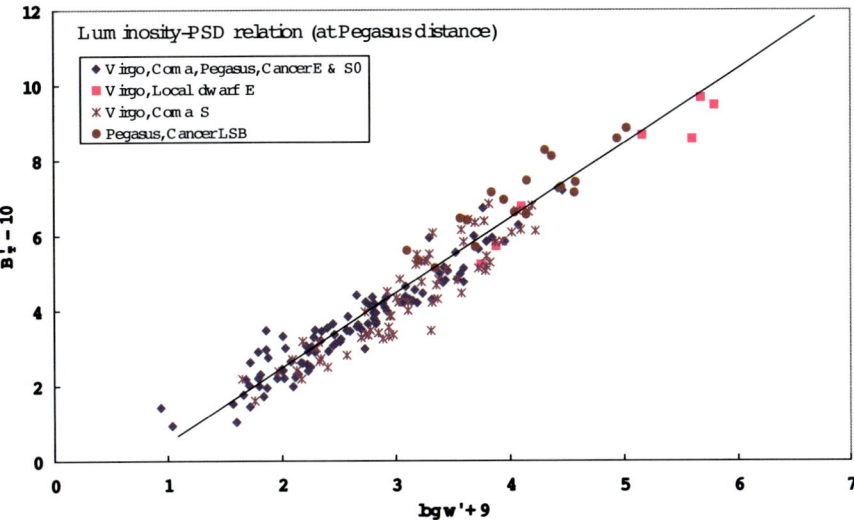

Figure 1: $B_T - \log w$ diagram for all the galaxies adopted in KKM and given in Appendix A. The primes of B' and w' indicate that these are scaled to the Pegasus Cluster distance; see Appendix A. The reference strait line has a gradient 2.0.

We may conclude that the $L - w$ relation, which is another way of edge-on projection of the basic plane, has an expression insensitive to the morphology, and suitable for treating galaxies of different morphological types collectively, though its details may be varying in individual cluster environment. Gibbons, Fruchter, & Bothun (2001) reported that the quality of fit of E galaxies to the average FP actually varies among clusters of galaxies, and that all X-ray-bright clusters in their sample fitted well to the average FP, suggesting the possible dependence of the details of FP on an intrinsic cluster property, probably related to the presence of significant substructure within a cluster. The properties of clusters may be examined using the new concept of the "Cluster FP" based upon parameters derived from X-ray and optical observations (cf. Fujita & Takahara 1999a, 1999b; Fritsch & Buchert 1999; Fujita & Takahara 2000; Sato et al. 2000).

The fact that, the $L - w$ relation in a similar form, $\log L = \alpha \log w + \beta$, was valid both for S and E galaxies, seemed to me suggestive for the presence of a common background for all types of nearby galaxies. When we assume that the optical galaxies are the stellar systems formed from barions imbedded in dark halos and that their evolution and morphology were destined by the dynamic initial conditions of the original barion-dark-matter mixture as a particle ensemble, the common validity of the PSD parameter for nearby galaxies from E to S may be an indication of its special role for unified implication of galaxy structure and evolution, like the role of the central density of the stars in the equilibrium phase.

Within the framework of CDM cosmology, many numerical simulations of galaxy formation and evolution were undertaken using the SPH + N-body code (cf. Steinmetz 1999). A detailed model calculation was performed by Koda, Sofue, & Wada (2000b) using GRAPE-SPH code in order to elucidate the physical background of the TF Relation for S galaxies. They adopted fixed initial conditions, leaving only mass and spin parameters free for rigidly rotating proto-galaxy spheres. Their simulation succeeded in reproducing the typical observational properties such as the global structure of the bulge and the exponential disk with the flat rotation curve, and produced a model TF Relation which had the same gradient as observed but showed a zero-point shift of ~ 1.5 mag in the I band. They concluded that the main factor of the TF sequence was the total mass m_T, and that the diagonal scatter was caused by the difference in the spin parameter, λ, with increasing λ leading to lower rotation velocity and brighter I magnitude. This trend was explained by the role of a rapid spin to centrifugally dilute the matter, which resulted in slow rotation velocity and slow star-formation process leading to relatively young bright stellar population at the present epoch. They suggested that the zero-point shift of the simulated TF Relation might be alleviated by choice of a larger cosmological constant H_0 (or a larger value of m/L). They also noted that the shift might have been affected by the star formation model using a rather coarse model mesh of ~ 1 kpc, and that local stellar energy feedback should have been better treated in detail.

Koda, Sofue, & Wada (2000b) did not show the cases of negligible λ; these would have simulated the E galaxies. The evolution history of the simulated S galaxies indicates that the basic properties, which command the TF Relation, are already determined around the epoch of the bulge formation, that is, the initial burst-like star formation. The period of the disk formation follows that epoch under the influence of the main gravitational potential field of the dark halo and the bulge. Disk structures are side-effects which appear as the results of the formation of these basic building blocks. This line of consideration suggests that all the types of galaxies from E to S are to be investigated in the numerical simulation as a continuous sequence of the spin parameter, λ, (and the parameter representing the hotness of the initial system), to reveal underlying common factors which control the basic properties of galaxies.

3   Non-Homology of Galaxy Structure

Hjorth & Madsen (1995) cast a doubt about the simple interpretation of the FP by adopting the structural homology and the empirical inference of m/L as function of m. Their argument was based on the observational evidences for systematic deviation of luminosity profiles from the fiducial 1/4-law for E galaxies and on the theoretical consideration of statistical mechanics (cf. Hjorth & Madsen 1993). They pointed out that, instead of the structural homology as usually invoked, the structural non-homology in kinematical structure of E galaxies as function of the luminosity may account for the tilt of the

FP. This working hypothesis led to the inference that the central phase-space density, f_c, would behave proportional to $L^{-1.5}$, instead of $L^{-2.35}$ in the case of the homology. The same doubt was expressed by Pahre, Djorgovski, & Calvarho (1995, 1998) in analyzing the scatter around the near-infrared and optical FP. They stated that the scatter was not fully understood in terms of metallicity and age effects as usually assumed, and that the departure of the near-infrared FP from the pure virial form might be explained by "slight systematic departures of the structure and dynamics of elliptical galaxies from the homology."

This line of consideration was tested by Busarello et al. (1997) using high-quality observational data to elucidate the structural non-homology of E galaxies. They succeeded in showing that the basic plane was thinner when the velocity dispersion that was "derived from the relative kinematic energy inside the equivalent radius", σ, was adopted instead of the central velocity dispersion, σ_0. The newly introduced average kinematic parameter σ increases slower than σ_0 as the mass increases: $\log \sigma \simeq 0.78 \log \sigma_0 + 0.46$. They claimed also that m/L was not correlated with the mass that was actually deduced from the rotation curves.

These investigations (and Koda, Sofue, & Wada 2000b) suggest that the basic plane of the galaxy-parameter space is not reflecting the relation for stellar systems of structural-homology in the virial equilibrium, but that some global property of the phase-space structure of non-homologous dynamic systems plays a significant role in defining the principal-factor plane. To characterize the global property of the phase-space structure of a dynamic system in equilibrium, the first candidate of a suitable parameter may be the average PSD, whose definition, however, is not unique for a muliple-particle system.

The physical processes, such as mass loss, merging, or dissipation, involved in the galaxy formation and evolution may in principle be investigated on the mass versus PSD-parameter plane. When we regard a nearby galaxy to be a relaxed dynamic ensemble of stellar particles imbedded in a relaxed dark halo, the global properties of the galaxy and the dark halo may well be characterized by their respective masses and average PSDs. In practice, however, we directly observe the photometric size parameter and the velocity parameters for stars or gas, but not their masses, and indirectly estimate the effective mass including both stars and dark halo. The observed luminosity in a certain photometric band, L_λ, is related to the stellar mass, m_s, through $L_\lambda \equiv m_s \times (L/m_s) \times (L_\lambda/L)$. The latter two terms are generally dependent on the evolutionary history of galaxies. When we are concerned with normal HSB galaxies at the present epoch, m_s approximately represents the barion mass of a galaxy. When we take the total mass, including the mass of the dark halo, m_T, into account, we have to consider a relation $L_\lambda \equiv m_T \times (m_s/m_T) \times (L/m_s) \times (L_\lambda/L)$. In studying galaxies of various morphology collectively on mass or luminosity versus PSD-parameter plane, we need more detailed information about the density ratio ρ_s/ρ_T as a function of the radial distance.

4   Low-Surface-Brightness Galaxies

It was known by photographic surveys that there were low surface-brightness (LSB) objects among nearby galaxies (Caldwell 1983; Binggeli, Sandage, & Tarenghi 1984, 1985; Bothun et al. 1986; Ichikawa, Wakamastu, & Okamura 1986; Caldwell & Bothun 1987). Several surveys using modern techniques were conducted to investigate their nature (i.e., Romanishin, Strom, & Strom 1983; Impey, Bothun, & Malin 1988; Bothun et al. 1991; Schombert et al. 1992; Impey et al. 1996). Most of the LSB objects detected in these surveys for nearby clusters of galaxies turned out to be dwarf galaxies having exponential luminosity profiles with varying colors from blue to red, indicating stellar population of various metallicity. A survey for H I gas in the LSB galaxies by Impey, Bothun, & Malin (1988) failed without any detection but for Malin 1, setting an upper limit of $m(\text{H\,I}) = 3 \sim 5 \times 10^6 \ m_\odot$; they were H I-poor.

O'Neil, Bothun, & Cornell (1997) published detailed surface-photometric results of their CCD search for LSB galaxies in the sky fields including the Pegasus and the Cancer Cluster of galaxies, followed by a publication of the *UBVI* color data by O'Neil et al. (1997). Among the total of 121 LSB galaxies observed in these optical surveys, 43 objects were detected in their follow-up radio survey in H I 21 cm line (O'Neil, Bothun, & Schombert 2000), to yield radial velocity, linewidths, $W_{20}(\text{H\,I})$ and $W_{50}(\text{H\,I})$, and hydrogen mass, $m(\text{H\,I})$. The gas-rich LSB galaxies detected in their H I survey had parameters over wide ranges, to increase the overall parameter space occupied by nearby galaxies: $W_{20}(\text{H\,I}) = 60 \sim 430$ km/sec, $M_B = -14.5 \sim -20$, $\mu_{0,B} = 22.0 \sim 25.0$, $B - V = 0.37 \sim 1.28$, and $m(\text{H\,I}) = 1 \sim 100 \times 10^8 \text{m}_\odot$. Judging from the absolute magnitude, color, and morphology, it can be said that these LSB are similar to the dwarf galaxies in the general field (cf. Borazza, Binggeli, & Prugniel 2001). The morphological types the authors assigned to them ranged from Sbc to Im. They found that a subset of the detected LSB galaxies had rotational velocity higher than ~ 200 km/sec and yet were at least an order of magnitude fainter than the Schechter's L^* ($M_B^* \simeq -21$), showing extreme departures from the standard TF Relation for normal HSB galaxies. This finding was partly expected because the TF Relation is not the best edge-on projection of the basic plane spun by the eigen-vectors. Some part of the large scattering of the LSB galaxies in their $L - W(\text{H\,I})$ diagram, however, may have been caused by rather inhomogeneous constituents of their sample, a part of which had no precise distance estimates.

I have adopted their data of LSB galaxies in the Pegasus and the Cancer Cluster (Appendix A, Table 2) to test the applicability of the same $L - w$ relation as was found for HSB galaxies in KKM. The disk-like LSB samples were restricted to those having the inclination angle $i = 30° \sim 65°$ in order to avoid excessive uncertainty due to the inclination correction either in the surface photometry or in the estimate of rotational velocity, as was the case in KKM. In addition, 4 E galaxies in the Pegasus and 1 E galaxy in the Cancer Cluster (Appendix A, Table 4) were also adopted to prove the consistency of the assumed distances among different clusters (Appendix A, Table 3). The resulting $B_T - \log w$ diagram is shown in Figure 1, which is scaled to the distance of the Pegasus Cluster. Since the E galaxies in the Pegasus and the Cancer Cluster are adopted from the same source as other galaxies in KKM from Faber et al. (1989), and since they actually do not show any deviation from the E samples of the Coma and the Virgo Cluster, no different specific symbol is assigned to them in Figure 1. As is seen in Figure 1, the LSB galaxies comprise with the natural extension of the $B_T - \log w$ relation for HSB galaxies, distributing almost common even with dwarf E galaxies.

It is to be noted that 3 LSB galaxies with the King profile also follow the general trend of other LSBs with the exponential profile. One may find that the gradient of the regression line for S+LSB galaxies is slightly steeper than that for E and S0 galaxies in Figure 1, but this is not certain, because the photometric system used for the LSB observation might not be quite the same as that for the HSB cases (see Appendix B in KKM). It should be also pointed out that the $B_T - \log w$ diagram is rather insensitive to errors in distance because of the gradient ~ 2.0 which is close to that of distance error vector, 2.5. This is the reason why this kind of expression is not suitable for practice of distance estimation but suitable for obtaining physical insights about the empirical relation. When the gradient $a = 2.0$ is accepted, the $L - w$ relation is transformed into another simple form, $\log D = SB + \log(v_c^2) + C$, with SB being the surface brightness in magnitude averaged over the area within the photometric diameter D. This relation defines the basic plane in the Cartesian space with the 3rd axis of $\log v_c$ additional to BSBD, and may be used for distance estimation if the constant C is determined empirically.

5   Visible- and Dark-Mass Conspiracy?

The consistent behavior of LSB galaxies with that of HSB galaxies raises anew the issue of "conspiracy" between the visible and the dark mass as a function of the radial distance, because the stellar system in LSBs seems to be only the core part of a larger H I disk as discussed by O'Neil, Bothun, & Schombert (2000).

The idea of the "universal rotation curve" for HSB S galaxies was proposed by Persic & Salucci (1991), and was substantiated by Persic, Salucci, & Stel (1996). They found that the rotation curves of all the bright S galaxies could be expressed by the universal form when they were scaled by the luminosity, L_B, and the optical disk scale length, R_M:

$$v(R) = 200(L_B/L_B^0)^\alpha \, [1 + f(L_B/L_B^0)(R/R_M - 1)], \qquad (1)$$

with $\alpha \sim 0.25$, $R_M = 2.2R_D$, $\log L_B^0 = 10.4$, and $f(x) = 0.625 \log x - 0.25$. The above equation is valid for $0.5 \lesssim R/R_{opt} \lesssim 1$, with the optical radius $R_{opt} = 3.2R_D$ and R_D being the exponential scale length of the disk. In some of low-luminosity galaxies, the contribution of m_s to the virial mass was found to be negligible even at small radial distance, what implied a kind of cooperation of visible mass and dark mass to realize the universal rotation curve.

McGaugh & de Blok (1998) noted that the rotation curves of LSB galaxies became similar when the scaling was applied, but not necessarily identical, to those of HSB galaxies. Swaters, Nadore, & Trewhella (2000), however, reported that their analysis of high resolution rotation curves derived from $H\alpha$ observation indicated the identical behaviors of HSB and LSB galaxies. This finding underlines the suspect of the visible- and dark-mass "conspiracy" by Salucci, Frenk, & Persic (1993). Persic, Salucci, & Stel (1996) actually constructed a semi-empirical composite model for the universal rotation curve, which was composed of two contributions from the visible disk and the dark halo. The local ratio of the two contributions was solely defined by L_B.

Giraud (1998a) additionally devised an analogous dynamic rotation model for HSB S galaxies:

$$v(R)^2 = \{G\ [m(R)/R^2] + \Gamma\ [m(R)/R^2]^{1/2}\}R, \qquad (2)$$

with the first right term being the contribution from the visible mass, $m(R)$, and the second term being that from the dark mass, $m_d(R) = g \cdot m(R)^{1/2} \cdot R$ with $g \equiv G/\Gamma$. He was successful in reproducing the observed rotational curves by adopting stellar mass-to-luminosity ratio $(m/L)_s \sim 2.7$ and ~ 1.5 for Sb and Sc galaxies, respectively, and varying values of $g = 1 \sim 6$. The excessively large value $g = 6$ was only for one LSB galaxy (NGC 289) among his samples. He stated that "this relation of the structure coupling accounts for the disk-halo conspiracy in HSB galaxies and is also valid in LSB dwarf galaxies". He also claimed that "its validity implies the existence of a physical mechanism responsible for the continuity between visible and dark mass distributions". The scatter in the TF relation was attributed by him to varying g-value with small g-value leading to luminosity excess and vice versa.

Further he tested the validity of the model on low-density spiral and dwarf galaxies (Giraud 1998b), in particular on two gas-dominant galaxies, DDO 154 and UGC 2684, with $m_{gas}/L \sim 7$, to conclude that the visible mass should include the gas mass which was "visible" in H I observation. Giraud (2000) additionally introduced invisible barion component which should have a gas-like distribution, representing cool components such as dust and molecules. For some LSB dwarf galaxies, $m_{gas-like}/m_{gas}$ amounted up to ~ 4. His results may mean that the "conspiracy" of the visible and the dark mass was made up already before the star formation took place in the disk. The distributions of barion and dark matter might have been "dictated by the dynamic history and the relative physical properties of the barion and the dark matter," to produce the universal rotation curve, and the resulting distribution further controlled the star formation history.

In summarizing the present status of investigations on E $\sim$ S galaxies, especially LSB and dwarf galaxies, I may suggest that the global properties of a galaxy as a dynamic system composed of barion and dark matter are primarily defined by its total mass, but its luminosity (star formation history) is strongly correlated with the global properties of the phase-space structure which the barion and the dark matter work out in "conspiracy". Until now the "conspiracy" was discussed more about S galaxies than about E galaxies in which the visible mass is dominant, but I may assume a kind of continuity between S and E galaxies with varying spin and hotness parameters of the barion-dark-matter dynamic systems. The interactions with outer world like merging, in-fall, or encounter influence the galaxy evolution, but the "conspiracy" may be again attained after the dynamic system has settled down in equilibrium. Merging of evolved galaxies may manifest itself by decreasing PSD through phase mixing (Lynden-Bell 1967; Carlberg 1986). Secular evolution of galaxy morphology due to gas inflow via angular momentum transfer and viscous transport, as suggested by Courteau, de Jong, & Broeils (1996), might be accompanied by secular readjustment of the dynamic structure. Although the PSD parameter I adopted is a mixture of the velocity parameter for total mass, $v_{\rm c}$, and the size parameter for visible mass, D, the "conspiracy" between the visible and the dark mass may allow me to adopt this kind of mixed combination of parameters in order to characterize the average properties of the phase-space structure of the composite system.

I dare not to claim that these data and arguments are sufficient to verify my adoption of the specific PSD parameter, w, to define the basic plane for E $\sim$ S galaxies. It is, however, desirable that theoretical researchers working on numerical simulations may pay more attention to the phase-space structure of dynamic galaxy models. On the other hand, the definition of the PSD parameter still shall be elaborated to better represent the global properties of the phase-space structure of galaxies as particle ensembles.

6   Microscopic View of Nearby Galaxies

Provided that the global dynamic structure of a galaxy may be represented by the mass and the average phase-space properties of the self gravitating system, the observational properties are bound to the barion, which is subject to dissipative processes, typically the gas contraction and the following star formation activities. The observed star light of nearby galaxies are interpreted in terms of stellar population synthesis (Kurth, Fritze-von Alvensleben, & Fricke 1999, and others), or in terms of modeling the chemical and luminosity evolution of galaxies (Arimoto, Yoshii, & Takahara 1992, and others). Several elaborate galaxy evolution models are available, in which conventionally users tune various model parameters to reproduce the observed stellar population. Although the available evolution models are now tested using high-z galaxies (Kodama, Bower, & Bell 1999, and others), most of the model parameters are still to be empirically adjusted. So far as we regard a galaxy to be an isolated system, the major parametric functions are the star formation rate (SFR) and the initial mass function (IMF), for which conventionally the Schmidt law (Schmidt 1959) and the Salpeter function (Salpeter 1955) are adopted and applied to macroscopic volumes within a model galaxy. Recent empirical approaches to estimate local IMF in a galaxy scale were undertaken by analyzing photometric data of star complexes in nearby galaxies, revealing the intricate nature of local IMFs (cf. Bresolin & Kennicut 1997; Sakimov & Smirnov 2000). In the actual physical processes, they are affected by large-scale dynamics such as spiral density waves and bar potential, or by local energy feed-back from massive stars into the interstellar media. Microscopic views of nearby galaxies based upon high-spatial-resolution data are indispensable for physical understanding of these varying and episodic elementary processes.

While the Magellanic Clouds, LMC and SMC, have been the targets of detailed observational studies as outstanding examples for dwarf satellite galaxies (cf. Chu et al. 1999), high-resolution studies of the two nearest standard spiral galaxies, M31 and M33, are handicapped because of their larger distances which are more than ten times far compared to LMC and SMC. In particular, M31 has a morphological type close to that of the Milky Way Galaxy and is most suitable to provide supplementary data to the Galactic ones to understand the local dynamic structures such as spiral arms and star-forming regions (cf. Hodge 1992). Systematic detailed optical surveys of M31 were carried out using KPNO 4 m and Palomar 5 m telescopes under seeing condition of $1''$–$2''$, to provide overviews of star clusters in the galaxy (Sargent et al. 1977; Hodge 1981; hereafter H81, among others). The ground-based observation improved the seeing up to $0''.6$–$0''.7$, which enabled us to produce color-magnitude diagram (CMD) of low-density stellar fields even at the distance of M31 or M33 (cf. Durrell, Harris, & Pritchet 2001; Cuillandre et al. 2001). When adaptive-optics techniques were applied to near-infrared observation, image photometry of $0''.3$–$0''.4$ resolution became possible for rather dense stellar fields of small areas ($\lesssim 1'$) (cf. Kodaira et al. 1998, 1999; Davidge 2000).

All these were competed by the HST observation using WFPC2, starting with Magnier et al. (1995) and Magnier et al. (1997) among others, followed by recent works by Joblanka et al. (2000) and Williams & Hodge (2001a, b) who succeeded in producing CMD for various classes of star clusters in M31. The impact of the high-resolution observation using HST-WFPC2 was demonstrated by Pleuss, Heller, & Fricke (2000) in a study of H II regions and star clusters in M101, which is located again about ten times farther than M31 or M33. Since the field of view of HST-WFPC2 is small ($80'' \times 80''$), a kind of survey was possible for M101, but it was not easy to cover a large portion of M31. At present, therefore, it is more effective to use prime-focus wide-field cameras of ground-based large telescopes in detailed optical surveys of M31 and M33 if their image quality is excellent.

7 Suprime-Cam M31 Images

With the intention to examine the capability of the Subaru Telescope in investigating the local dynamics and the star formation processes in the M31 disk, several test frames were secured using the prime-focus wide-field camera

(Suprime-Cam) during its verification period in July 1999. The camera was equipped with a mosaic 8K×8K CCD with a pixel size of $0''.2$/pix. The B, V, R, and Hα frames which we processed, with exposure time of 2 min × 5 for each, showed median seeing of $0''.5$–$0''.6$ after stacking. Our initial inspection of the field shown in Figure 2 has revealed many new details as follows, while the photometric work is still under way.

(1) Dark Clouds:

Most of the prominent absorption features show continuous dark patterns along the spiral-like large structure, and it is rather difficult to designate specific domains by separating one from another as was done in H81. The continuous dark lanes show thin fibrous substructures in many regions of moderate absorption (Figure 3), indicating that this kind of fine structures might be densely stacked in regions of high absorption. Many of large features of low absorption have flocculent appearance (Figure 2 and 3), suggesting that these may be drifting out of the disk plane as "cirrus" (cf. Howk 1999). We do not, however, see any clear sign of dark-cloud "chimneys" which are standing out from the disk plane, such as were reported for some of the edge-on galaxies (cf. Sofue 1987: Sofue, Wakamatsu, & Malin 1994). Further inspection reveals that the fibrous lanes often break up on their thin ends into chains of dark knots, and that even very small dark knots ($\lesssim 7$ pc) are present in isolation (Figure 4). These small features might be artifacts, which are accidentally produced by compilation of stars in front of dark clouds, but they might possibly be physical globules of dark clouds, and need to be investigated more in detail in connection to star formation process.

(2) OB Associations and Star Clusters:

Bright blue stars are distributed along the spiral-like large structures with intermittences by dark clouds. To assign OB association as in H81 is easier than to assign dark clouds, but designation of their substructures may be subject to some arbitrariness (cf. Efremov, Ivanov, & Nikolov 1987; Gouliermis 2000).

The preliminary results of the present survey for star clusters have been compared with H81, Battistini et al. (1987; hereafter B87), Battistini et al. (1993), and Barmby et al. (2001). The following discussion shall be focused only onto the prominent objects of special interests readily found in the comparison. All the globular clusters in H81 are identified in the field of Figure 2, but some of them are found to show loose appearances similar to compact open clusters. These are listed in Appendix B Table 5, a, together with $B - V$ colors adopted from B87. Among the 6 G-objects, two objects (K154, K288) are doubted to have cluster nature in B87, and one object (K225) is listed with a comment "very diffuse" in their Table D which contains "miscellaneous objects". The other 3 are listed in B87 as globular clusters but have relatively blue colors. There are another category of 6 objects (Appendix B Table 5, b) which appear merely as aggregations of stars, or globular-like objects with star aggregations around them. The objects of the latter category are identified as globular clusters in B87. Among these 6 objects, 4 have very blue colors indicating young population.

Actually Williams & Hodge (2001a) made CMDs of 4 G-clusters (G38, G44, G94, G293; only G44 = K291 is in our present field) using HST-WFPC2 data, to find out that these compact clusters were "well populated with upper-main-sequence and various numbers of supergiants", in analogous to blue globulars in LMC (cf. Brocato et al. 1989). Williams & Hodge (2001b) made a survey for young star cluster candidates in outer spiral arms of M31 by applying an automated technique to HST-WFPC2 images, to identify 79 new objects in 13 fields of 60 arcmin^2. Most of them are "small ($\lesssim$ 5pc) and young (10 $\sim$ 200 Myr) star groups, located within large OB associations" and "showing excess in mean surface density of bright blue stars" ($M_V < 0.1$, $B - V < 0.45$). One object in Appendix Table 5 (G42 = K289b) was identified by them as one of such objects to have age of $\log t = 7.751$. On the other hand, we found 3 C-objects in H81 Table B (open cluster) to show globular-like compactness (Appendix B Table 5, c). One of them, C88 = K299 was identified with high certainty as globular cluster in B87.

All of these findings strongly indicate that intermediate clusters between the fiducial open and the globular clusters may abundantly reside in the M31 disk, and that a part of them have a compact globular-like appearance. We have also identified numerous very small, compact objects ($\lesssim$ 7pc), which may be young star clusters similar to those found by Williams & Hodge (2001b). A full photometric analysis of the Suprime-Cam frames will enable us to make a statistical study of cluster sizes in M31, as was done for M101 by Pleuss, Heller, & Fricke (2000) using HST-WFPC2, and to evaluate the local SFR and IMF across the M31 disk, although accurate stellar photometry in the crowded field needs special cares and techniques (cf. Kodaira et al. 1998). A few distinct objects are added in Appendix B Table 5, d, which were listed in the previous catalogues, but cannot be surely classified. Some of them are suspected to be background galaxies.

(3) H II Regions and nebulous Stars:

When our images and identification list of H II regions are compared with those of Pellet et al. (1978), it is found that almost all emission features were well detected by them, except for faint star-like objects. We notice that Pellet et al. (1978) identified some diffuse extended emission regions and lanes of emission sources we do not recognize. These features are apparently connected to the regions of high stellar density, and we suspect that these features might be artifacts; their Hα plates might have suffered under worse image quality due to substantially longer exposure than their R plates between which a blink-comparator was operated for the search.

Figure 5 shows the south-western extent of the prominent open cluster C179 imbedded in a large H II bubble, which contacts with a dark cloud D118. This sort of composite images is a useful tool to obtain global insights about

Figure 2: False-color picture of the south-western quadrant of the M31 disk ($25' \times 18'$) produced from B (blue), V (yellow), and Hα (red) frames taken with Suprime-Cam. The picture is centered at $0^h 38^m_{\cdot}1$, $+40°28'$ (1950.0), with north at the top and east to the left.

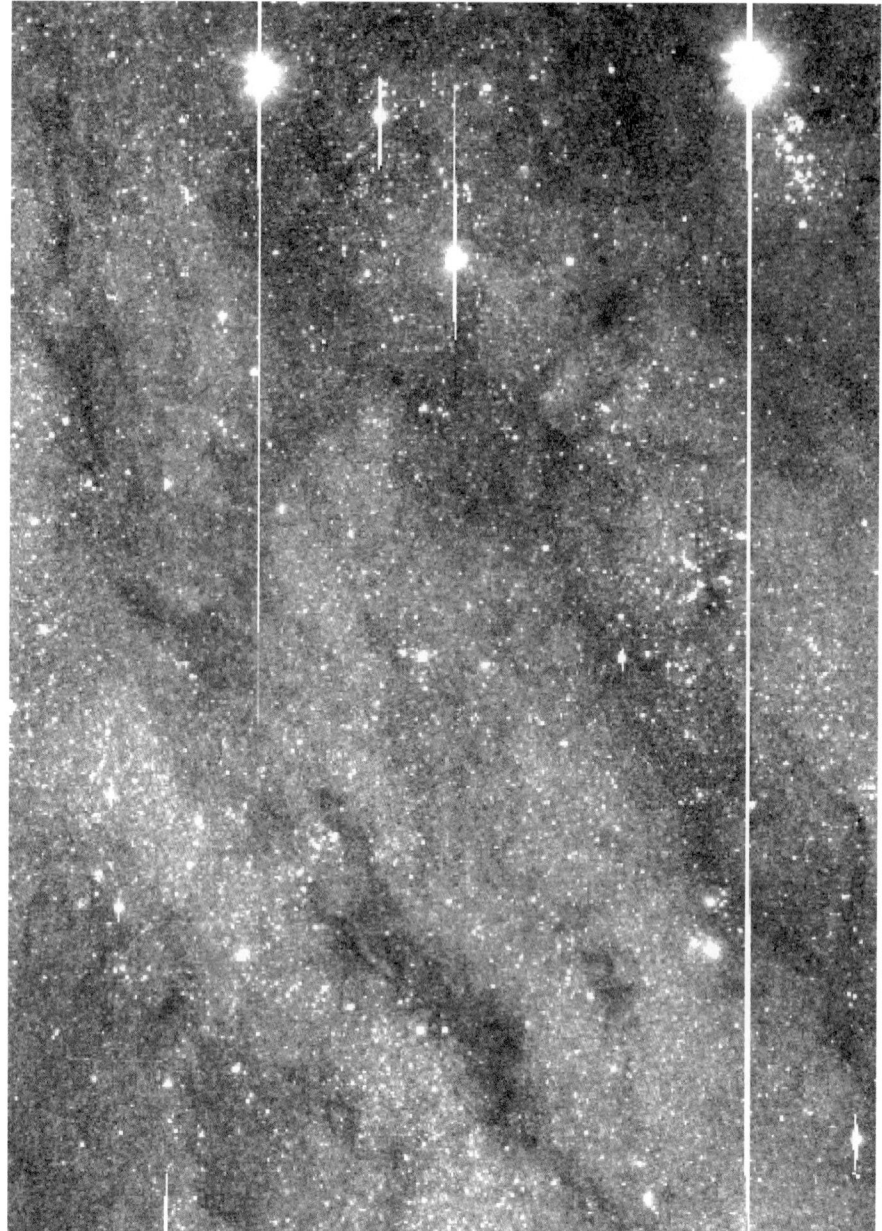

Figure 3: *B*-band image, showing examples of fibrous dark clouds in a $6\rlap{.}'0 \times 4\rlap{.}'3$ region north of N206 centered around $0^h37\rlap{.}^m80$, $+40°33\rlap{.}'9$ (1950.0), with north at the top and east to the left.

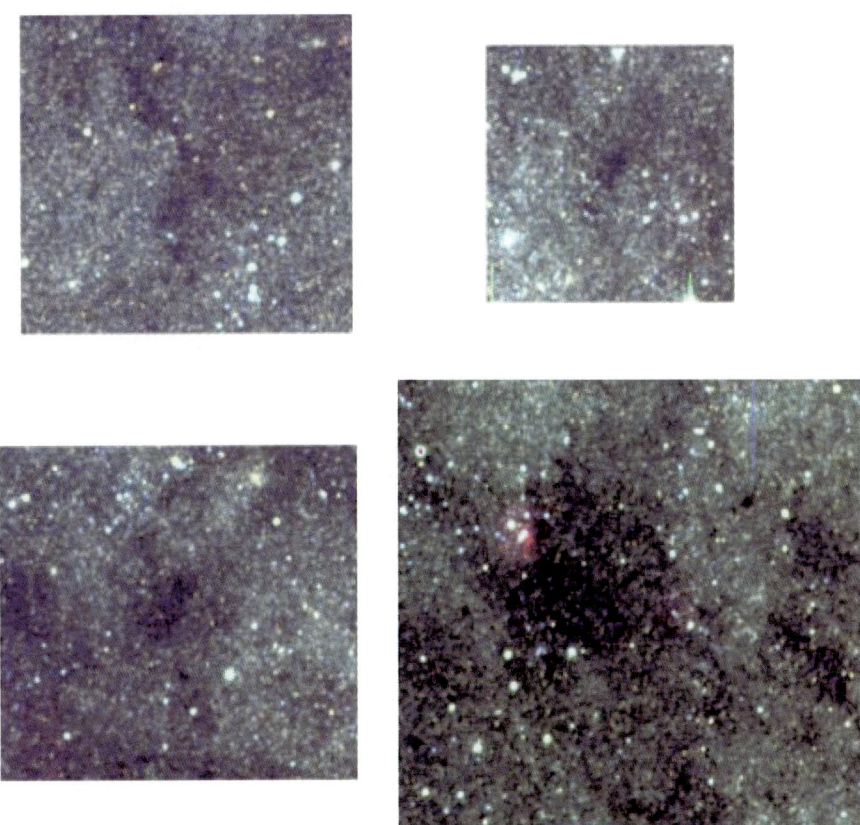

Figure 4: Examples of globule-like dark features around $0^h37^m_.85$, $+40°39'_.3$ (top, left) ($95'' \times 101''$); $0^h37^m_.75$, $+40°35'_.7$ (top, right) ($70'' \times 77''$); $0^h37^m_.93$, $+40°31'_.7$ (bottom, left) ($54'' \times 54''$); and $0^h37^m_.72$, $+40°31'_.1$ (bottom, right) ($65'' \times 71''$). North is at the top and east is to the left. The linear scale is the same for the 4 frames.

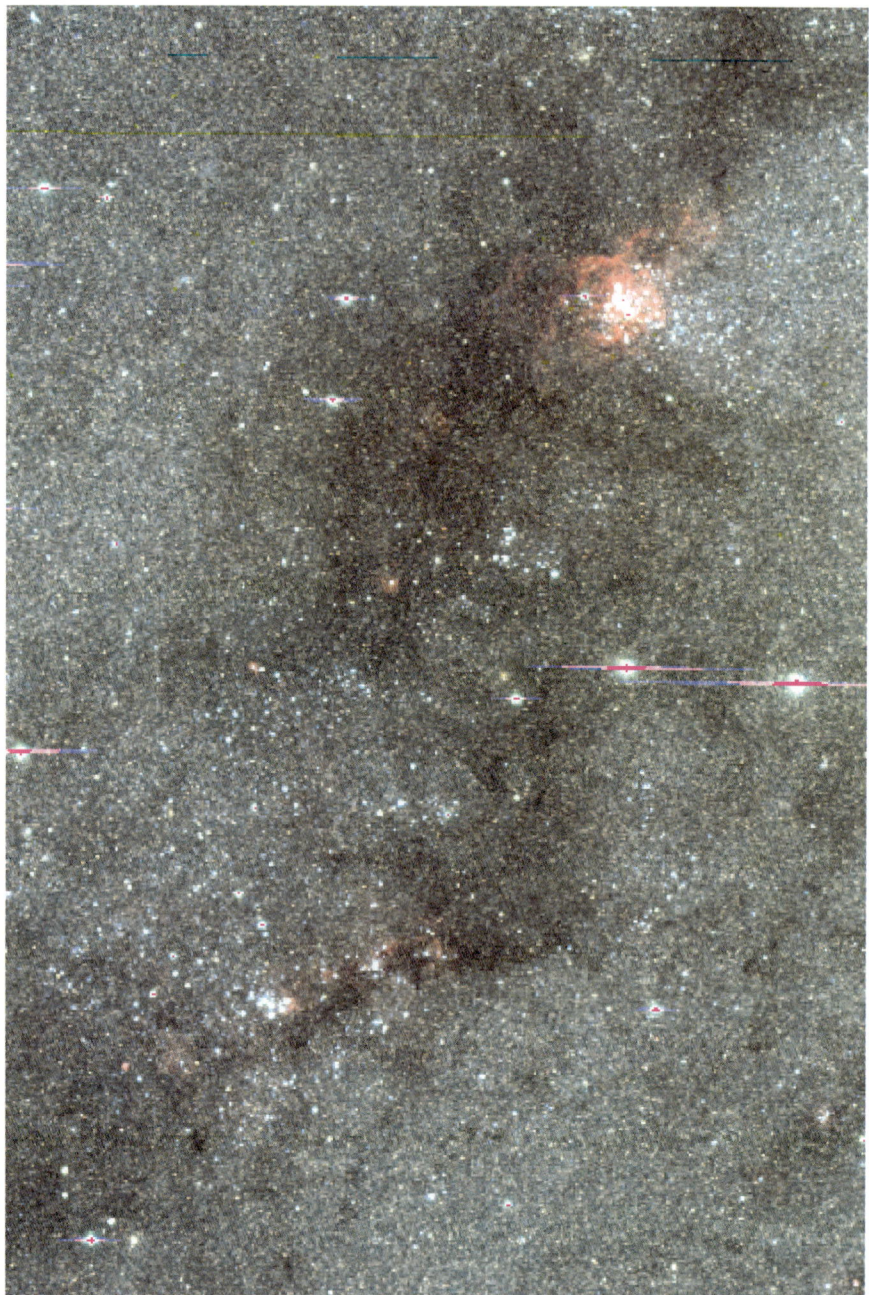

Figure 5: False-color picture of the south-western extent of open cluster C179, showing examples of H II regions around dark clouds D118 and D74 (C- and D-designations are according to H81). The frame (7.'2 × 5.'0) is centered at $0^h 38^m 59$, $+40° 33.'5$ (1950.0) with east at the top and north to the right.

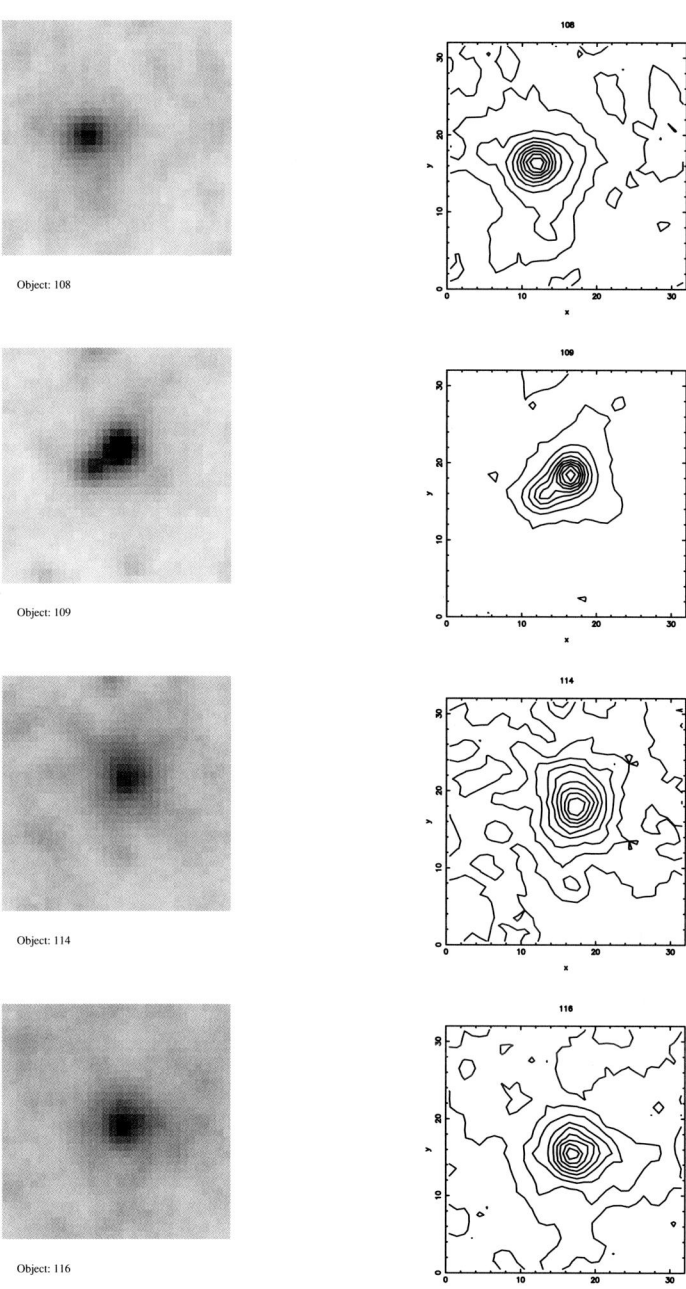

Figure 6: Examples of nebulous stellar images. From the top: K108 ($0^h 37^m.394$, $+40°33'.63$), K109 ($0^h 37^m.508$, $+40°33'.62$), K114 ($0^h 38^m.641$, $+40°33'.52$), and K116 ($0^h 38^m.763$, $+40°33'.51$). K-numbers are preliminary IDs in the present work. Each frame covers $15'' \times 15''$, and the isophotes are in a linear scale dividing the surface-brightness range in the frame by ten steps.

the dynamic situation around star-forming regions. As are seen in Figure 5, there are numerous small emission features, down to the star-image scales ($\lesssim$ 3pc); some stellar images are surrounded by emission ring, while some of stellar images themselves are dominated by Hα emission. Further, we detect many nebulous stellar images in and along dark clouds, which are probably early-type stars imbedded in scattering dust layers. Some examples are shown in Figure 6 with their isophote maps. In order to gain physical pictures of the interaction between young stars and surrounding interstellar media, we further need spectroscopic approach, for example using FOCAS/IRCS in the case of the Subaru Telescope.

These preliminary results demonstrate that Suprime-Cam will be a powerful guiding instrument for the microscopic researches of nearby galaxies, especially when it is combined with high-spatial-resolution spectrographs and radio interferometers such as ALMA. In connection to these topics, and at the end of my talk, I may express my strong wish that the Japanese astronomy community may soon join with the European and the American in enhancing the ALMA, which will be the most powerful facility not only to explore the world of the primeval galaxies but also to establish microscopic views of nearby galaxies, as well.

Acknowledgement

I wish to express my sincere thanks to all of my colleagues and friends who supported me and collaborated with me in my research activities, particularly in the Subaru-Telescope project. My special gratitude is due to Dr. Satoshi Miyazaki who developed the Suprime-Cam and is collaborating with me in the evaluation of the M31 data.

References

Arimoto, N., Yoshii, Y., & Takahara, F. 1992, A&A 253, 21

Barazza, F. D., Binggeli, B., & Prugniel, P. 2001, A&A 373, 12

Barmby, P. et al. 2000, AJ 119, 727

Battistini, P. et al. 1987, A&AS 67, 447 (B87)

Battistini, P. et al. 1993, A&A 272, 77

Bender, R., Burstein, D., & Faber, S. M. 1992, ApJ 399, 462

Bender, R., Burstein, D., & Faber, S. M. 1993, ApJ 411, 153

Binggeli, B.,Sandage, A., & Tarenghi, M. 1984, AJ 89, 64

Binggeli, B., Sandage, A., & Tarenghi, M. 1985, AJ 90, 1681

Binney, J. & Tremaine, S. 1987, Galactic Dynamics (Princeton: Princeton Univ. Press)

Bothun, G. D., Mould, J. R., Caldwell, N., & MacGillivary, H. T. 1986, AJ 92, 1007

Bothun, G.D. et al. 1991, ApJ 376, 404

Bresolin, F. & Kennicutt, R. C. Jr. 1997, AJ 113, 975

Brocato, E., Buonanno, R., Castellani, V., & Walker, A. R. 1989, ApJS 71, 25

Brosche, P. 1973, A&A 23, 259

Burstein, D., Bender, R., Faber, S. M., & Northenius, R. 1997, AJ 114, 1365

Busarello, G., Capaccioli, M., Capozziello, S., Lengo, G., & Puddu, E. 1997, A&A 320, 415

Caldwell, N. 1983, Ph. D. Thesis, University of Cambridge

Caldwell, N. & Bothun, G. D., 1987, AJ 94, 1126

Carlberg, R. G. 1986, ApJ 310, 593

Cuillandre, J.-Ch., Lequeux, J., Allen, R. J., Mellier, Y., & Bertin, E. 2001, ApJ 554, 190

Chiba, M. & Yoshii, Y. 1995, ApJ 442, 82

Chu, Y.-H., Suntzeff, N. B., Hesser, J., & Bohlender, D. A. eds. 1999, IAU Symp. No. 190

Courteau, S., De Jong, R. S., & Broeils, A. H. 1996, ApJ 457, L73

Courteau, S. & Rix, H.-W. 1999, ApJ 513, 561

Davidge, T. J. 2000, AJ 119, 748

Djorgovski, S. 1995, ApJ 438, L29

Djorgovski, S. & Davis, M. 1987, ApJ 313, 59

Dressler, A., et al. 1987, ApJ 313, 42

Durrell, P. R., Harris, W. E., & Pritchet, Ch. J. 2001, AJ 121, 2557

Efremov, Yu. N., Ivanov, G. R., & Nikolov, N. S. 1987, Ap&SS 135, 119

Faber, S. M. & Jackson, R. E. 1976, ApJ 204, 668

Faber, S. M. et al. 1987, in: Nearly Normal Galaxies, from the Planck Time to the Present, ed. S. M. Faber (New York: Springer), 175

Faber, S. M. et al. 1989, ApJS 69, 763

Fritsch, Ch. & Buchert, Th. 1999, A&A 344, 749

Fujita, Y. & Takahara, F. 1999a, ApJ 519, L51

Fujita, Y. & Takahara, F. 1999b, ApJ 519, L55

Fujita, Y. & Takahara, F. 2000, PASJ 52, 317

Gibbons, R. A., Fruchter, A. S., & Bothun, G. D. 2001, AJ 121, 649

Giraud, E. 1998a, AJ 116, 1125

Giraud, E. 1998b, AJ, 116 2177

Giraud, E. 2000, ApJ 531, 701

Gouliermis, D. et al. 2000, AJ 119, 1737

Han, J.-L., Deng, Z., Zou, Zh., Wu, X.-B., & Jing, Y. 2001, PASJ 53, 853

Hjorth, J. & Madsen, J. 1993, MNRAS 265, 237

Hjorth, J. & Madsen, J. 1995, ApJ 445, 55

Hodge, P. W. 1981, Atlas of the Andromeda Galaxy (Seatle and London: Univ. Washington Press) (H81)

Hodge, P. W. 1992, The Andromeda Galaxy (Dordorecht: Kluwer)

Howk, J. Ch. 1999, ASS 269–270, 293

Ichikawa, S.-I., Wakamatsu, K.-I., & Okamura, S. 1986, ApJS 60, 475

Illingworth, G., Kelson, D., Van Dokkun, P., & Franx, M. 1999, ASS 269–270, 485

Impey, C., Bothun, G. D., & Malin, D. 1988, ApJ 330, 634

Impey, C. et al. 1996, ApJS 105, 209

Jablonka, P. et al. 2000, A&A 359, 131

Kelson, D. D. et al. 2000a, ApJ 529, 768

Kelson, D. D., Illingworth, G. D., Van Dokkum, P. G., & Franx, M. 2000b, ApJ 531, 184

Koda, J., Sofue, Y., & Wada, K. 2000a, ApJ 531, L17

Koda, J., Sofue, Y., & Wada, K. 2000b, ApJ 532, 214

Kodaira, K. 1989, ApJ 342, 122

Kodaira, K., Okamura, S., & Watanabe, M. 1983, ApJ 274, L49

Kodaira, K., Watanabe, M., & Okamura, S. 1986, ApJS 62, 703

Kodaira, K., Okamura, S., & Ichikawa, S.-I. 1990, ed. Photometric Atlas of Northern Bright Galaxies (Tokyo: Univ. Tokyo Press)

Kodaira, K. et al. 1998, ApJS 118, 177

Kodaira, K., Vansevicius, V., Tamura, M., & Miyazaki, S. 1999, ApJ 519, 153

Kodaira, K., Kashikawa, N., & Misawa, T. 2000, ApJ 531, 665 (KKM)

Kodama, T., Bower, R. G., & Bell, E. F. 1999, MNRAS 306, 561

Kurth, O. M., Fritze-von Alvensleben, U., & Fricke, K. J. 1999, A&AS 138, 19

Lynden-Bell, D. 1967, MNRAS 167, 101

Magnier, E. A. et al. 1995, A&AS 114, 215

Magnier, E. A., Hodge, P., Battinelli, P., Lewin, W. H.,& van Paradijis, J. 1997, MNRAS 292, 490

McGaugh, S. S. & de Blok, W. J. G. 1998, ApJ 499, 41

Okamura, S., Kodaira, K., & Watanabe, M. 1984, ApJ 280, 7

O'Neil, K., Bothun, G. D., & Cornell, M. E. 1997, AJ 113, 1212

O'Neil, K., Bothun, G. D., Schombert, J., Cornell, M. E., & Impey, C. D. 1997, AJ 114, 2448

O'Neil, K., Bothun, G. D., & Schombert, J. 2000, AJ 119, 136

Pahre, M. A., Djorgovski, S. G., & De Carvalho, R. R. 1995, ApJ 453, L17

Pahre, M. A., Djorgovski, S. G., & De Carvalho, R. R. 1998, AJ 116, 1591

Pellet, A. et al. 1978, A&AS 31, 439

Persic, M. & Salucci, P. 1991, ApJ 368, 60

Persic, M., Salucci, P., & Stel, F. 1996, MNRAS 281, 27

Pleuss, P. O., Heller, C. H., & Fricke, K. J. 2000, A&A 361, 913

Romanishin, K., Strom, K., & Strom, S. 1983, ApJS 53, 105

Salpeter, E. E. 1955, ApJ 121, 161

Sakhibov, F. & Smirnov, M. 2000, A&A 354, 802

Salucci, P., Frenk, C. S., & Persic, M. 1993, MNRAS 262, 392

Sargent, W. L. W., Kowal, Ch. T., Hartwick, F. D. A., & van den Bergh, S. 1977, AJ 82, 947

Sato, S. et al. 2000, ApJ 537, L73

Schmidt, M. 1959, ApJ 129, 243

Schombert, J. et al. 1992, AJ 103, 1107

Shimasaku, K. 1993, ApJ 413, 59

Sofue, Y. 1987, PASJ 39, 547

Sofue, Y., Wakamatsu, K., & Malin, D. F. 1994, AJ 108, 2102

Steinmetz, M. 1999, ASS 269–270, 513

Swaters, R. A., Madore, B. F., & Trewhella, M. 2000, ApJ 531, L107

Takase, B., Kodaira, K., & Okamura, S. 1984, ed: An Atlas of Selected Galaxies (Tokyo: Univ. Tokyo Press)

Tully, R. B. & Fisher, J. R. 1977, A&A 54, 661

Watanabe, M. 1983, Ann. Tokyo Astron. Obs. 19, 121

Watanabe, M., Kodaira, K., & Okamura, S. 1982, ApJS 50, 1

Watanabe, M., Kodaira, K., & Okamura, S. 1985, ApJ 292, 72

Whitemore, B. C. & Kirshner, R. P. 1981, ApJ 250, 43

Williams, B. F. & Hodge, P. W. 2001a, ApJ 548, 190

Williams, B. F. & Hodge, P. W. 2001b, astro-ph/0105506

Zou, Z.-L. & Han, J.-L. 2000, Chin. Phys. Lett. 17, 935

Appendix A:
Observational Data of Additionally Adopted Galaxies

The observational data are summarized in this Appendix for the galaxies which are discussed in section 2 and 4 and included in Figure 1, in addition to those galaxies in the list of KKM. Table 1 presents the dwarf galaxies in the Virgo Cluster and the Local Group which are adopted from Bender, Burstein, & Faber (1993) and discussed in section 2. The photometric diameter D at the surface-brightness $\mu_B = 25$ is given in units of arcsec ($''$). The velocity parameter is given in units of km/sec. The total B magnitude B_T and the PSD parameter w in the last two columns are provided with a prime which indicates that the parameters are scaled to the distance of the Pegasus Cluster according to Table 3, as are plotted in Figure 1. When $H_0 = 75$ km/sec is adopted, $B_T' = 10$ corresponds to $M_B = -23.4$ and $\log w' = 0$ to phase-space density of 0.013 m$_\odot$ pc^{-3} (km/sec)$^{-3}$.

The observational data in Table 1 are adopted from Bender, Burstein, & Faber (1993), but the original distance to the Virgo cluster (20.7 Mpc) was modified to be consistent with that in KKM as is given in Table 3, while the distances to the 4 dwarf galaxies in the Local Group are kept as are in Bender, Burstein, & Faber (1993) (0.7 Mpc). The data for the latter are brought into the common distance scale with those for the former by using the distance of the Virgo Cluster in Table 3, $R = 1333$ km/sec with $H_0 = 75$ km/sec/Mpc.

Table 2 presents the LSB galaxies in the Pegasus and the Cancer Cluster which are adopted from O'Neil, Bothun, & Schombert (2000) and discussed in section 4. The adopted inclination angle i is the average of those given in O'Neil, Bothun, & Cornell (1997) and in O'Neil, Bothun, & Schombert (2000).

Table 4 lists the E galaxies in the Pegasus and the Cancer Cluster which are adopted from the same source as in KKM, Faber et al. (1989). These E galaxies are utilized to support the consistency of the adopted distances among the HSB and the LSB galaxies in different clusters in section 4.

Table 1: Virgo and Local dwarf E galaxies

ID	B_T	$\log D_{25}('')$	$\log \sigma_0$	$\log w' + 9$	$B_T' - 10$
N4431	13.58	2.08	1.830	3.880	5.73
N4515	13.09	1.98	1.954	3.748	5.24
I3393	14.63	1.91	1.740	4.102	6.78
N147	9.63	2.89	1.369	5.675	9.68
N185	9.43	2.84	1.352	5.792	9.48
N205	8.63	3.02	1.623	5.161	8.68
N221	8.53	2.66	1.903	5.601	8.58

Table 2: LSB galaxies in the Pegasus (p) and the Cancer (c) Cluster

ID	B_T	$W_{20}(HI)$	i	$r_{25,B}$	$\log w'+9$	B'_T-10	$B-V$	profile	type
p1-2	18.27	268	61°	8.9	4.313	8.27	0.72	exp.	Sc
p1-3	17.15	391	42	11.1	3.842	7.15	0.64	exp.	Sc
p1-4	16.56	104	30	12.9	4.146	6.56	0.86	exp.	Sm
p1-7	15.14	135	30	27.0	3.348	5.14	0.90	exp.	Sc
p2-3	17.42	112	44	9.0	4.583	7.42	0.44	exp.	Sbc
p3-3	17.29	108	47	10.9	4.455	7.29	-	exp.	Im
p5-1	17.46	170	60	13.4	4.152	7.46	0.48	exp	Sm
p5-4	18.11	168	65	10.7	4.372	8.11	0.70	exp	Sc
p5-5	18.58	71	43	7.4	4.943	8.58	-	King	Im
p6-4	15.61	215	66	41.1	3.100	5.61	0.67	exp.	Sbc
p9-1	17.14	84	37	9.8	4.572	7.14	-	exp.	Sc
p9-4	18.86	91	63	6.8	5.025	8.86	-	exp.	Sc
c1-6	16.77	173	49	12.7	3.701	5.70	0.88	King	Sc
c3-6	16.41	218	67	22.2	3.202	5.34	0.52	exp.	Sbc
c4-2	17.69	103	49	11.1	4.043	6.62	-	King	Im
c4-3	17.48	170	53	14.3	3.630	6.41	0.49	exp.	Sm
c6-1	17.53	385	45	9.6	3.568	6.46	-	exp.	Sc
c8-1	18.02	144	55	10.9	3.949	6.95	0.58	exp.	Sc

Table 3: Distance of clusters

	Virgo	Pegasus	Cancer	Coma	Local Group
R (km/sec)	1333	3581	5874	7461	–
note		R:	adopted	from Faber et al. (1989)	0.7 Mpc

Table 4: Pegasus and Cancer E galaxies

ID	B_T	$\log D_{25}('')$	$\log \sigma_0$	$\log w' + 9$	$B'_T - 10$
N7562	12.32	2.138	2.385	2.142	2.32
N7613	14.38	1.788	2.142	3.085	4.38
N7619	12.00	2.238	2.528	1.799	2.00
N7626	11.99	2.168	2.369	2.098	1.99
N2563	12.84	2.148	2.416	1.661	1.77

Appendix B: Prominent Star-Clusters of Special Interests in the South-Western Quadrant of the M31 Disk

The prominent star-cluster candidates in the south-western quadrant of the M31 disk, which are of special interests and discussed in section 7, are listed in Table 5. They are classified into 4 groups of different interests, a ~ d. The members of the first group (a) have an open-cluster-like appearance although they were suspected to be globular clusters in previous studies such as H81. The members in group (b) appear to be small star aggregations or globular-like objects combined with star aggregations, rather than simple globular clusters as previously classified in H81, and those in group (c) look like globular clusters in spite of former classification as open clusters in H81. The objects in last group (d) show similarity in their appearance to galaxies in the background.

The K-numbers are for the preliminary identification list which is used by Kodaira & Miyazaki for the present Suprime-Cam survey. The common bases of 0^h and $+40°$ are omitted from the measured equatorial coordinates (1950.0). The appearances of the objects on the false-color image are briefly described as "compact cluster", "globular-like + stars", "star-aggregation-like", "globular-like", or "non-circular". The size and the brightness found in the present survey are indicated as large (l), medium (m), or small (s), and bright (b), medium (m), or faint (f), respectively. The G- and C-numbers are according to Table A (globular cluster) and B (open cluster) of H81. The identification numbers in B87 are supplemented by their assurance grade, $A \sim E$, with A being the best, and by the comments about the appearance, "very diffuse", "diffuse", "irregular", or "cluster". Magnitude V and color $B - V$ are adopted from B87 whenever available.

Table 5: Prominent star-clusters of special interests in the south-western quadrant of the M31 disk

Present ID	R.A.	Dec.	form	size, bright.	H81	B87	V	$B-V$
a (open-cl.-like Gs)								
K154	38^m884	$30'95$	comp. cl.	l, b	G99	E galaxy?	–	–
K216	38.097	25.08	comp. cl.	m, m	G71	18B	17.83	0.55
K225	37.534	23.53	star aggr.	s, f	G47	189D v. dif.	–	–
K254	38.675	20.33	comp. cl.	m, m	G94	342B dif.	17.78	0.39
K288	38.488	17.83	comp. cl.	m, m	G85	E H II/galaxy	–	–
K291	37.327	17.51	comp. cl.	m, m	G44	319A	17.71	0.19
b (star-aggr.-like Gs)								
K193	37.615	27.51	gl.+star?	l, b	G52	5A	15.67	1.02
K227	37.569	22.86	star aggr.	s, m	G48	190D irreg.	18.38	0.15
K232	37.563	22.60	star aggr.	s, m	G49	322B	17.92	-0.08
K267	38.427	19.42	gl.+star?	m, b	G81	341A	16.34	0.93
K274	38.080	18.63	star aggr.	s, m	G69	452C cl.?	17.87	0.22
K289b	37.290	17.67	gl.+star?	m, m	G42	318A	16.97	0.23
c (globular-like Cs)								
K36	38.145	37.48	globular	m, m	C168	246C dif.	–	–
K261	37.766	19.77	globular	l, b	C107?	195D	15.19	0.20
K299	37.580	16.28	globular	m, m	C88	323A	17.74	0.93
d (galaxy-like objects)								
K241	37.688	21.36	non-circ.	l, b	–	192D	16.85	1.27
K242	37.708	21.22	non-circ.	m, m	C112	–	–	–

Astronomische Gesellschaft: Reviews in Modern Astronomy **15**, 27–56 (2002)

Ludwig Biermann Award Lecture

X-ray Evidence for Supermassive Black Holes at the Centers of Nearby, Non-active Galaxies

Stefanie Komossa

Max-Planck-Institut für extraterrestrische Physik
Postfach 1312, 85741 Garching, Germany
skomossa@mpe.mpg.de, http://www.xray.mpe.mpg.de/~skomossa/

Abstract

We first present a short overview of X-ray probes of the black hole region of active galaxies (AGN) and then concentrate on the X-ray search for supermassive black holes (SMBHs) in optically non-active galaxies.

The first part focuses on recent results from the X-ray observatories Chandra and XMM-Newton which detected a wealth of new spectral features which originate in the nuclear region of AGN.

In the last few years, giant-amplitude, non-recurrent X-ray flares have been observed from several non-active galaxies. All of them share similar properties, namely: extreme X-ray softness in outburst, huge peak luminosity (up to $\sim 10^{44}$ erg/s), and the absence of optical signs of Seyfert activity. Tidal disruption of a star by a supermassive black hole is the favored explanation of these unusual events. The second part provides a review of the initial X-ray observations, follow-up studies, and the relevant aspects of tidal disruption models studied in the literature.

1 Introduction

1.1 The search for SMBHs at the centers of galaxies

The study of supermassive black holes and their cosmological evolution is of great interest for a broad range of astrophysical topics including facets of galaxy formation and general relativity. In the last few decades, a number of different methods were developed to search for supermassive black holes (SMBHs) in external galaxies. Their detection in large numbers would clarify our understanding of the early phases of the evolution of galaxies. In *active* galactic nuclei SMBHs are now generally believed to be the prime mover of the non-stellar activity. X-ray observations in the near future are expected to offer the opportunity of detecting some of the distinctive features of strong field gravity, thereby also providing the ultimate proof for the existence of black holes in AGN.

There is now strong evidence for the presence of massive dark objects at the centers of many galaxies. Does this hold for *all* galaxies? If so, why are some SMBHs 'dark'? Questions of particular interest in the context of galaxy/AGN evolution are: When and how did the first SMBHs form and how do they evolve? What fraction of galaxies have passed through an active phase, and how many now have non-accreting and hence unseen supermassive black holes at their centers (e. g. Lynden-Bell 1969, Rees 1989)?

Several approaches were followed to study these questions. Much effort has concentrated on the determination of central object masses from measurements of the *dynamics of stars and gas* in the nuclei of nearby galaxies. Earlier (ground-based) evidence for central quiescent dark masses in galaxies (Kormendy & Richstone 1995) has been strengthened by recent HST results (see Kormendy & Gebhardt 2001 for a review).

A quite accurate determination of black hole mass was enabled by the detection of water vapor maser emission from the mildly active galaxy NGC 4258 (Miyoshi et al. 1995). The water masers, whose motion can be precisely mapped with VLBI, are located in a very compact disk in Keplerian rotation around the central SMBH. The fortunate geometry of the disk, nearly edge-on, allows to obtain the BH mass with high accuracy: $M_{\mathrm{BH}} = 3.6\,10^7$ $M_\odot$ (Neufeld & Maloney 1995, Greenhill et al. 1995).

Still closer to the SMBH, in *active* galaxies with broad line region (BLR hereafter), the technique of BLR reverberation mapping (e. g. Peterson 2001) provides a powerful tool to estimate the BH mass via the clouds' distance from the center and their velocity field (e. g. Peterson & Wandel 2000, Ferrarese et al. 2001).

1.2   X-ray probes of the black hole region of AGN

Whereas the dynamics of stars and gas probe rather large distances from the SMBH, high-energy *X-ray emission* originates from the immediate vicinity of the black hole. In *active* galaxies, excellent evidence for the presence of SMBHs is provided by the detection of luminous hard power-law like X-ray emission, rapid variability, and the discovery of evidence for relativistic effects in the iron-K line profile. X-ray observations currently provide the most powerful way to explore the black hole region of AGN.

X-rays at the centers of AGN arise in the accretion-disk–corona system (e. g. Mushotzky et al. 1993, Svensson et al. 1994, Collin et al. 2000, and references therein). On larger scales, but still within the central region, X-rays might be emitted by a hot intercloud medium at distances of the broad or narrow-line region (e. g. Elvis et al. 1990).

The X-rays which originate from the accretion-disk region are reprocessed in form of absorption and partial re-emission (e. g. George & Fabian 1991, Netzer 1993, Krolik & Kriss 1995, Collin-Souffrin et al. 1996, Komossa & Fink 1997b) as they make their way out of the nucleus. The reprocessing bears the disadvantage of veiling the *intrinsic* X-ray spectral shape, and the spectral disentanglement of many different potentially contributing components is not always easy. However, reprocessing also offers the unique chance to study the physical conditions and dynamical states of the reprocessing material (see Komossa 2001 for a review), like: the outer parts of the accretion disk; the ionized absorber; the torus, which plays an important role in AGN unification schemes (Antonucci 1993); and the BLR and NLR. Detailed modeling of the reprocessor(s) is also necessary to recover the shape and properties of the *intrinsic* X-ray spectrum.

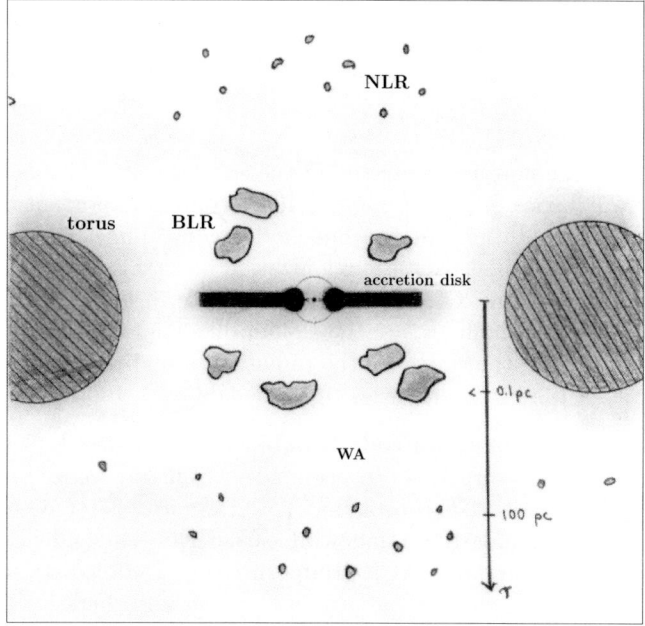

Figure 1: Sketch of the central region of Seyfert galaxies. The black hole and accretion disk region is surrounded by two systems of gas clouds, the broad line region (BLR) and narrow-line region (NLR). These show up by their characteristic line emission in optical spectra of AGN, and their presence is usually used to identify and classify AGN. The molecular torus, and variants of it, are thought to play an important role in unification schemes of Seyfert galaxies by blocking the direct view on the BLR for certain viewing directions of the observer (Antonucci 1993). Somewhere outside the bulk of the BLR, a relatively recently discovered component of the active nucleus is located, the so-called 'warm' or ionized absorber (WA).

A modification of this picture was recently proposed by Elvis (2000, 2001). In his model, the BLR clouds arise from a flow of gas which rises vertically from a narrow range of radii from the accretion disk. The flow then bends and forms a conical wind moving radially outwards. The BLR clouds are identified with the cool phase of this two-phase medium. Warm absorbers appear if we view the continuum source through the wind (Elvis 2000, see his Fig. 1).

Recent progress has been made based on the improved spectral resolution of the new generation of X-ray observatories, *Chandra* and *XMM-Newton*. Both missions have imaging detectors and grating spectrometers aboard. Their energy sensitivity bandpass covers ~(0.1–10) keV. Below, a short review of results from these observatories is given, starting at relatively large distances from the SMBH (NLR), and then moving further inward (warm absorber and accretion disk region).

X-ray emission lines, X-ray narrow-line region. The detection of a high-temperature, narrow-line, X-ray emitting plasma in NGC 4151 was reported by Ogle et al. (2000), confirming earlier evidence for extended X-ray emission from this galaxy (Elvis et al. 1983). The X-ray gas is spatially coincident with the NLR and extended narrow-line region. For the first time, numerous emission-lines were detected in the X-ray spectrum of NGC 4151 with the HETG (High Energy Transmission Grating Spectrometer) aboard *Chandra*.

The X-ray emission lines detected in the spectrum of NGC 4151 and several Seyfert 2 galaxies (e.g. Mrk 3, NGC 1068) contain important information on the physical conditions in the line-emitting medium, like temperature, density, and the main gas excitation/ionization mechanism – photoionization or collisional ionization. Of particular importance in determining the main power mechanism of the lines, are the Helium-like triplets (Gabriel & Jordan 1969; see our Fig. 2), the widths of the radiative recombination continua, and the strengths of the Fe-L complexes (e.g. Liedahl et al. 1990).

X-ray absorption lines, ionized absorber. With *ROSAT*, the signatures of so-called 'warm' absorbers, absorption edges of highly ionized oxygen ions at $E_{O\,VII} = 0.74$ keV and $E_{O\,VIII} = 0.87$ keV, were first detected in MCG−6-30-15 (Nandra & Pounds 1992), following earlier *Einstein* evidence for highly ionized absorbing material in AGN (Halpern 1984). Detailed studies of many other AGN followed, and the signatures of warm absorbers have now been seen in about 50 % of the well-studied Seyfert galaxies (see Komossa 1999 for a review). First constraints place the bulk of the ionized material outside the BLR, and depending on its covering factor and location, the warm absorber may be one of the most massive components of the active nucleus. Evidence for ionized absorption was also found in some very high-redshift quasars, starting with observations of PKS 2351–154 (Schartel et al. 1997). Some (but not all) warm absorbers were suggested to contain dust, based on otherwise contradictory optical–X-ray observations (e.g. Brandt et al. 1996, Komossa & Fink 1997b, Komossa & Bade 1998). The first possible detection of Fe-L dust features in the X-ray spectra of MCG−6-30-15 and Mrk 766 was recently reported by Lee et al. (2001) and Lee (2001).

The high-resolution spectrum of the Seyfert galaxy NGC 5548 obtained with the *Chandra* Low Energy Transmission Grating Spectrometer (LETGS) shows many narrow absorption lines of highly ionized metal ions of oxygen, neon, iron, etc. (Fig. 2) confirming the presence of a warm absorber in this galaxy (Kaastra et al. 2000). Similar signatures of ionized material have

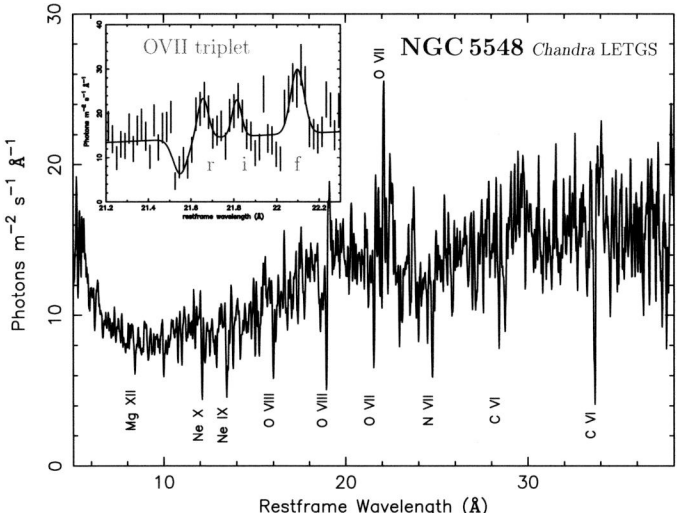

Figure 2: *Chandra* LETGS X-ray spectrum of NGC 5548 (Kaastra et al. 2000). The inset shows a zoom of the O VII triplet to which a resonance line, two intercombination lines (unresolved), and a dipole-forbidden line contribute.

been detected with *Chandra* and *XMM-Newton* in several AGN including NGC 3783 (Kaspi et al. 2000), IRAS 13349+2438 (Sako et al. 2001), NGC 4051 (Collinge et al. 2001), Mrk 509 (Yaqoob et al. 2002), and MCG −6-30-15 (see next paragraph). First results show that the ionized absorption is complex with a range in ionization states.

Assuming that the ionized absorber outflow is driven by radiation pressure of the central continuum sources, Morales & Fabian (2001) demonstrated that observations can then be used for an estimate of black holes masses in AGN. They derive masses of $M_{BH} \simeq 10^{6.5-7}$ $M_\odot$ for the galaxies of their sample.

Accretion-disk region, Fe-K line. The most direct probe of the black hole region and particularly of special and general relativistic effects is emission from the inner part of the accretion disk (see Fabian 2001 for a review). Tanaka et al. (1995) reported the detection of a broadened FeKα line in MCG−6-30-15. The line profile is well explained by the special relativistic effects of beaming and transverse Doppler effect, and the general relativistic effect of gravitational redshift. Depending on details of modeling the continuum, broad-winged Fe lines may also be present in several XMM spectra of AGN (Nandra 2001). At certain times, the red wing of MCG−6-30-15 is very broad extending down to very soft energies (Wilms et al. 2001).

With *XMM* it has also become clearer that the Fe-line profiles are complex and the line has several sides of formation including likely the BLR (NGC 5548), the torus (NGC 3783, Mrk 205), the X-ray ionization cone of NGC 1068, and a contribution from the outer parts of the accretion disk (MCG−6-30-15).

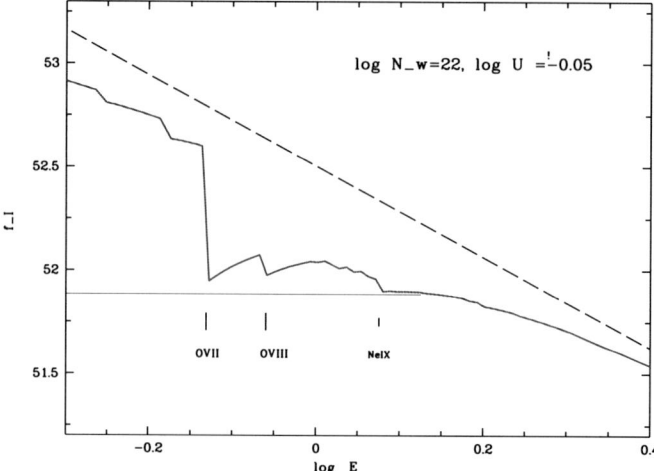

Figure 3: Soft X-ray spectrum of an AGN, plotted as log flux [a. u.] versus log Energy [keV]. The dashed line shows the input continuum spectrum. The thick solid line gives the spectrum after passage of a warm absorber. The calculation was carried out with the photoionization code *Cloudy* (Ferland 1993). Input parameters (ionization parameter U, column density N_w) were chosen similarly to those obtained from a Beppo-SAX observation of Mrk 766 (Matt et al. 2000) except that U was lowered. Only the absorption edges are shown, and are labeled in the graph (absorption lines were omitted). If this theoretical absorption spectrum is now re-fit, without knowledge of the intrinsic continuum (its shape and level), t wo fundamentally different solutions are possible: (i) a high-level steep continuum in which case the absorption-solution is recovered, or (ii) a low-level flat continuum (the horizontal thin solid line in the graph) in which case the presence of a strong soft excess (at lower energies than the O VII edge) plus some very broad emission lines are inferred.

The case of MCG −6-30-15, X-ray spectral complexity. Whereas the signature of an ionized absorber in the form of narrow absorption lines was detected in this galaxy with both, XMM-Newton (Branduardi-Raymont et al. 2001) and *Chandra* (Lee et al. 2001), new interpretations of some of the spectral features were put forward: Branduardi-Raymont et al. suggested that the dominant soft X-ray features, so far interpreted as metal absorption edges of the warm absorber, can be better understood in terms of relativistically broadened emission lines which originate in the accretion disk. On the other hand, Matsumoto & Inoue (2001) noted that the *ASCA*-detected broad wing of the iron K line – so far interpreted in terms of relativistic broadening due to the line's origin in the inner parts of the accretion disk – could be successfully modeled by invoking a t wo-component warm absorber.

Fig. 3 summarizes and visualizes one of the basic underlying ideas in the discussion of emission versus absorption features at soft X-ray energies (for many additional details see Branduardi-Raymont et al. 2001 and Lee et al. 2001): Depending on where the continuum is placed in Fig. 3, one would either infer the presence of *a huge soft excess plus broad emission lines*, or a powerlaw spectrum modified by *absorption edges*. We do not discuss these ideas further here except for noting that a disk-line interpretation of the bulk of the soft X-ray features of MCG−6-30-15 would leave the puzzle of the discrepant optical and X-ray absorption of this galaxy unanswered, which could be solved by introducing a dusty warm absorber (Reynolds et al. 1997).

Deep high-resolution X-ray observations of, and a search for, variability of the spectral features will be a very important next step in disentangling all components of the complex X-ray spectrum of this galaxy.

1.3 The X-ray search for SMBHs in ULIRGs and LINERs

It is now generally believed that *active* galactic nuclei (AGN) are powered by accretion onto SMBHs. The search for heavily obscured SMBHs in ultraluminous infrared galaxies (ULIRGs), and for low-luminosity AGN (LLAGN) in LINER galaxies is another interesting topic. It will only be briefly touched here, since the emphasis of this review will be on recent evidence for SMBHs in non-active 'normal' galaxies (next Section).

1.3.1 ULIRGs

ULIRGs, characterized by their huge power-output in the infrared which exceeds $10^{12} L_\odot$ (Sanders & Mirabel 1996), are powered by massive starbursts or SMBHs. The discussion, which one actually dominates received a lot of attention in recent years (e.g. Joseph 1999, Sanders 1999). In particular, only a small fraction of ULIRGs show AGN signatures in their optical and infrared spectra. Do the remaining ones nevertheless harbor AGN? X-ray variability and luminous hard X-ray emission are excellent indicators of obscured AGN activity.

With a redshift $z = 0.024$ and a far-infrared luminosity of $\sim 10^{12} L_\odot$, NGC 6240 is one of the nearest members of the class of ULIRGs. Whereas X-rays from distant Hyperluminous IR galaxies, HyLIRGs, were not detected by Wilman et al. (1999), and the ULIRGs in the study of Rigopoulou et al. (1996) were X-ray weak, NGC 6240 turned out to be exceptionally X-ray luminous. Starburst-driven superwinds are the most likely interpretation of the extended emission (see Schulz & Komossa 1999 for alternatives), albeit being pushed to their limits to explain the huge power output (Schulz et al. 1998). The hard spectral component present in the *ROSAT* energy band was interpreted as scattered emission from an obscured AGN (Schulz et al. 1998, Komossa et al. 1998) which shows up more clearly at higher energies, up to 100 keV (e.g. Vignati et al. 1999, Mitsuda 1995, Ikebe et al. 2000). The intrinsically luminous AGN ($L_x \approx 10^{44}$ erg/s), can account for at least a substantial fraction of the FIR power output of NGC 6240.

Using *ASCA*, Nakagawa et al. (1999) studied the hard X-ray properties of a sample of 10 ULIRGs. Among these 50 % have hard X-ray detections.

The most stringent upper limit for the presence of any hard X-ray emission was reported for Arp 220. The possibility that it is a Compton-thick source cannot be excluded, though.

1.3.2 LINERs

LINER (Low-Ionization Nuclear Emission-Line Region) galaxies are characterized by their optical emission line spectrum which shows a lower degree of ionization than AGN. Their major power source and line excitation mechanism have been a subject of lively debate ever since their discovery. LINERs manifest the most common type of activity in the local universe. If powered by accretion, they probably represent the low-luminosity end of the quasar phenomenon, and their presence has relevance to e.g. the evolution of quasars, the faint end of the Seyfert luminosity function, the soft X-ray background, and the presence of SMBHs in nearby galaxies.

The X-ray properties of LINERs are inhomogeneous. Spectra of a sample of objects studied by Komossa et al. (1999) are best described by a composition of soft thermal emission and a powerlaw with varying relative contributions of the two components from object to object. Several studies of individual objects are consistent with these results (e.g. Mushotzky 1982, Koratkar et al. 1995, Cui et al. 1997, Ptak et al. 1999, Roberts et al. 1999). X-ray luminosities are in the range $\sim 10^{38-41}$ erg/s; below those typically observed for Seyfert galaxies. The general absence of short-time scale (hours–weeks) X-ray variability (Ptak et al. 1998, Komossa et al. 1999) is consistent with the suggestion that LINERs accrete in the advection-dominated mode (e.g. Yi & Boughn 1998, 1999, and references therein). However, clear positive X-ray detections of LLAGNs in LINERs are still rare. One potential problem is to distinguish powerlaw emission of the X-ray binary population of the host galaxy from that of a genuine LLAGN. First *Chandra* results on LINERs show that few, if any, are obscured by absorbers of high column density (Ho et al. 2001). Four out of eight LINERs of that study possess compact nuclear cores, consistent with AGNs.

1.4 The X-ray search for SMBHs in *non-active* ('normal') galaxies and tidal disruption flares as probes

How can we find *dormant* SMBHs in *non-active* galaxies? Lidskii & Ozernoi (1979) and Rees (1988) suggested to use the flare of electromagnetic radiation produced when a star is tidally disrupted and accreted by a SMBH as a means to detect SMBHs in nearby *non-active* galaxies.

A star on a near-radial 'loss-cone' orbit gets tidally disrupted once the tidal gravitational forces exerted by the black hole exceed the self-gravitational force of the star (e.g. Hills 1975, Lidskii & Ozernoi 1979, Diener et al. 1997). The tidal radius is given by

$$r_{\mathrm{t}} \simeq 7\,10^{12}\,\Big(\frac{M_{\mathrm{BH}}}{10^6\,M_\odot}\Big)^{\frac{1}{3}}\Big(\frac{M_*}{M_\odot}\Big)^{-\frac{1}{3}}\frac{r_*}{r_\odot}\ \mathrm{cm}\,. \tag{1}$$

The star is first heavily deformed, then disrupted. About 50 %–90 % of the gaseous debris becomes unbound and is lost from the system (e. g. Young et al. 1977, Ayal et al. 2000). The rest will eventually be accreted by the black hole (e. g. Cannizzo et al. 1990, Loeb & Ulmer 1997). The stellar material, first spread over a number of orbits, quickly circularizes (e. g. Rees 1988, Cannizzo et al. 1990) due to the action of strong shocks when the most tightly bound debris interacts with other parts of the stream (e. g. Kim et al. 1999). Most orbital periods will then be within a few times the period of the most tightly bound matter (e. g. Evans & Kochanek 1989; see also Nolthenius & Katz 1982, Luminet & Marck 1985).

Explicit predictions of the emitted spectrum and luminosity during the disruption process and the start of the accretion phase are still rare (see Sect. 3 for details). The emission is likely peaked in the soft X-ray or UV portion of the spectrum, initially (e. g. Rees 1988, Kim et al. 1999, Cannizzo et al. 1990; see also Sembay & West 1993).

2   Giant-amplitude X-ray flares from *non-active* galaxies

2.1   Summary of the original X-ray observations

With the X-ray satellite *ROSAT* some rather unusual observations have been made in the last few years: the detection of giant-amplitude, non-recurrent X-ray outbursts from a handful of *optically non-active* galaxies, starting with the case of NGC 5905 (Bade et al. 1996, Komossa & Bade 1999). Based on the huge observed outburst luminosity the observations were interpreted in terms of tidal disruption events. Below, we first give a brief review of the properties of all published X-ray flaring non-active galaxies and then discuss the favored outburst scenario. A Hubble constant of $H_0 = 50$ km/s/Mpc is adopted throughout the paper.

So far, four X-ray flaring *non-active* galaxies have been detected: NGC 5905 (Bade et al. 1996, Komossa & Bade 1999; see Fig. 4), RX J1242−1119 (Komossa & Greiner 1999), RX J1624+7554 (Grupe et al. 1999), and RX J1420 +5334 (Greiner et al. 2000)[1]; first results on a fifth candidate were presented by Reiprich & Greiner (2001). Based on the position they report we refer to this source as RX J1331−3243. All these galaxies show similar properties:

- huge X-ray peak luminosity (up to $\sim 10^{44}$ erg/s),

- giant amplitude of variability (up to a factor ~ 200),

- ultra-soft X-ray spectrum ($kT_{bb} \simeq 0.04$–0.1 keV when a black body model is applied).

---

[1] The X-ray position error circle of RX J1420+53 contains a second galaxy for which an optical spectrum is not yet available. Based on the galaxy's morphology, Greiner et al. argue that it is likely non-active.

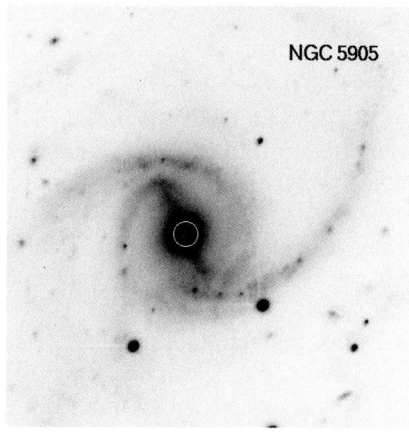
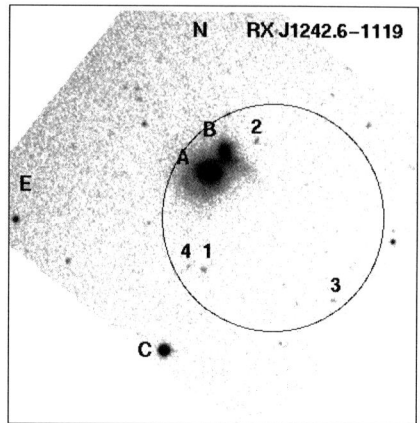

Figure 4: Optical images of NGC 5905 (right) and RX J1242–1119 (left). The circles mark the positional uncertainty of the X-ray emission.

A summary of the observations is provided in Table 1. In Fig. 7 X-ray lightcurves of NGC 5905 and RX J1420+53 are overplotted, shifted in time to the same date of outburst to allow direct comparison. So far, the best sampled lightcurve is that of NGC 5905. The 'merged' lightcurve is consistent with a fast rise and a decline on a time scale of months to years. We performed a preliminary analysis of an archival *ASCA* observation of NGC 5905 carried out in 1999. The flux of NGC 5905 did not drop further compared to the last *ROSAT* observations.

2.2   Optical observations

Given the unusual X-ray properties of these galaxies a very important question was: what is the optical classification of the flaring galaxies. Particularly: are there any hints of weak permanent Seyfert activity? This question is of great interest when discussing outburst scenarios[2].

Optical spectra were taken at different times and at different telescopes starting several years after the X-ray high-states. The optical spectrum of NGC 5905 turned out to be of H II-type consistent with its classification prior to the observed X-ray outburst. Neither broad wings in the Balmer lines nor AGN-typical emission-line ratios, nor high-ionization lines, which usually indicate the presence of an AGN, were detected. The spectra of the other galaxies only show absorption lines from the host galaxies.

---

[2] For instance, theorists working on tidal disruption of stars repeatedly argued that it would be important to be certain about the (optically) *non-active* nature of the flaring galaxy, to exclude AGN-related variability mechanisms (changes in the accretion disk).

Table 1: Summary of the X-ray and optical properties of the flaring normal galaxies during outburst. z gives the redshift, T_{bb} is the black body temperature derived from a black body fit to the X-ray high-state spectrum (cold absorption was fixed to the Galactic value in the direction of the individual galaxies), 'no emi.' means: no optical emission lines were detected. $L_{x,bb}$ gives the intrinsic luminosity in the (0.1–2.4) keV band based on the black body fit. This is a lower limit to the actual peak luminosity since we most likely have not caught the sources exactly at maximum light, since the spectrum may extend into the EUV, and since it was conservatively assumed that no additional X-ray absorption occurs intrinsic to the galaxies.

galaxy name	z	opt. type	kT_{bb} [keV]	$L_{x,bb}$ [erg/s]
NGC 5905	0.011	H II	0.06	$3\ 10^{42}*$
RX J1242−1119	0.050	no emi.	0.06	$9\ 10^{43}$
RX J1624+7554	0.064	no emi.	0.097	$\sim 10^{44}$
RX J1420+5334	0.147	no emi.	0.04	$8\ 10^{43}$
RX J1331−3243	0.051	no emi.		

*Mean luminosity during the outburst; since the flux varied by a factor $\sim$3 during the observation, the peak luminosity is higher.

2.3   Radio observations

Radio observations are important for two reasons: Firstly, they allow the search for a peculiar, optically hidden AGN at the center of each flaring galaxy (as already discussed by Komossa & Bade (1999) this possibility is a very unlikely explanation for the X-ray flares. It is very important, though, to safely *exclude* exotic AGN scenarios). Besides hard X-ray observations, compact radio emission is a good indicator of AGN activity because radio photons can penetrate even high-column density dusty gas which is not transparent to optical or soft X-ray photons. Secondly, radio emission could possibly be produced in relation to the X-ray flare itself.

2.3.1   NVSS and FIRST search
for radio emission from the X-ray outbursters

A search for radio emission from the X-ray flaring galaxies was performed utilizing the NRAO VLA Sky Survey (NVSS) catalogue (Condon et al. 1998), which contains the results of a 1.4 GHz radio sky survey north of δ=−40°. Except NGC 5905 no flaring galaxy has a NVSS detection. At 1.4 GHz the emission of NGC 5905 appears extended and is thus related to the galaxy instead of the nucleus (see also next Section).

Table 2: Coordinates (J 2000) of the optical centers of the galaxies identified as counterparts to the X-ray flares (NGC 5905, RX J1242−1119A, RX J1624+7554) or X-ray position (RX J1420+5334, RX J1331−3243).

galaxy name	coordinates	
	RA	DEC
NGC 5905	$15^h15^m23.4^s$	$+55°31'02''$
RX J1242−1119A	$12^h42^m38.5^s$	$-11°19'21''$
RX J1624+7554	$16^h24^m56.5^s$	$+75°54'56''$
RX J1420+5334	$14^h20^m24.2^s$	$+53°34'11''$
RX J1331−3243	$13^h31^m57.6^s$	$-32°43'20''$

NGC 5905 is also detected in the FIRST VLA sky survey at 20cm (e. g. Becker et al. 1995). No radio emission from RX J1420+53 was found. The FIRST catalogue detection limit at the source position is 0.96 mJy/beam. None of the other outbursters is located within a FIRST survey field.

2.3.2   Radio emission from NGC 5905

21 cm neutral hydrogen line emission was detected by van Moorsel (1982; see also Staveley-Smith & Davies 1987) using the Westerbork Synthesis Radio Telescope (WSRT). The emission is spatially resolved (Fig. 3 of van Moorsel 1982) with an extent of diameter 7.3′. Peaks in the H I emission closely follow the spiral arms. Whereas the bulk of the radio emission detected at the frequency of the 21cm line is unrelated to the nucleus, van Moorsel also briefly mentions the presence of unresolved continuum emission of 13.2 mJy.

Extended radio emission was also found by Hummel et al. (1987) at 1.49 GHz, whereas Brosch & Krumm (1984) reported upper limits at 5 GHz (for both, extended emission and a nuclear source). Israel & Mahoney (1990) give an upper limit at 57.5 MHz (see our Fig. 5).

Hummel et al. (1987) reported the presence of an unresolved ($\lesssim 2''$) core source at a frequency of 1.49 GHz. Similar sources were found in 41 % of the H II galaxies of the 'complete sample' of Hummel et al. (1987; their Tab. 1).

Finally, we note that the NVSS and FIRST surveys were performed after the X-ray outburst of NGC 5905. However, the NVSS value is consistent with previous measurements of extended radio emission at that frequency. Similarly, the FIRST value is consistent with the pre-flare measurement of Hummel et al. (1987) at the same frequency.

In order to search for a radio source at the nucleus of NGC 5905 after the X-ray outburst (besides a permanent AGN, radio emission could be produced in relation to the X-ray flare) a radio observation with the VLA A array at 8.46 GHz was carried out by M. Dahlem in 1996. No radio source was detected

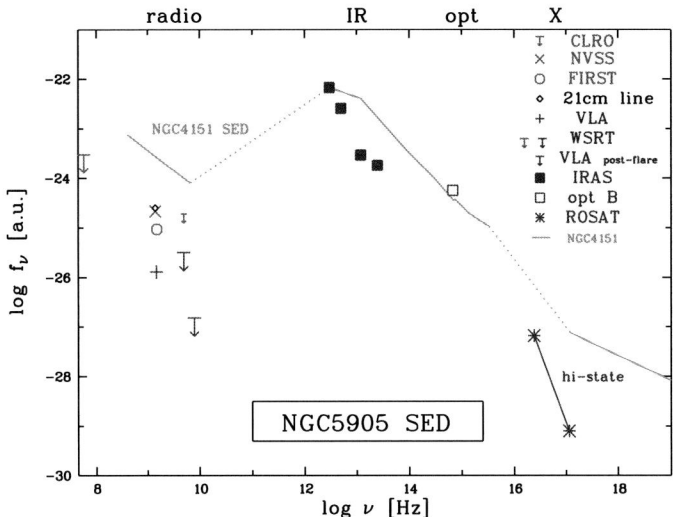

Figure 5: Multi-wavelength continuum spectrum (SED) of NGC 5905 (symbols). The SED of the active galaxy NGC 4151 (Komossa 2001) is shown for comparison by the solid/dotted line. Arrows denote upper limits. Radio data, from left to right: arrow: Clark Lake Radio Observatory TPT array (Israel & Mahoney 1990); lozenge: 21 cm line (van Moorsel 1982); cross: NVSS survey; open circle: FIRST survey; plus: VLA (Hummel et al. 1987); upper limits: WSRT (upper point: total emission within 2.4′, lower point: nuclear emission within 10″; Brosch & Krumm 1984); upper limit: VLA post-flare observation (Komossa & Dahlem 2001); filled squares: IRAS (taken from NED data base), open square: optical B-magnitude (taken from NED), asterisks: X-ray high-state emission (Komossa & Bade 1999). Note: data were taken with different aperture sizes and resolution and at different times.

within the central field of view of $100″×100″$ with a 5σ upper limit for the presence of a central point source of 0.15 mJy (Komossa & Dahlem 2001). Assuming a distance of 75.4 Mpc of NGC 5905 this translates into an upper limit of $L_{8.46GHz} \leq 1.0\,10^{20}$ W/Hz.

The radio measurements of NGC 5905 are summarized in Fig. 5 together with multi-wavelength observations.

2.3.3   Origin of the radio emission of NGC 5905

A large contribution to the radio emission of NGC 5905 comes from the 21cm line of neutral hydrogen. In addition there is some radio continuum emission at the same frequency and at 1.49 GHz (FIRST). No other radio detection was reported.

The radio emission of large samples of spiral galaxies was examined by e.g. Brosch & Krumm (1984), Hummel et al. 1987, Giuricin et al. (1990), Israel & Mahoney (1990), Sadler et al. (1995), and Falcke (2001). Radio emis-

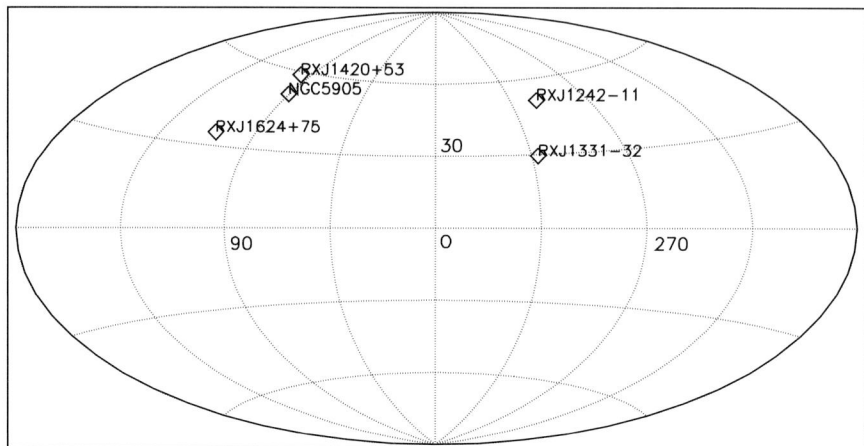

Figure 6: Locations of the X-ray flaring galaxies on the sky (Galactic II coordinates in aitoff projection).

sion (extended and from the inner few arcseconds) was generally detected from a number of the non-active spiral galaxies in the samples. E.g. Hummel et al. find unresolved ($\lesssim 2''$) core sources in 41 % of the H II galaxies of their 'complete sample' at 1.49 GHz. At 57.5 MHz, Israel & Mahoney (1990) detected 68 out of 133 observed galaxies. Trends were repeatedly reported in the literature, that paired H II galaxies (like NGC 5905) and interacting galaxies show enhanced total and central radio emission compared to isolated galaxies (e. g. Giuricin et al. 1990, and references therein). Enhanced star-formation activity was considered a possible explanation of this effect. In some cases nuclear radio sources in spirals could possibly be explained by radio supernovae (Sadler et al. 1995).

In summary, the radio emission of NGC 5905 is not unusual for its class and does not indicate the presence of a luminous, optically hidden AGN[3].

3   Favored outburst scenario:
tidal disruption of a star by a SMBH

3.1   (Rejected or unlikely) alternatives to tidal disruption: stellar sources, lensing, GRB, hidden AGN

Most examined outburst scenarios do not survive close scrutiny (Komossa & Bade 1999), because they cannot explain the huge maximum luminosity (e. g. X-ray binaries within the galaxies, or a supernova in a dense medium), are

---

[3]The radio observations do not exclude the presence of some low-level nuclear radio activity related to a low-luminosity AGN, like it may be present for instance in many LINERs (e. g. Falcke 2001, and references therein).

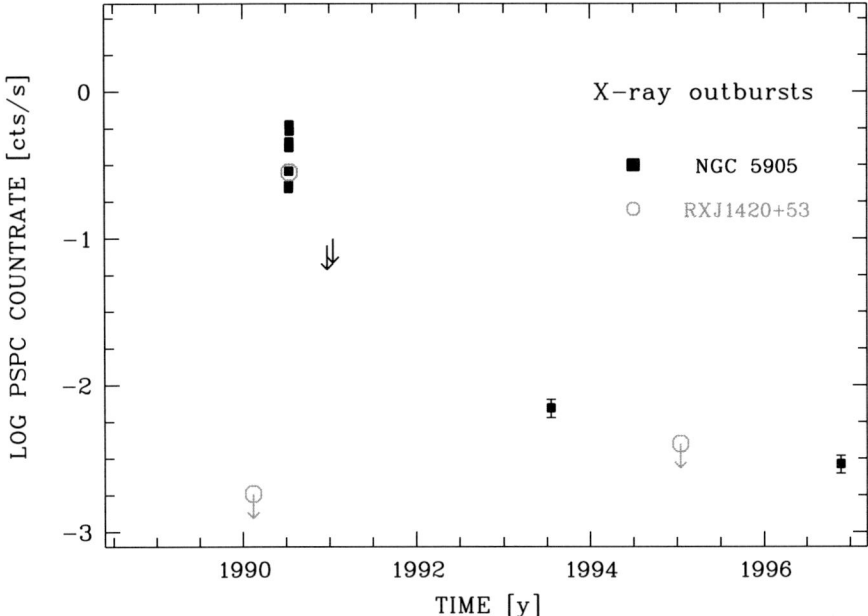

Figure 7: X-ray light curve of NGC 5905 (black squares), and RX J1420+53 (open circles; shifted in time to match outburst date of NGC 5905). Arrows denote upper limits. A preliminary analysis of an archival *ASCA* observation of NGC 5905 carried out in 1999 shows that the source flux did not drop further compared to the last *ROSAT* observations.

inconsistent with the optical observations (gravitational lensing), or predict a different temporal behavior (X-ray afterglow of a Gamma-ray burst; see e. g. Fig. 2 of Bradt et al. 2001). Standard AGN scenarios cannot account for the soft X-ray flares and the absence of optical AGN-like emission lines (see the discussion by Komossa & Bade 1999 and Komossa & Voges 2001 for more details).

3.2   Tidal disruption model

Except possibly for some GRB-related emission mechanisms, the huge peak outburst luminosity nearly inevitably calls for the presence of a SMBH[4]. This, in combination with the complete absence of any signs of AGN activity at all wavelengths, makes tidal disruption of a star by a SMBH the favored outburst mechanism.

After a short overview of aspects of tidal disruption models discussed in the literature, we apply these models to the X-ray flare observations.

---

[4] In fact, even most GRB scenarios involve the presence of a black hole as ultimate energy reservoir

3.2.1   Tidal disruption, short overview

Historically, tidal disruption of captured stars by black holes was first considered in relation to star clusters and galactic nuclei (e. g. Frank & Rees 1976),
and was applied to the nuclei of active galaxies where it was suggested as a
means of fueling AGN (e. g. Hills 1975, Sanders 1984), or to explain UV–X-ray
variability of AGN (e. g. Kato & Hoshi 1978). Shields & Wheeler (1978) then
argued that tidal disruptions of captured stars cannot provide an effective
source of fuel of AGN, basically because of two problems: Firstly, the disruption rate is not high enough to sustain a permanent gas flow, and low-angular
momentum orbits are quickly depleted of stars. If, however, the stellar density
close to the black hole is high enough, stellar collisions would dominate the
gas supply over tidal disruption. Secondly, it is difficult to account for the
luminosity of the most luminous quasars since these have masses where the
tidal radius is inside the Schwarzschild radius.

A star will only be disrupted if its tidal radius lies outside the Schwarzschild
radius of the black hole, else it is swallowed as a whole. This happens for black
hole masses larger than $\simeq 10^8$ $M_\odot$; in case of a Kerr black hole, tidal disruption
may occur for even larger BH masses if the star approaches from a favorable
direction (Beloborodov et al. 1992). More massive black holes may still strip
the atmospheres of giant stars.

Due to the complexity of the problem, theoretical work focused on different subtopics of the complete problem, and on stars of solar mass and
radius. Calculations and numerical simulations of the disruption process, the
stream-stream collision, the accretion phase, the changes in the stellar distribution of the surroundings and the depopulation and refilling of low-angular
momentum orbits, and the disruption rates have been studied in the literature (e. g. Nolthenius & Katz 1982, 1983, Carter & Luminet 1985, Luminet &
Marck 1985, Evans & Kochanek 1989, Laguna et al. 1993, Diener et al. 1997,
Ayal et al. 2000, Ivanov & Novikov 2001, Kochanek 1994, Lee et al. 1995,
Kim et al. 1999, Hills 1975, Gurzadyan & Ozernoi 1979, 1980, Cannizzo et al.
1990, Loeb & Ulmer 1997, Ulmer et al. 1998, Frank & Rees 1976, Lightman &
Shapiro 1977, Norman & Silk 1983, Sanders & van Oosterom 1984, Rauch &
Ingalls 1998, Rauch 1999, Syer & Ulmer 1999, Magorrian & Tremaine 1999).
Di Stefano et al. (2001) recently considered the case of $M > M_\odot$, and sug-
gested that remnants of tidally stripped stars might be detected as supersoft
X-ray sources at the centers of nearby galaxies.

3.2.2   Tidal disruption in *active* galaxies (?)

During the last several years, tidal disruption was occasionally considered as
explanation of some properties of *active* galaxies (either AGN as a class, or
individual peculiar observations) although alternative interpretations existed
in each case: Tidal disruption was applied by Eracleous et al. (1995) in a duty
cycle model to explain the UV brightness/darkness of LINERs. Roos (e. g.

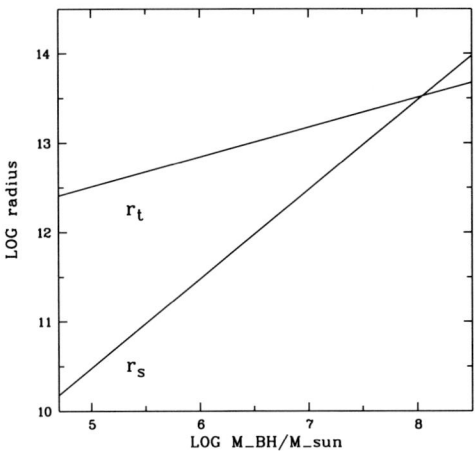

Figure 8: Run of tidal radius and Schwarzschild radius in dependence of black hole mass, for a star of solar mass and radius. For large SMBH masses, stars are swallowed whole, since the tidal radius is no longer outside the Schwarzschild radius.

1992) suggested an origin of the BLR clouds of AGN in terms of the gaseous debris of tidally disrupted stars. Peterson & Ferland (1986) proposed tidal disruption as a possible explanation for the transient brightening and broadening of the He II line observed in the Seyfert galaxy NGC 5548. Brandt et al. (1995) and Grupe et al. (1995) reported the detection of an X-ray outburst from the active galaxy Zwicky 159.034 (IC 3599). Besides other outburst mechanisms, tidal disruption was briefly mentioned as a possibility. Based on high-resolution post-outburst optical spectra Komossa & Bade (1999) classified IC 3599 as Seyfert type 1.9. In the UV spectral region t wo UV spikes were detected at and near the center of the elliptical galaxy NGC 4552. The central flare, although rather weak, was interpreted by Renzini et al. (1995) as accretion event (the tidal stripping of a star's atmosphere by a SMBH, or the accretion of a molecular cloud). There are several indications (e. g. from radio observations), that NGC 4552 shows permanent low-level activity.

3.2.3   Emission of radiation, temporal evolution, and model uncertainties

Intense electromagnetic radiation will be emitted in three phases of the disruption and accretion process: Firstly, during the stream-stream collision when different parts of the bound stellar debris first interact with themselves (Rees 1988). Kim et al. (1999) have carried out numerical simulations of this process and find that the initial burst due to the collision may reach a luminosity of 10^{41} erg/s, under the assumption of a BH mass of 10^6 M$_\odot$ and a star of solar mass and radius. Secondly, radiation is emitted during the accretion of the stellar gaseous debris. Finally, the unbound stellar material leaving the system may shock the surrounding interstellar matter and cause intense emission like in a supernova remnant (Khokhlov & Melia 1996).

Many details of the tidal disruption and the related processes are still unclear. In particular, the flares cannot be standardised. Observations would depend on many parameters, like the type of disrupted star, the impact parameter, the spin of the black hole, effects of relativistic precession, and the complication of radiative transfer by effects of viscosity and shocks (Rees 1990). Uncertainties also include the amount of the stellar material that is accreted (part may be ejected as a thick wind, or swallowed immediately). Related to this is the duration of the flare-like activity, which may be months or years to tens of years (e. g. Rees 1988, Cannizzo et al. 1990, Gurzadyan & Ozernoi 1979) followed by a decline on a longer time scale. The flare duration depends on how fast the stellar debris circularizes and how fast it accretes. If both time scales are short and the material accretes at the same rate as it falls back towards the black hole, the decline time scale of the flare scales as

$$\frac{dm}{dt} \propto t^{-\frac{5}{3}} \tag{2}$$

(e. g. Evans & Kochanek 1989, Rees 1990). Otherwise energy output will be spread more evenly in time (Cannizzo et al. 1990).

Another uncertainty in predicting the flare spectrum arises from the mode of accretion: does it proceed via a thin disk (e. g. Cannizzo et al. 1990) or a thick disk (Ulmer et al. 1998), and, if in a thick disk, under which angle do we view it?

3.3 Order of magnitude estimates and consistency checks

3.3.1 Inferences from X-ray observations

Although many details of the actual tidal disruption process are still unclear, some basic predictions have been repeatedly made in the literature how a tidal disruption event should manifest itself observationally:

- (1) the event should be of finite duration (a 'flare'),

- (2) it should be very luminous (up to $L_{\max} \approx 10^{45}$ erg/s in maximum), and

- (3) it should reside in a galaxy which is otherwise perfectly *non*-active (to be sure to exclude an upward fluctuation in gaseous accretion rate of an *active* galaxy).

All three predictions are fulfilled by the X-ray flaring galaxies; particularly by NGC 5905 and RXJ1242−1119, which are the two best-studied cases so far.

In addition, we can do some further order of magnitude estimates and consistency checks. The luminosity emitted if the black hole is accreting at its Eddington luminosity can be estimated by

$$L_{\mathrm{edd}} = \frac{4\pi G M m_{\mathrm{p}} c}{\sigma_{\mathrm{T}}} \simeq 1.3 \times 10^{38} M/M_{\odot} \ \mathrm{erg/s}. \tag{3}$$

In case of NGC 5905, a BH mass of at least a few $\sim 10^4$ M$_{\odot}$ would be required to produce the observed luminosity. This is a *lower limit* on the black hole mass since we likely did not observe L_x at its peak value due to observational gaps in the lightcurve, and since other conservative assumptions were made[5]. A comparison with the SMBH mass of NGC 5905 using indirect *optical* methods is performed in Sect. 3.3.2. For the other galaxies, using again L_{edd}, we infer BH masses reaching up to a few 10^6 $M_\odot$. This is, again, a lower limit. Alternative to a complete disruption event the atmosphere of a giant star could have been stripped.

In a simple black body approximation, the temperature of the accretion disk scales with black hole mass as

$$T \simeq 8\,10^4 \left(\frac{M_{BH}}{M_\odot}\right)^{\frac{1}{12}} \text{K} \ \ (\text{at } r_t), \quad T \simeq 2\,10^7 \left(\frac{M_{BH}}{M_\odot}\right)^{-\frac{1}{4}} \text{K} \ \ (\text{at } 3\,r_S). \quad (4)$$

This gives $T_{r_{tidal}} \simeq 3\,10^5$ K, $T_{3r_S} \simeq 7\,10^5$ K for M=10^6 $M_\odot$ where r_S is the Schwarzschild radius. Using black body fits of the X-ray flare spectra we find temperatures in a similar range; $T_{obs} \simeq (4\text{-}10)\,10^5$ K. Like in AGN, X-ray powerlaw tails could develop. They might have escaped detection during the observations since they are weak or they may develop only after a certain time after the start of the accretion phase. We soon expect first results from a *Chandra* and *XMM* observation of RX J1242−1119, which will give valuable constraints on the post-flare evolution.

The Eddington time scale for the accretion of the stellar material is given by

$$t_{edd} \simeq 4\,\eta_{0.1}(M_{BH}/10^6 M_\odot)(M_*/0.1 M_\odot) \text{ yrs}. \quad (5)$$

Uncertainties in estimating the total duration of the tidal disruption event arise from questions like: how much material is actually accreted or expelled, does a strong wind develop, etc. (see Sect. 3.2.3). The events are expected to last for months to years (e. g. Rees 1988). Observationally, the duration of the events was at least several days, followed by gaps in the observations. The source fluxes were then significantly down several years later (e. g. Fig. 9 of Komossa & Bade 1999). Apart from theoretical uncertainties in the prediction of the decline time scale in the total luminosity output (Sect. 3.2.3) the emission will also likely shift outside the *ROSAT* band, from the X-ray to the EUV-UV band, with increasing time.

Finally, we note that the redshift distribution of the few sources observed so far is consistent with the predicted tidal disruption rate, in the sense that the events are sufficiently distant to define a large volume of space, in which the detection of a few events would be expected.

3.3.2 Non X-ray estimates of the black hole mass in NGC 5905

An optical rotation curve was obtained by Komossa & Bade (1999), which allowed an estimate of the mass enclosed within 0.7 kpc: $M \approx 10^{10}$ $M_\odot$. This

---

[5]E. g. the amount of absorption was fixed to the lowest possible value, N_{Gal}

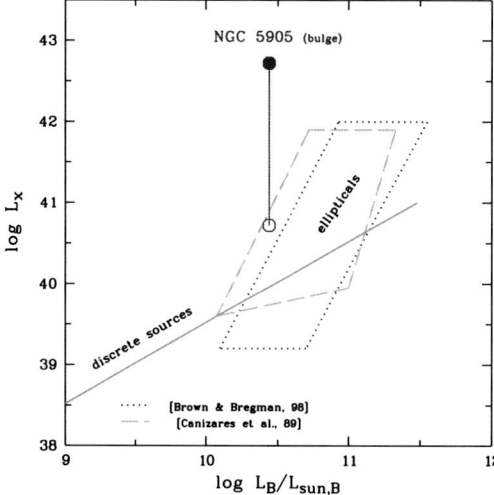

Figure 9: Position of NGC 5905 in the $L_x - L_B$ diagram in outburst and low-state. The dashed lines mark the region populated by some samples of elliptical galaxies.

immediately provides an upper limit on BH mass, but the volume sampled is still too large to estimate the actual BH mass. An HST-based rotation curve would significantly improve the above limit and thus the constraints on a central dark mass.

In order to get a better (non-X-ray) estimate of the BH mass of NGC 5905, we used the correlation between bulge properties and BH mass.

NGC 5905 has a total blue magnitude of $m_{B,0} = 12.1^m$. Using the bulge-to-disk luminosity ratio generally valid for galaxies of the Hubble-type of NGC 5905 (SBb), $k = 0.25$ (Salucci et al. 2000), gives the absolute bulge blue magnitude, $B_{T,0}^{bulge} = -20.5$. We then compared with t wo recent studies that correlate BH mass and bulge blue luminosity: (i) the work of Ferrarese & Merritt (2000; FM00), which concentrates mostly on elliptical galaxies, and (ii) the work of Salucci et al. (2000; S00) on spiral galaxies.

Using the relation between $B_{T,0}^{bulge}$ and M_{BH} of FM00 (their 'sample A', their Tab. 2; see also e. g. Franceschini et al. 1998) gives a BH mass of a few times 10^8 $M_\odot$ for NGC 5905. This is close to the limiting BH mass for tidal disruption of a solar-type star to work. However, it has to be kept in mind that the $B_{T,0}^{bulge}$ - M_{BH} relation shows a very large scatter (in contrast to the M_{BH} - σ relation) and that the few spirals in the sample of FM01 tend to be located below the relation followed by ellipticals. Therefore, in a second step, we used the results of S00 on BH masses in (late-type) spiral galaxies (their Fig. 6). In that case we obtain an upper limit for the BH mass of NGC 5905 of $M_{BH} \lesssim 10^7$ $M_\odot$.

Finally, the relation between radio luminosity and BH mass of Franceschini et al. (1998, their Fig. 3; see also Wu & Han 2000) was employed. The measured radio upper limit for the nucleus of NGC 5905 then translates into an upper limit on black hole mass of $M_{BH} \lesssim 2.5 \, 10^8$ $M_\odot$. Results are summarized in Tab. 3.

Table 3: Summary of mass estimates of the black hole at the center of NGC 5905, employing different methods (see the text for details).

energy band	method	BH mass
X-rays	$L_{Eddington}$	$>$ few 10^4 $M_\odot$
optical	$M_{B,bulge} - M_{BH}$ correlation, spirals	$\approx 10^7$ $M_\odot$
	$M_{B,bulge} - M_{BH}$ correlation, ellipticals	$\approx$ few 10^8 $M_\odot$
	rotation curve	$\ll 10^{10}$ $M_\odot$
radio	nuclear radio power $- M_{BH}$ correlation	$< 2.5\,10^8$ $M_\odot$

3.3.3   Properties of the host galaxies

The expected rate of tidal disruption events is about one event in *at least* $\sim10^4$ years per galaxy (e. g. Magorrian & Tremaine 1999), and the whole *ROSAT* data base has to be employed for a systematic search for further tidal disruption events (for first results see Komossa & Bade 1999, and below).

The disruption rate depends on the efficiency with which the loss-cone orbits are re-filled (Frank and Rees 1976, Lightman and Shapiro 1977, Shields & Wheeler 1978). This can be done by perturbations of the stellar orbits e. g. by merging events (Roos 1981) or a triaxial gravitational potential in the galaxy's core (Norman & Silk 1983).

Are the observed flaring galaxies special in this context (i. e., did any process aid in re-populating the loss-cone orbits)? So far, not much is known about the host galaxies of the few flaring galaxies. The best studied case is NGC 5905. It is interesting to note that this galaxy possesses multiple triaxial structures with a secondary bar (Friedli et al. 1996, Wozniak et al. 1995) which might aid occasional tidal disruption events by disturbing the stellar velocity fields. NGC 5905 is in a pair with NGC 5908.

RX J1242−11 is actually a pair of galaxies at similar redshift, and it is well possible that both galaxies are interacting. The X-ray error circle of another outburster, RX J1420+53, also includes two galaxies. However, a redshift is so far only available for the brighter of the two.

The Hubble types of the flaring galaxies are not known in all cases. NGC 5905, of type SB (Keenan 1937), is one of the largest spiral galaxies known (e. g. Romanishin 1983), whereas some of the other host galaxies look like ellipticals. Deeper optical imaging is presently in progress.

4 Search for further X-ray flares

While we wait for the next-generation of X-ray all-sky surveys, we can still make use of an existing data base which has not yet been fully exploited, the *ROSAT* data base (Voges et al. 1999).

In a first step to search for further cases of strong X-ray variability the sample of nearby galaxies of Ho et al. (1995) and *ROSAT* all-sky survey (RASS) and archived pointed observations were used. The sample of Ho et al. has the advantage of the availability of optical spectra of good quality, which are necessary when searching for 'truly' non-active galaxies. 136 out of the 486 galaxies in the catalogue were detected in pointed observations. The source countrates were then compared with those measured during the RASS.

4.1 Non-active galaxies

No other X-ray flaring, optically in-active galaxy was found. The absence of any further flaring event among the sample galaxies is entirely consistent with the expected tidal disruption rate of one event in $\sim 10^{4-5}$ years per galaxy (e. g. Magorrian & Tremaine 1999).

The next step, presently in progress, will be an extended search for X-ray flaring events based on the whole *ROSAT* all-sky survey database. This approach will allow statistical inferences on the abundance of SMBHs in non-active galaxies (Sembay & West 1993).

4.2 AGN

Several of the sample galaxies show variability by a factor 10–30. All of these are well-known AGN.

Many active galactic nuclei are variable in X-rays with a range of amplitudes, typically a factor 2–3 (e. g. Mushotzky et al. 1993, Ulrich et al. 1997). The cause of variability is usually linked in one way or another to the central engine; for instance by changes in the accretion disk (e. g. Piro et al. 1988, 1997), or by variable obscuration (e. g. Komossa & Fink 1997a, Komossa & Janek 2000). None of the well-studied X-ray variable AGN of the present sample are candidates for tidal disruption events since their lightcurves show recurrent variability on different time scales.

An example for a highly variable AGN among the present sample galaxies is NGC 4051. Its long-term *ROSAT* X-ray lightcurve exhibits variability in countrate by a factor ~ 30. Only a small part of the variability of NGC 4051 can be explained with a variable warm absorber, the rest is likely intrinsic (Komossa & Fink 1997a, Komossa & Janek 2001, and references therein).

Even higher total amplitude of variability is detected in two subsequent *ROSAT* observations of NGC 3516. The X-ray countrate varies by a factor ~ 50 (Komossa & Bade 1999, Komossa & Halpern 2001, in prep.); variable cold absorption likely plays a major part in explaining the observations.

5   Future perspectives

X-ray outbursts from non-active galaxies provide important information on the presence of SMBHs in these galaxies, and the link between active and normal galaxies. One advantage of this method compared to other approaches to search for central dark masses – like HST-based galaxy rotation curves – is that the X-ray flare emission originates from the *very vicinity* of the SMBH. Therefore, potentially it provides much tighter constraints on the black hole mass. Flares can also be detected out to larger cosmological distances.

Follow-up studies: future X-ray surveys.   Future X-ray surveys, like those planned with the *LOBSTER* ISS X-ray all-sky monitor (Fraser 2001), *MAXI* (Yuan et al. 2001, Mihara 2001), and *ROSITA* (Predehl 2001) will be valuable in finding more of these outstanding sources. In addition, a number of flare events are expected to be detected (Yuan et al. 2002) in pointings of the *XMM-Newton* and *Chandra* missions.

On the one hand, all-sky surveys will be important in detecting the brightest events, due to their large areal sky coverage. Those surveys with sensitivity at the softest energies will be most efficient. On the other hand, deeper pointings on limited areas, or 'pencil beam' surveys will increase the number of (more distant, on average fainter) events (see Yuan et al. 2002 for $\log N$–$\log S$ estimates of the expected number of flares detectable with *XMM-Newton*, and Sembay & West 1993 for a general discussion). Concerning deep surveys in limited fields of view, two effects are important. Firstly, there is a limiting maximal brightness the events can reach in the context of the tidal disruption scenario: the most luminous events are those with black hole masses around 10^8 M$_\odot$, with high accretion rate, and the maximal possible fraction emitted in the X-ray band (definite upper limit: $L_{\text{flare}} \lesssim 10^{45-46}$ erg/s). Secondly, there is a limiting distance out to which flare events are detectable because the flare spectra are very soft, and for increasing redshift more and more of the (black-body-like) emission is shifted out of the observable energy band (Fig. 10). In addition, distant galaxies may show a higher intrinsic fraction of cold gas which will heavily absorb at soft X-rays.

After the discovery of new X-ray flare events, *rapid follow-up* multi-wavelength observations will be essential. Apart from valuable new constraints on the favored outburst scenario which would also allow a refinement of model calculations, these observations would enable us to address a number of important topics:

Absorption-line spectroscopy of the IGM/ISM.   As the flare emission travels through the ISM of the host galaxy, and the IGM, absorption features will be imprinted on the X-ray spectrum. These can then be used to study the properties of the absorbing material.

Emission-line spectroscopy of the circumnuclear material.   If the soft X-ray flare emission has an extension into the EUV, which is highly likely, then optical observations will be important in order to detect potential emission

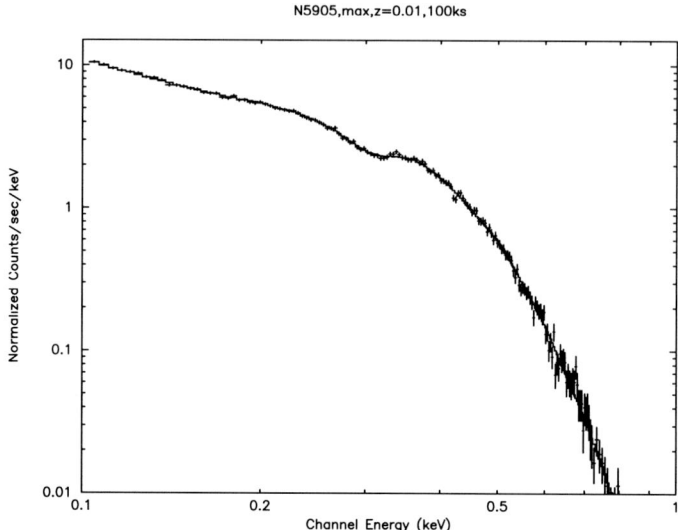

Figure 10: Simulation of an XMM spectrum (EPIC pn instrument) of an X-ray flare similar to the one observed from NGC 5905, assuming a medium-deep exposure of 100 ksec and a single black body spectrum of $kT = 0.05$ ke V absorbed by $N_{Gal} = 1.5\,10^{20}$ cm^{-2} at a redshift of $z = 0.01$. At $z = 0.1$ the very soft black-body-like emission is partly redshifted out of the instrument's energy bandpass, causing a further drop in the observed countrate ($\sim$30 % at $z = 0.1$) in addition to the flux decrease because of the larger distance. The effect increases with redshift such that very distant flares can no longer be detected since shifted out of the observable energy range. Among the X-ray flares, that of NGC 5905 was of lowest luminosity. More luminous flares would therefore be detectable out to much higher redshifts. In addition to the continuum spectrum, we may expect to detect emission features which arise in the accreted material, or absorption features from the ISM of the host galaxy.

lines that were excited by the outburst emission. Firstly, any gaseous material close to the nucleus is expected to show an emission-line response. The time variability of these lines will allow a reverberation mapping of the circumnuclear gas; line profiles, and line-ratios will allow to estimate the velocity structure and physical conditions (density, abundances) of this gas. In particular, this would also enable us to search for the presence of a BLR in these optically in-active galaxies.

Probing the realm of strong gravity. If observed with high spectral and temporal resolution with the next generation of X-ray telescopes, like the *XEUS* mission sensitive between 0.05–30 ke V, the flare spectra may allow to probe the realm of strong gravity, since the temporal evolution of the stellar debris, and of potential spectral features, will depend on relativistic precession effects around the Kerr metric.

Acknowledgements:

It is a pleasure to thank the Astronomische Gesellschaft for awarding the Ludwig-Biermann price and for inviting me to give this talk.

I would like to thank Weimin Yuan, Michael Dahlem, Jules Halpern, Norbert Bade, Andrew Ulmer, Niel Brandt, Martin Elvis, David Meier, and the participants of the CAS-MPG Workshop on High-Energy Astrophysics at Ringberg Castle for very useful discussions and comments on the subject of X-ray flares from non-active galaxies, and Norbert Bade for taking the optical image of NGC 5905 which is shown in Fig. 4. I gratefully remember Henner Fink for introducing me to the work with X-ray data, for discussions and helpful advice. Henner Fink passed away in December 1996. The *ROSAT* project was supported by the German Bundesministerium für Bildung, Wissenschaft, Forschung und Technologie (BMBF/DLR) and the Max-Planck-Society. Preprints of this and related papers can be retrieved at http://www.xray.mpe. mpg.de/~skomossa/

References

Antonucci, R. 1993, ARA&A 31, 473

Ayal, S., Livio, M., Piran, T. 2000, ApJ 545, 772

Bade, N., Komossa, S., Dahlem, M. 1996, A&A 309, L35

Becker, R. H., White, R. L., Helfand, D. J. 1995, ApJ, 450, 559

Beloborodov, A. M., Illarionov, A. F., Ivanov, P. B., Polnarev, A. G. 1992, MNRAS 259, 209

Bradt, H., Levine, A. M., Marshall, F. E., et al. 2001, to appear in ESO Astrophysics Symposia, F. Frontera et al. (eds), [astro-ph/0108004]

Brandt, W. N., Pounds, K. A., Fink, H. H. 1995, MNRAS 273, L47

Brandt, W. N., Fabian, A. C., Pounds, K. A. 1996, MNRAS 278, 326

Branduardi-Raymont, G., Sako, M., Kahn, S., et al. 2001, A&A 365, L140

Brosch, N., Krumm, N. 1984, A&A 132, 80

Brown, B. A., Bregman, J. N., 1998, ApJ 495, L75

Canizares, C. R., Fabbiano, G., Trinchieri, G. 1987, ApJ 312, 503

Cannizzo, J. K., Lee, H. M., Goodman J. 1990, ApJ 351, 38

Carter, B., Luminet, J. P. 1985, MNRAS 212, 23

Collin, S., Abrassart, A., Czerny, B., Dumont, A.-M., Mouchet, M., EDPS Conf. Series in Astron. & Astrophys., in press [astro-ph/0003108]

Collin-Souffrin, S., Czerny, B., Dumont, A.-M., Zycki, P. T. 1996, A&A 314, 393

Collinge, M. J., Brandt, W. N., Kaspi, S., et al. 2001, ApJ 557, 2

Condon, J. J., Cotton, W. D., Greissen, E. W., et al. 1998, AJ 115, 1693

Cui, W., Feldkun, D., Braun, R. 1997, ApJ 477, 693

Diener, P., Frolov, V. P., Khokhlov, A. M., Novikov, I. D., Pethick, C. J. 1997, ApJ 479, 164

Di Stefano, R., Greiner, R., Murray, S., Garcia, M. 2001, ApJ 551, L37

Elvis, M. 2000, ApJ 545, 63

Elvis, M. 2001, in press, [astro-ph/0109513]

Elvis, M., Briel, U., Henry, J. P. 1983, ApJ 268, 105

Elvis, M., Fassnacht, C., Wilson, A. S., Briel, U. 1990, ApJ 361, 459

Eracleous, M., Livio, M., Binette, L. 1995, ApJ 445, L1

Evans, C. R., Kochanek, C. S. 1989, ApJ 346, L13

Fabian, A. 2001, in: TEXAS Symposium, in press [astro-ph/0103438]

Falcke, H. 2001, in: Reviews in Modern Astronomy 14, R. Schielicke (ed.), 15

Ferland, G. J. 1993, University of Kentucky, Physics Department, Internal Report

Ferrarese, L., Merritt D. 2000, ApJ, 539, L9

Ferrarese, L., Pogge, R. W., Peterson, B. M., et al. 2001, ApJ 555, L79

Franceschini, A., Vercellone, S., Fabian, A. C. 1998, MNRAS 297, 817

Frank, J., Rees, M. J. 1976, MNRAS 176, 633

Fraser, G. 2001, in: MAXI workshop on AGN variability, in press

Friedli, D., Wozniak, H., Rieke, M., Martinet, L., Bratschi, P. 1996, A&AS 118, 461

Gabriel, A. H., Jordan, C. 1969, MNRAS 145, 241

George, I. M., Fabian, A. C. 1991, MNRAS 249, 352

Giuricin, G., Bertotti, G., Mardirossian, F., Mezzetti, M. 1990, MNRAS 247, 444

Greenhill, L. J., Henkel, C., Becker, R., Wilson, T. L., Wouterloot, J. G. A. 1995, A&A 304, 21

Greiner, J., Schwarz, R., Zharikov, S., Orio, M. 2000, A&A 362, L25

Grupe, D., Beuermann, K., Mannheim, K., et al. 1995, A&A 299, L5

Grupe, D., Leighly, K., Thomas, H. 1999, A&A 351, L30

Gurzadyan, V. G., Ozernoi, L. M. 1979, Nature 280, 214

Gurzadyan, V. G., Ozernoi, L. M. 1980, A&A 86, 315

Hills, J. G. 1975, Nature 254, 295

Halpern, J. P. 1984, ApJ 281, 90

Ho, L. C., Filippenko, A. V., Sargent, W. L. W. 1995, ApJS 98, 477

Ho, L. C., et al. 2001, ApJ, in press

Hummel, E., van der Hulst, J. M., Keel, W. C., Kennicutt, R. C. Jr. 1987, A&AS 70, 517

Ikebe, Y., Leighly, K., Tanaka, Y., et al. 2000, MNRAS 316, 433

Israel, F. P., Mahoney, M. J. 1990, ApJ 352, 30

Ivanov, P. B., Novikov, I. D. 2001, ApJ 549, 467

Joseph, R. D. 1999, Ap&SS 266, 321

Kaastra, J., Mewe, R., Liedahl, D. A., Komossa, S., Brinkman, A. C. 2000, A&A 354, L83

Kaspi, S., Brandt, W. N., Netzer, H., et al. 2001, ApJ 554, 216

Kato, M., Hoshi, R. 1978, Prog. Theor. Phys. 60/6, 1692

Keenan, P. C. 1937, ApJ 85, 325

Khokhlov, A., Melia, F. 1996, ApJ 457, L61

Kim, S. S., Park, M.-G., Lee, H. M. 1999, ApJ 519, 647

Kochanek, C. 1994, ApJ 422, 508

Komossa, S. 1999, in: ASCA/ROSAT Workshop on AGN and the X-ray Background, T. Takahashi, H. Inoue (eds), ISAS Report, p. 149; [also available at astro-ph/0001263]

Komossa, S. 2001, A&A 371, 507

Komossa, S. 2001, in: IX. Marcel Grossmann Meeting on General Relativity, Gravitation and Relativistic Field Theories, V. Gurzadyan et al. (eds), World Scientific, in press

Komossa, S., Bade, N. 1998, A&A 331, L49

Komossa, S., Bade, N. 1999, A&A 343, 775

Komossa, S., Böhringer, H., Huchra, J. 1999, A&A 349, 88

Komossa, S., Dahlem, M. 2001, in: MAXI workshop on AGN variability, ISAS Report, in press

Komossa, S., Fink, H. 1997a, A&A 322, 719

Komossa, S., Fink, H. 1997b, A&A 327, 483

Komossa, S., Greiner, J. 1999, A&A 349, L45

Komossa, S., Janek, M. 2000, A&A 354, 411

Komossa, S., Schulz, H., Greiner, J. 1998, A&A 334, 110

Komossa, S., Voges, W. 2001, in: MPG-CAS workshop on high-energy astrophysics, preprint at: http://www.xray.mpe.mpg.de/~skomossa/publrev.html

Koratkar, A., Deustua, S. E., Heckman, T., et al. 1995, ApJ 440, 132

Kormendy, J., Richstone, D. O. 1995, ARA&A 33, 581

Kormendy, J., Gebhardt, K. 2001, in press [astro-ph/0105230]

Krolik, J. H., Kriss, G. A. 1995, ApJ 447, 512

Laguna, P., Miller, W. A., Zurek, W. H., Davies, M. B. 1993, ApJ 410, L83

Lee, H. M., Kang, H., Ryu, D. 1995, ApJ 464, 131

Lee, J., Ogle, P., Canizares, C., et al. 2001, ApJ 554, L13

Lee, J. 2001, talk given at: Workshop on X-ray spectroscopy of active galactic nuclei with Chandra and XMM-Newton (Garching, Dec. 2001)

Lidskii, V. V., Ozernoi, L. M. 1979, Sov. Astron. Lett. 5(1), 16

Liedahl, D. A., Kahn, S. M., Osterheld, A. L., Goldstein, W. H. 1990, ApJ 350, L37

Lightman, A. P., Shapiro, S. L. 1977, ApJ 211, 244

Loeb, A., Ulmer, A. 1997, ApJ 489, 573

Luminet, J. P., Marck, J.-A. 1985, MNRAS 212, 57

Lynden-Bell, D. 1969, Nature 223, 690

Magorrian, J., Tremaine, S. 1999, MNRAS 309, 447

Matsumoto, C., Inoue, H. 2001, in: MAXI workshop on AGN variability, in press

Matt, G., Perola, G. C., Fiore, F., et al. 2000, A&A 363, 863

Mihara, T. 2001, in: MAXI workshop on AGN variability, in press

Mitsuda, K. 1995, Ann. N. Y. Acad. Sc. 759, Proc. 17th Texas Symp. Relat. Ap. and
       Cosm., H. Böhringer, G. E. Morfill, J. E. Trümper (eds), 213

Miyoshi, M., Moran, J., Herrnstein, J., et al. 1995, Nature 373, 127

Morales, R., Fabian, A. C. 2001, MNRAS, in press [astro-ph/0109050]

Mushotzky, R. F. 1982, ApJ 256, 92

Mushotzky, R. F., Done, C., Pounds, K. A. 1993, ARA&A 31, 717

Nakagawa, T., Kii, T., Fujimoto, R., et al. 1999, Ap&SS 266, 43

Nandra, P. 2001, talk given at: Workshop on X-ray spectroscopy of active galactic
       nuclei with Chandra and XMM-Newton (Garching, Dec. 2001)

Nandra, K., Pounds, K. A. 1992, Nature 359, 215

Netzer, H. 1993, ApJ 411, 594

Neufeld, D. A., Maloney, P. R. 1995, ApJ 447, L17

Nolthenius, R. A., Katz, J. I. 1982, ApJ 263, 377

Nolthenius, R. A., Katz, J. I. 1983, ApJ 269, 297

Norman, C., Silk, J. 1983, ApJ 266, 502

Ogle, P. M., Marshall, H. L., Lee, J. C., Canizares, C. R. 2000, ApJ 545, L81

Peterson, B. M., Ferland, G. J. 1986, Nature 324, 345

Peterson, B. M. 2001, in: Advanced lectures on the starburst-AGN connection, I.
       Aretxaga et al. (eds), World Scientific, 3

Peterson, B. M., Wandel, A. 2000, ApJ 540, L13

Piro, L., Massaro, E., Perola, G. C., Molteni, D. 1988, ApJ 325, L25

Piro, L., Balucinska-Church, M., Fink H., et al. 1997, A&A 319, 74

Predehl, P. 2001, in: MAXI workshop on AGN variability, in press

Ptak, A., Yaqoob, T., Mushotzky, R., Serlemitsos, P., Griffiths, R. 1998, ApJ 501,
       L37

Ptak, A., Serlemitsos, P., Yaqoob, T., Mushotzky, R. 1999, ApJS 120, 179

Rauch, K. P. 1999, ApJ 514, 725

Rauch, K. P., Ingalls, B. 1998, MNRAS 299, 1231

Rees, M. J. 1988, Nature 333, 523

Rees, M. J. 1989, in: Reviews in Modern Astronomy 2, G. Klare (ed.), 1

range 1.5–3 keV is radically different from the prediction, with a peak at a redshift in the range 0.5–0.7. This is still the case, if the objects belonging to the large scale structures around $z = 0.7$ in the CDFS are removed. The total number of objects at redshift less than 1 is significantly higher than the model predictions, even ignoring the 40 % spectroscopic incompleteness. The peak at redshifts below 1 is also significant, if the normal star forming galaxies in the sample are removed. This clearly demonstrates that the population synthesis models will have to be modified to incorporate different luminosity functions and evolutionary scenarios for intermediate-redshift, low-luminosity AGN.

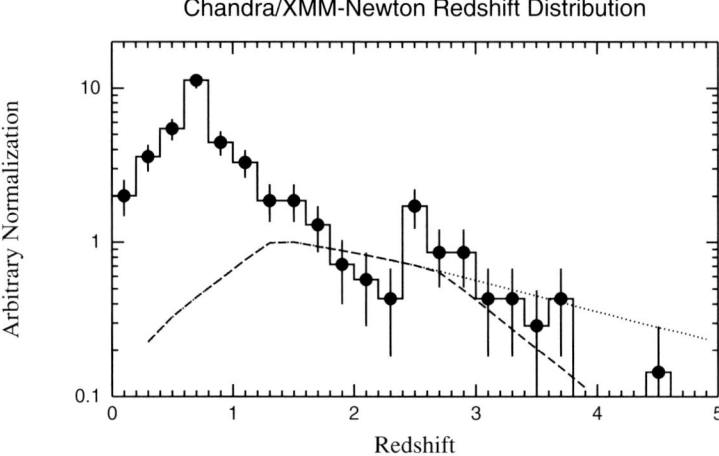

Figure 10: Redshift distribution of ~ 300 X-ray selected AGN and galaxies in the deep *Chandra* and *XMM-Newton* survey samples given in table 1 (solid circles and histogram), compared to model predictions from population synthesis models (Gilli et al. 2001). The dashed line shows the prediction for a model, where the commoving space density of high-redshift QSO follows the decline above $z = 2.7$ observed in optical samples (Schmidt, Schneider & Gunn, 1995; Fan et al. 2001). The dotted line shows a prediction with a constant space density for $z > 1.5$. The two model curves have been normalized to their peak at $z = 1$, while the observed distribution has been normalized to roughly fit the models in the redshift range 1.7–3.

7   The AGN evolution at high redshift

The comparison between the observed and predicted $N(z)$ distributions at high redshifts is complicated by the possible existence of large-scale structure in the pencil beam survey (there is e. g. a possibly significant excess of objects around $z = 2.5$ in the CDFS), but also by redshift-dependent selection effects and in general by the still relatively small volume sampled and therefore poor counting statistics in the number of objects. In addition, the overall normalization of the curves is uncertain because of the significant mismatch of the distribution at low z. Nevertheless, the observed distribution is roughly consistent with both predictions in the redshift range $z = 1.6$–3.8. There is, however, a significant discrepancy between the observed distribution and the constant space density model (dotted line) at redshifts above 4, where only one object was detected, while about 8 objects would be predicted from the constant space density model. From Figure 9 it becomes apparent, that the dearth of X-ray selected AGN is probably not due to optical spectroscopic selection effects. The one object detected at $z = 4.45$ already in the ROSAT data of the Lockman Hole (Schneider et al. 1998) has an optical magnitude of $R = 23$ and is therefore not at the spectroscopic limit of the samples. Also the Ly_α and C IV lines for QSOs in the redshift range 4–5 fall well into the optical range. The observed redshift distribution therefore gives a strong indication for a decline of the QSO space density beyond a redshift of 3.8.

A similar conclusion about a decline of the X-ray selected AGN space density at high redshifts can be obtained from the absence of QSOs with $z > 5$ in all X-ray survey samples so far. (There was a recent announcement of a QSO at $z = 5.2$ in the *Chandra* observation of the HDF-N, but this does not change the conclusions discussed below). Figure 11 shows a prediction of number counts for high-redshift QSO from Haiman & Loeb (1999), according to which a large number of $z > 5$ AGN should be detected in any deep survey with *Chandra*. This theoretical model assumes the X-ray luminosity function at $z = 3.5$ determined from the ROSAT surveys and extrapolates it backwards in time assuming a simple hierarchical CDM model. The figure also shows limits for the number counts of $z > 5$ AGN from X-ray surveys at varying flux limits. The most distant QSO among ~ 2000 objects in the ROSAT Bright Survey (RBS, Schwope et al. 2000) has a redshift of 2.8, the lack of higher redshift objects is, however, not constraining given the high flux limit of this survey. The lack of $z > 5$ AGN in the ROSAT Deep and Ultradeep Surveys (Schmidt et al. 1998, Lehmann et al. 2000, 2001) is still just consistent with the Haiman & Loeb predictions, the highest-redshift object in the UDS is RX J105225.9+571905 at $z = 4.45$ (Schneider et al. 1998). The *Chandra* Deep survey, while only about 60 % spectroscopically identified, still provides an upper limit for the number counts of $z > 5$ AGN significantly lower than the prediction, using the conservative assumption that less than half of the unidentified objects are at redshifts larger than 5. Finally, the 400 ksec *Chandra* observation in the Hubble Deep Field proper, providing 100 % identifications for 12 sources in the field and their highest redshift object at $z = 4.42$ just outside the HDF-N also gives an upper limit about a factor of three lower than the Haiman & Loeb prediction.

The information about the space density of X-ray selected AGN is still limited by the small number statistics in the deep X-ray surveys which cover too small a solid angle. More and wider fields have been surveyed by both *Chandra* and *XMM-Newton*. As soon as the tedious and time consuming optical follow-up work in these fields is completed, we will be able to learn more about the decline of the X-ray AGN and therefore their formation at early redshifts. The possible discrepancy between a declining space density

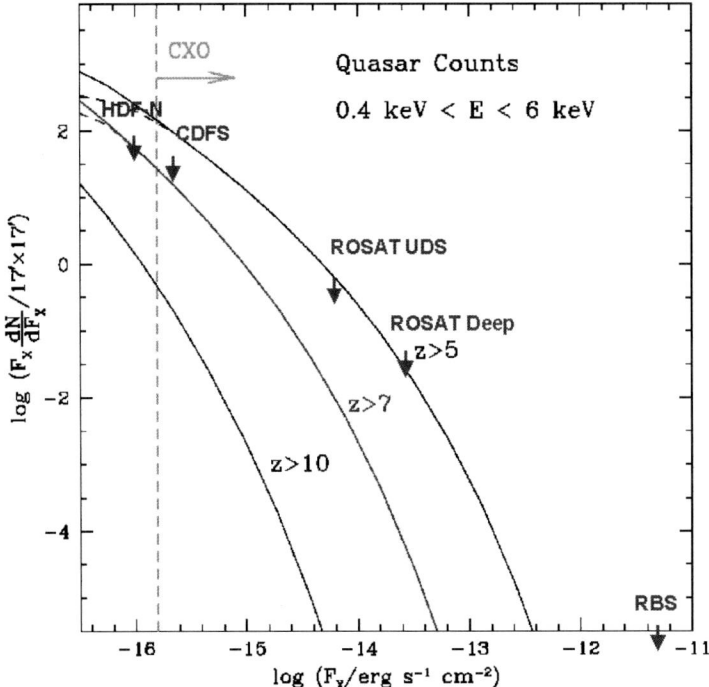

Figure 11: Prediction of the number density of AGN with redshits larger than 5, 7 and 10, respectively as a function of flux in a typical 17×17 arcmin *Chandra* field of view from Haiman & Loeb (1999). Upper limits measured in X-ray surveys at various flux limits are indicated.

of optical and radio-selected QSOs above a redshift of 2.7 and an apparently constant space density of X-ray selected AGN with a decline beyond a redshift of ~ 4 could still be understood in terms of the different luminosity and therefore different black hole mass of the objects involved. The optical and radio surveys cover a large solid angle to a modest flux limit and therefore pick up only the most luminous and therefore most massive objects at high redshift. The deep pencil beam surveys, on the other hand, sample a much smaller volume to much fainter flux limits and therefore select high-redshift AGN which are intrinsically more than a factor of 10 less luminous and therefore probably less massive than the objects selected in wide-angle surveys. In the hierarchical large scale structure formation the smaller cold dark matter halos collapse earlier than the larger ones. Given the correlation between black hole mass and galaxy mass (and presumably dark matter mass), it is expected that the lower mass black holes are formed earlier than the most massive objects and thus that lower luminosity AGN appear earlier than the most luminous QSOs. This concept can be tested with more optical identifications of *Chandra* and *XMM-Newton* surveys and with future, even more sensitive X-ray telescopes, like the ESA/ISAS XEUS mission.

References

Barger, A.J., Cowie, L.L., Mushotzky, R.F., Richards, E.A. 2001, AJ 121, 662

Barger, A.J., Cowie, L.L., Bautz, M.W., et al. 2001, AJ 122, 2177

Böhringer, H., Voges, W., Fabian, A.C., et al. 1993, MNRAS 264, L25

Böhringer, H., Matsushita, K., Churazov, E., et al. 2002, A&A 382, 804

Brandt, W.N., Hornschemeier, A.E., Alexander, D.M., et al. 2001, AJ 122, 1

Brandt, W.N., Alexander, D.M., Hornschemeier, A.E., et al. 2001, AJ 122, 2810

Briel, U.G., Henry, J.P. 1995, A&A 302, L9

Cen, R., Ostriker, J.P. 1999, ApJ 514, 1

Churazov, E., Brüggen, M., Kaiser, C.R., et al. 2001, ApJ 554, 261

Comastri, A. et al. 1995, A&A 296, 1

Davé, R., Cen, R., Ostriker, J., et al. 2001, ApJ 552, 473

den Herder, J.W., Brinkman, A.C., Kahn, S.M., et al. 2001, A&A 365, L7

Fabian, A.C. 1994, ARA&A 32, 277

Fabian, A.C., Barcons, X., Almaini, O., Iwasawa, K. 1998, MNRAS 297, L11

Fabian, A.C., Sanders, J.S., Ettori, S., et al. 2000, MNRAS 318, L65

Fabian, A.C., Mushotzky, R.F., Nulsen, P.E.J., Peterson, J.R. 2001, MNRAS 321, L20

Fan, X. et al. 2001, AJ 121, 54

Fiore, F., LaFranca, F., Giommi, P., et al. 1999, MNRAS 306, 55

Fiore, F., LaFranca, F., Vignali, C., et al. 2000, NewA 5, 143

Gebhardt, K., Bender, R., Bower, G., et al. 2000, ApJ 539, 13

Giacconi, R., Rosati, P., Tozzi, P., et al. 2001, ApJ 551, 624

Giacconi, R., Zirm, A., Wang, P., et al. 2002, ApJS 139, 369

Gilli, R., Salvati, M., Hasinger, G. 2001, A&A 366, 407

Granato, G.L., Danese, L., Francheschini, A. 1997, ApJ 486, 147

Haiman, Z., Loeb, A. 1999, ApJ 519, 479

Hashimoto, Y., Hasinger, G., Arnaud, M., et al. 2002, A&A 381, 841

Hasinger, G., Burg, R., Giacconi, R., et al. 1993, A&A 275, 1

Hasinger, G., Burg, R., Giacconi, R., et al. 1998, A&A 329, 482

Hasinger, G., Giacconi, R., Gunn, J.E., et al. 1999, A&A 340, 27

Hasinger, G., Altieri, B., Arnaud, M., et al. 2001, A&A 365, 45

Hasinger, G., Lehmann, I. 2001, Proc. "Where's the Matter?", Marseille, France, 25–29 June 2001, eds. L. Tresse & M. Treyer, in press

Hornschemeier, A.E., Brandt, W.N., Garmire, G.P., et al. 2000, ApJ 541, 49

Jansen, F., Lumb, D., Altieri, B., et al. 2001, A&A 365, L1

Lehmann, I., Hasinger, G., Schmidt, M., et al. 2000, A&A 354, 35

Lehmann, I., Hasinger, G., Schmidt, M., et al. 2001, A&A 371, 833

Lehmann, I., Hasinger, G., Murray, S.S., Schmidt, M. 2002, Proc. "High Energy Universe at Sharp Focus" (astro-ph/0109172)

Mainieri, V., Bergeron, J., Rosati, P., et al. 2002, Proc. Symp. "New Visions", astro-ph/0202211

Mason, K.O., Breeveld, A., Much R. et al. 2001, A&A 365, L36

McNamara, B.R., Wise, M.W., Nulsen, P.E.J., et al. 2001, ApJ 562, L149

McHardy, I., Jones, L.R., Merrifield, M.R., et al. 1998, MNRAS 295, 641

Miyaji, T., Hasinger, G., Schmidt, M. 2000, A&A 353, 25

Miyaji, T., Hasinger, G., Schmidt, M. 2001, A&A 369, 49

Murray, S.S. et al. 2002, in prep.

Mushotzky, R.F., Cowie, L.L., Barger, A.J., Arnaud, K.A. 2000, Nature 404, 459

Nicastro, F., Zezas, A., Drake, J., et al. 2002, astro-ph/0201058

Norman, C., Hasinger, G., Giacconi, R., et al. 2001, ApJ, in press (astro-ph/0103198)

Owen, F.N., Eilek, J.A., Kassim, N.E. 2000, ApJ 543, 611

Rasmussen, A., Paerels, F., Kahn, S.M. 2002, in prep.

Peterson, J.R., Paerels, F.B.S., Kaastra, J.S., et al. 2001, A&A 365, L104

Phillips, L.A., Ostriker, J., Cen, R., et al. 2001, ApJ 554, L9

Rosati, P., Tozzi, P., Giacconi, R., et al. 2002, ApJ 566, 667

Schmidt, M., Schneider, D.P., Gunn, J.E. 1995, AJ 114, 36

Schmidt, M., Hasinger, G., Gunn, J.E., et al. 1998, A&A 329, 495

Schneider, D.P., Schmidt, M., Hasinger, G., et al. 1998, AJ 115, 1230

Schwope, A., Hasinger, G., Lehmann, I., et al. 2000, AN 321, 1

Shaver, P.A. et al. 1996, Nature 384, 439

Sliwa, W., Soltan, A.M., Freyberg, M.J. 2001, A&A 380, 397

Soltan, A.M., Hasinger, G., Egger, R., Snowden, S., Trümper, J. 1996, A&A 305, 17

Stern, D., Moran, E.C., Coil, A.L., et al. 2002a, ApJ 568, 71

Stern, D., Tozzi, P., Stanford, S.A., et al. 2002b, astro-ph/0203392

Strüder, L., Briel, U., Dennerl, K., et al. 2001, A&A 365, L18

Szokoly, G., Hasinger, G., Rosati, P., et al. 2002, in prep.

Tamura, T., Kaastra, J.S., Peteson, J.R., et al. 2001, A&A 365, L87

Thompson, D., Pozzetti, L., Hasinger, G., et al. 2001, A&A 377, 778

Tozzi, P., Rosati, P., Nonino, M., et al. 2001, ApJ 562, 42

Tripp, T.M., Giroux, M.L., Stocke, J.T., et al. 2001, ApJ 563, 724

Turner, M.J.L., Abbey, A., Arnaud, M., et al. 2001, A&A 365, L27

Weisskopf, M.C. 1999, Proc. NATO-ASI held in Crete, Greece, 7–18 June, 1999, astro-ph/9912097

Wu, X.-P., Xue, Y.-J. 2001, ApJ 560, 544

Zamorani, G., Mignoli, M., Hasinger, G., et al. 1999, A&A 346,731

Seeing the Universe
in the Light of Gravitational Waves

Karsten Danzmann and Albrecht Rüdiger

Max-Planck-Institut für Gravitationsphysik – Albert-Einstein-Institut
and Universität Hannover
Callinstraße 38, 30167 Hannover, Germany
E-mail: kvd@mpq.mpg.de

Abstract

*The existence of gravitational waves is the most prominent of Einstein's
predictions that has not yet been directly verified. The space project
LISA is approved by ESA as a cornerstone mission in the field of "Fundamental Physics", and is currently the object of a joint ESA/NASA
study aimed at launch in 2011. This space project shares its goal and
principle of operation with the ground-based interferometers currently
under construction: the detection and measurement of gravitational
waves by laser interferometry. Ground and space detection differ in
their frequency ranges, and thus the detectable sources. At low frequencies, ground-based detection is limited by seismic noise, and yet more
fundamentally by 'gravity gradient noise', thus covering the range from
a few Hz to a few kHz. On five sites worldwide, detectors of armlengths
from 0.3 to 4 km are nearing completion. They will progressively be put
in operation between 2001 and 2003. Future enhanced versions are being planned, with scientific data not expected until 2008, i. e. near the
launch of the space project LISA. It is only in space that detection of
signals below, say, 1 Hz is possible, opening a wide window to a different class of interesting sources of gravitational waves. The project LISA
consists of three spacecraft in heliocentric orbits, forming a triangle of
5 million km sides. A technology demonstrator, designed to test vital
LISA technologies, is to be launched, aboard a SMART-2 mission, in
2006.*

1 Introduction

This JENAM 2001 conference will highlight many new developments and dis-
coveries in astronomy, most of them in the field of electromagnetic waves, from
radio to gamma rays. There are other radiations, such as the flux of particles

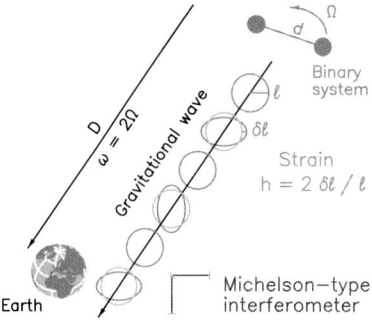
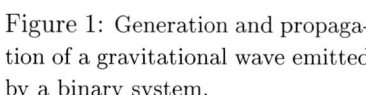
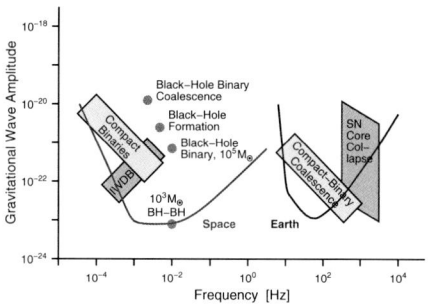

Figure 1: Generation and propagation of a gravitational wave emitted by a binary system.

Figure 2: Some sources of gravitational waves, with sensitivities of *Earth* and *Space* detectors.

from the sun and outer space, and in particular neutrinos, again from our sun (although fewer than we understand) and also from cosmic events far out in the universe.

This talk will be about a new window in astronomical observation presently being opened: the detection and measurement of gravitational waves. These waves share their elusiveness with the neutrinos: they have very little interaction with the measuring device, which is why these gravitational waves have not yet directly been detected. But that same feature also is a great advantage: due to their exceedingly low interaction with matter, gravitational waves can give us an unobstructed view into astrophysical and cosmological events that will forever be obscured in the electromagnetic window.

The price we have to pay is that, in order to detect and measure these gravitational waves, we will require the most advanced technologies in optics, lasers, and interferometry.

Efforts to observe these gravitational waves with ground-based interferometers have gone into their final phase of commissioning, and the international collaboration on placing a huge interferometer into an interplanetary orbit is close to reaching final approval.

We will briefly discuss the characteristics of large GW detectors being built right now. In this talk we will learn how the detectors on ground and in space differ, where aims and technologies overlap, and what can scientifically be gained from the complementarity of these researches.

2   Gravitational waves

In two publications [1, 2], Albert Einstein has predicted the existence and estimated the strength of gravitational waves. They are a direct outcome of his Theory of General Relativity, but they would be a necessary consequence of all theories having a finite velocity of light. Good introductions to the nature of gravitational waves, and on the possibilities of measuring them were given in two publications of Kip Thorne's [3, 4].

It can be shown that gravitational waves of measurable strengths are emitted only when large cosmic masses undergo strong accelerations, for instance – as shown schematically in Figure 1 – in the orbits of a (close) binary system. The effect of such a gravitational wave is an apparent strain in space, transverse to the direction of propagation, that makes distances ℓ between test bodies shrink and expand by small amounts $\delta\ell$, at twice the orbital frequency: $\omega = 2\,\Omega$. The strength of the gravitational wave, its "amplitude", is generally expressed by $h = 2\,\delta\ell/\ell$. An interferometer of the Michelson type, typically consisting of two orthogonal arms, is an ideal instrument to register such differential strains in space.

But what appears so straighforward turns out to be an almost insurmountable problem. It lies in the magnitude, or rather: the smallness, of the effect.

2.1   Strength of gravitational waves

With a linearized approximation, the so-called "quadrupole formula", the strength of the gravitational wave emitted by a mass quadrupole can be estimated, and for a binary with components of masses M_1 and M_2, or their respective Schwarzschild radii R_1, R_2, the strain h to be expected is of the order

$$h \approx \frac{R_1 R_2}{d\,D} \tag{1}$$

where d and D are the distances between the partners and from binary to the observer (see Figure 1). For neutron stars, and even better for black holes, the distance d can be of the order of a Schwarzschild radius, which then would further simplify the estimate.

From such an in-spiral of a neutron star binary out at the Virgo cluster (a cluster of about 2000 galaxies, $D \sim 10\,\text{Mpc}$ away), we could expect a strain of something like $h \approx 10^{-22}$. Similar (or even lower) strengths might be expected from supernovae out at Virgo cluster distances. That we insert such a large distance as the Virgo cluster is to have a reasonable rate of a few events per year. Inside a single galaxy (as ours), we would not count more than a few supernovae per century.

So all we have to do is to measure – in a Michelson interferometer of kilometer dimensions – path changes in the order of 10^{-19} m. *Hopeless?* The sensitivities obtained with prototypes of ground-based interferometers bear evidence that it is within reach.

And yet, despite the smallness of the interaction, gravitational waves are by no means a "weak" phenomenon. On the contrary, they are linked with cosmic events of high energy transfers. Some examples show this clearly.

The *binary system* containing the Hulse-Taylor pulsar PSR 1913+16, much publicised through the 1993 Nobel prize, loses its orbital energy primarily through the emission of gravitational radiation; no other loss mechanism comes anywhere near it. It is a very convincing, albeit indirect, evidence of the existence of gravitational waves, as observation and prediction agree to much better than 1 %.

A *supernova*, on the other hand, in its final milliseconds of collapse, emits more power than the (visible) luminescence of all the stars of the universe combined. Although most of this is in neutrinos, an appreciable part is also emitted in gravitational waves, from the rebound of the core.

The formation, and the interaction, of massive and supermassive black holes, believed to be at the centers of most galaxies, are perhaps the most violent events in the Universe, and the gravitational waves emitted by them will be our best chance to investigate the physics involved.

2.2 What makes gravitational waves so special ?

Let us see what can be learned about the nature of gravitational waves in a comparison with electromagnetic waves. What they do have in *common* is the velocity of propagation: the speed of light c. And also, from conservation of energy, the decline in amplitude with the inverse of the distance travelled, see Eq. (1).

But there are fundamental *differences* between electrical and gravitational forces. While in electromagnetism we have two opposite charges, gravity is governed by mass having only one sign (and the gravitational forces being attractive).

The most basic radiation mode in electromagnetism is thus the dipole, in the case of gravitation, however, the quadrupole. One result of that is that the frequency ω of the emitted gravitational wave is twice the orbital frequency Ω, as an identical mass distribution is already reached after half an orbital period. The quadrupole radiation also implies a different polarization pattern: In electromagnetism, the two polarizations are off by 90°, in gravitational waves by 45°. This will mean that the Michelson interferometer indicated in Figure 1 is optimized for a wave as the one shown $(+)$, whereas it would be insensitive to a wave 45° off $(\times)$.

Electrical attraction between, say, electron and proton is nearly 10^{40} times bigger than the gravitational one. But this does not justify considering gravity negligibly weak. The strong electrical attraction leads to mutual saturation, and thus globally to a cancelling: there are practically no large agglomerations of free charges in nature. What we observe in electromagnetic radiation are mainly the non-coherent motions of individual charges on microscopic scales.

For gravitation, on the other hand, we have masses of only one sign, governed by attraction. Thus in gravitational radiation we see collective, coherent motions of large masses. And huge masses are required to counteract the "small" value of the gravitational constant.

The electromagnetic radiation is easily obscured and absorbed by matter between source and observer; not so the gravitational waves: their interaction with matter is extremely weak, it reaches us effectively unblemished by any obstacles.

This is the great advantage: gravitational waves allow us to observe events that remain hidden "in the light" of electromagnetic radiation. Thus completely different types of information, and new information in astronomy, astrophysics, fundamental physics can be reaped.

2.3   Complementarity of ground and space observation

Figure 2 shows some typical sources of gravitational radiation. They range in frequency over a vast spectrum, from the kHz region of supernovae and final mergers of compact binary stars down to mHz events due to formation and coalescence of supermassive black holes. Indicated are sources in two clearly separated regimes: events in the range from, say, 5 Hz to several kHz (and only these will be detectable with terrestrial antennas), and a low-frequency regime, 10^{-5} to 1 Hz, accessible only with a space project such as LISA. In the following sections we will see how the sensitivity profiles of the detectors come about. No detector covering the whole spectrum shown could be devised.

Clearly, one would not want to miss the information of either of these two (rather disjoint) frequency regions. The upper band ("Earth"), with supernovae and compact binary coalescence, can give us information about relativistic effects and equations of state of highly condensed matter, in highly relativistic environments. Binary inspiral is an event type than can be calculated to high post-newtonian order, as shown, e. g., by Buonanno and Damour [5]. This will allow tracing the signal, possibly even by a single detector, until the final merger, a much less predictable phase. The ensuing phase of a ring-down of the combined core does again lend itself to an approximate calculation, and thus to an experimental verification. Chances for detection are reasonably good, but not by wide margins.

The events to be detected by LISA, on the other hand, may have extremely high signal-to-noise ratios, and failure to find them would shatter the very foundations of our present understanding of the universe. The strongest signals will come from events involving (super-)massive black holes, their formation as well when galaxies with their BH cores collide. But also the (quasi-continuous) signals from neutron-star and black-hole binaries are among the events to be detected ('Compact Binaries' in Figure 2). Interacting white dwarf binaries inside our galaxy ('IWDB' in Figure 2) may turn out to be so numerous that they cannot all be resolved as individual events. Catastrophic events such as the Gamma-ray bursts are not yet well enough understood to estimate their emission of gravitational waves, but there is a potential of great usefulness of GW detectors mainly at low frequencies. The combined observation with electromagnetic and gravitational waves could lead to a deeper understanding of the violent cosmic events in the far reaches of the universe [6].

3   Ground-based interferometers

The underlying concept of all our ground-based detectors is the Michelson interferometer (see schematic in Figure 3), in which an incoming laser beam is divided into t wo beams travelling along different (perpendicular) arms. On their return, these t wo beams are recombined, and their interference (measured with a photodiode PD) will depend on the difference in the gravitational wave effects that the t wo beams have experienced.

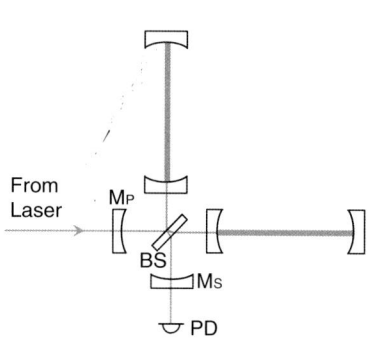

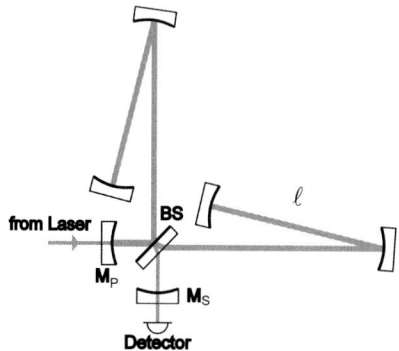

Figure 3:   Advanced Michelson interferometer with Fabry-Perots in the arms and extra mirrors M_P, M_S for power and signal recycling.

Figure 4:   The DL4 configuration with dual recycling to be used in GEO 600

The changes δL in optical path become the larger the longer the optical paths L are made, optimally about half the wavelength of the gravitational wave: e. g. to a seemingly unrealistic 150 km for a 1 kHz signal. Schemes were devised to make the optical path L significantly longer than the geometrical arm length ℓ, which is limited on Earth to only a few km. One way is to use 'optical delay lines' in the arms, the beam bouncing back and forth between t wo concave mirrors (a modest version of this is shown in Figure 4). The other scheme is to use Fabry-Perot cavities (Figure 3), again with the aim of increasing the interaction time of the light beam with the gravitational wave. For GW frequencies beyond the inverse of the storage time τ, the response of the interferometer will, however, roll off with frequency, as $1/f\tau$.

3.1   The detector prototypes

It is a fortunate feature that on our way to the large-scale detectors we were able to go through generations of ever-improving prototypes. It was only with their positive results that the proposals for large-scale proposals received sufficient credibilty.

After pioneering work by Rai Weiss [7] at MIT (1972), other groups at Munich/Garching, at Glasgow, then Caltech, Paris-Orsay, Pisa, and later in Japan and Australia, also entered the scene. Their prototypes range from a few meters up to 30, 40, and even 100 m. They all use an optical scheme modelled after the Michelson interferometer. An alternative detection scheme, a Sagnac configuration, is being investigated at Stanford.

Based on the idea of Weiss, Garching had also adopted the scheme of the optical delay-line. The advantage was a very rapid attainment of sensitivities that were limited only by shot noise [8], by gradually reducing newly discovered noise mechanisms. It was only in the later 1990's that the (technologically more challenging) prototypes with Fabry-Perot cavities reached similar, meanwhile even better, sensitivities.

Even though some of these prototypes reached the sensitivities of cryogenic resonant-mass antennas, they were never meant to be used as detectors, but rather as test-beds for verifying new schemes and configurations devised to overcome otherwise limiting noise effects.

The "phase noise" reduction achieved in these prototypes approaches that required in full-fledged *terrestrial* interferometers, and it is by many orders of magnitude better than required (at low frequencies) for a *space mission*.

3.2 The large-scale projects

Table 1 gives an impression of the wide international scope of the interferometer efforts, listed according to size of detector. All of the large-scale projects will use low-noise Nd:YAG lasers ($\lambda = 1.064\,\mu$m), pumped with laser diodes for high overall efficiency. A wealth of experience has accumulated on highly stable and efficient lasers, and also the space mission will profit from that. More details about the laser source in subsection 4.1.

Table 1: Current and future projects of ground-based GW detectors

Country:	USA		FRA	ITA	GER	GBR	JPN
Institute:	MIT,	Caltech	CNRS	INFN	AEI	Glasgow	NAO, U-Tokyo, ICRR

Large Interferometric Detectors: the current generation

Project name:	LIGO		VIRGO	GEO 600	TAMA 300
Arm length ℓ:	4 km 2 km	4 km	3 km	600 m	300 m
Site (State)	Hanford (WA)	Livingston (LA)	Pisa ITA	Hannover GER	Mitaka JPN

Large Interferometric Detectors: the future generation

Planning (start):	1995		1999	1998
Arm length ℓ:	4 km	4 km	3 km	3 km
Site (State)	Hanford (WA)	Livingston (LA)	EUROPE	Kamioka JPN
Project name:	Advanced LIGO		EURO	LCGT
special features:	active isolation, suspension, RSE		high seismic rejection; cryogenic, diffractive optics, tunable	cryogenic, underground

LIGO   The largest is the US project named LIGO [9, 10]. It comprises *two* facilities at two widely separated sites, in the states of Washington and Louisiana. Both will house a 4 km interferometer, Hanford an additional 2 km one. At both sites construction has long been completed, installation of the optics in the vacuum enclosures is done, and locking of the interferometers has now become routine. These three interferometers are designed for coincidence operation, allowing autonomous measurements inside the US project LIGO.

VIRGO   Next in size (3 km) is the French-Italian project VIRGO [11], being built near Pisa, Italy. An elaborate seismic isolation system, with seven-stage pendulums (see Section 4.3), will allow measurement down to GW frequencies of 10 Hz or even below, but still no overlap with the space interferometer LISA. Currently, a short-arm interferometer, with all mirrors inside the central building, is being used for performance testing.

GEO 600   For the detector of the British-German collaboration, GEO 600 [12, 13], with a de-scoped length of 600 m, construction and installation of the optics in the vacuum system are finished, and locking of the full power-recycled Michelson is routinely achieved. GEO 600 will employ the advanced optical technique of "signal recycling" [14] to make up for the shorter arms. This interferometric scheme, or its counterpart "resonant sideband extraction" [15], will later be transferred to the upgrades of LIGO and to the planned Australian detector.

TAMA 300   In Japan, on a site at the National Astronomical Observatory in Tokyo, construction, vacuum system, and optics installation of the detector called TAMA 300 [16] with 300 m armlength are completed, and several data runs of the Michelson have been successful. A recent run with a total of 1000 hours exhibited encouragingly long in-lock duty cycles   [17]. The sensitivity-enhancing scheme of power recycling will have, however, yet to be added. Separate tests with power recycling appeared promising. TAMA is, just as LIGO and VIRGO, equipped with standard Fabry-Perot cavities in the arms.

ACIGA   Australia (not included in table) also had to cut back from earlier plans of a 3 km detector, due to lack of funding. Currently a 80 m prototype detector is being built near Perth, Western Australia, with the aim of investigating new interferometry configurations [18], follow-ons to the GEO schemes of signal recycling and/or RSE. The design and the site will allow later extension to 3 km.

3.3   International collaboration

As it happens, these projects are scheduled such that first scientific operation can be expected around the years 2002/03. It is fortunate that the various projects are rather well in synchronism. For the received signal to be meaningful, coincident recordings from at least two detectors at well-separated sites are essential. A minimum of three detectors (at three different sites) is required to locate the position of the source, and there is general agreement that only with at least four detectors can we speak of a veritable gravitational wave *astronomy*, based on a close collaboration in the exchange and analysis of the experimental data.

3.4   First common data run   (Note added in proof)

At the turn of the year 2001/2002, a common data run between all three LIGO detectors and GEO 600 was undertaken, consisting of 17 days of mostly uninterrupted operation. This common effort can be considered quite a success.

It exhibited reasonable duty cycles of the interferometers being locked, in the GEO detector being improved considerably (up to better than 95 %) "on the fly", by upgrading the automatic mirror alignment. GEO was, however, not yet equipped with the 'signal-recycling' mirror, so not yet in its final high-sensitivity implementation.

With the detectors not yet being at the intended sensitivity level, the aim was not a search for gravitational waves, but rather to rehearse the activities of data acquisiton and, then, also of data analysis. The analysis of the data, just underway, can give further clues on the 'healthiness' of the interferometers.

4   Noise and sensitivity

The measurement of gravitational wave signals is a constant struggle against the many types of noise entering the detectors. The most prominent of such noise sources will be discussed below.

4.1   Laser noise

The requirements on the quality ('purity') of the laser light used for the GW interferometry are a great technological challenge. As it happens, the light sources for the ground-based and the space-borne interferometers will both be Nd:YAG lasers, in the form of non-planar ring oscillators [19], see Figure 5. Pumped by laser diodes, they exhibit a high overall efficiency. Their good tunabilty allows efficient stabilization schemes.

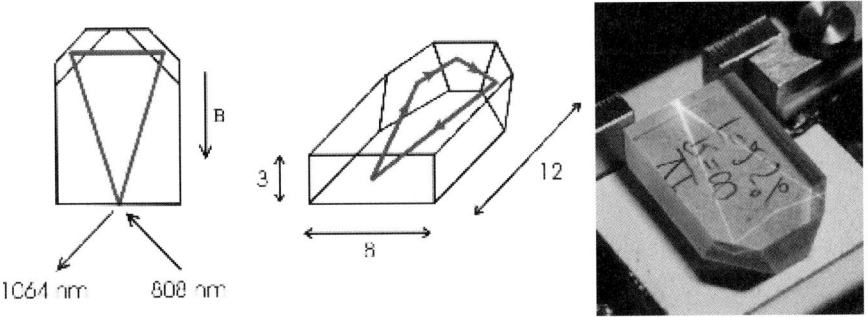

Figure 5: NPRO laser, scheme, dimensions in mm (left), photo (right)

Frequency stability A perfect Michelson interferometer (with exactly matching arms) would be insensitive to frequency fluctuations of the light used. The detectors will, however, by necessity have unequal arms, the ones on the ground due to civil engineering tolerances and a particular modulation scheme chosen, the space detector due to unavoidable imperfections in the orbits of the individual spacecraft.

Therefore, a very accurate control of the laser frequency is required, with (linear) spectral densities of the frequency fluctuations of the order $\widetilde{\delta\nu} = 10^{-4}\,\mathrm{Hz}/\sqrt{\mathrm{Hz}}$. Control schemes have been devised to reach such extreme stability, albeit only in the frequency band required, and not all the way down to DC (which would set an all-time record in frequency stability: $\widetilde{\delta\nu}/\nu = 3\times 10^{-19}/\sqrt{\mathrm{Hz}}$).

Power stability Again due to asymmetries of the interferometer, the incoming laser beam needs to be closely controlled as to its power, in the frequency band of interest. Here, however, a power stability in the order of 10^{-7} is seen to be sufficient [19].

Beam purity Any geometrical asymmetry of the Michelson interferometer will make it prone to noise from geometrical fluctuations of the laser beam. Thus the illumination of the Michelson is required to be an almost pure TEM_{00} mode. For small light powers, below 1 W as in the space project, this purity can be gotten by passing the light through a single-mode fiber. For the laser powers needed in the ground-based interferometers, however, a 'mode-cleaner' is used: a non-degenerate cavity that is tuned for the TEM_{00} mode, but suppresses the (time dependent) lateral modes that represent fluctuations in position, orientation, and width of the beam [20].

4.2   Thermal noise

All optical components – and in particular the mirrors – will cause fluctuations in the optical paths also due to their thermal vibrations, their *Brownian motion*. The noise coming from the pendulum modes of motion is most prominent at low frequencies, rolling off steeply towards higher frequencies. The noise due to the vibrational modes of the substrate rolls off less steeply and is thus a serious disturbance at intermediate frequencies.

By choice of materials (high mechanical Q) and appropriate shaping of the substrates (to keep their resonant frequencies above our kHz range) the effect of these thermal motions can be reduced.

Intensive research is going into the development and choice of appropriate materials for the mirror substrates (pure fused silica, sapphire), and the proper treatment for attaining the highest mechanical Q, e. g. several 10^7. Such high values can be maintained only if the bonding to the suspension 'wires' does not introduce losses. Special bonding techniques are required using fibers of material identical to the substrate (monolithic suspension). Efficient collaboration between the European groups under the leadership of the University of Glasgow has given very promising results.

4.3   Seismic noise

The mirrors between which the distances are to be monitored are suspended as pendulums in vacuum, to isolate them from extraneous vibrations: from seismic and acoustic noise.

Combinations of various schemes (pendulum suspension, lead-and-rubber stacks, even active position control) are used to reduce seismic noise by many powers of 10, which is relatively easy for frequencies above, say, 100 Hz. It is only with extreme effort that this lower frequency bound can be lowered to 10 Hz or less. Not only does the natural noise rise drastically towards low frequencies, but also the pendulum isolation becomes less effective. This causes the very steep rise to low frequencies in the righthand sensitivity curve in Figure 2.

VIRGO has developed an extremely powerful seimic isolation system, the "super-attenuator", consisting of a series of 6 successive pendulum stages, in conjunction with an 'inverted pendulum', and an active isolation stage. This suspension will allow VIRGO to extend GW search to lower frequencies than other terrestrial detectors. Figure 6 shows such an attenuator, with a total height of about 10 m. Preloaded cantilever springs in each stage provide excellent vertical isolation.

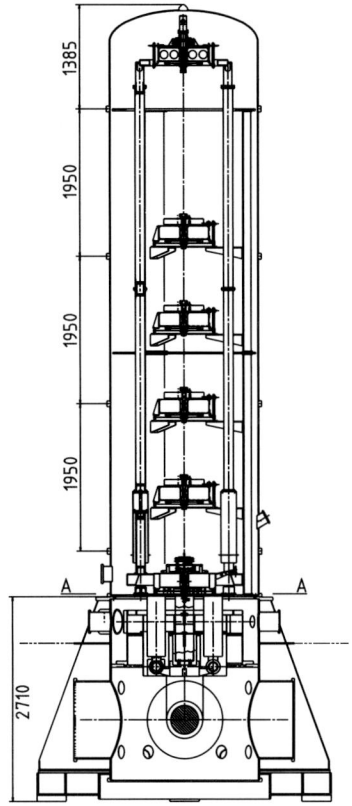

Figure 6: Super-attenuator suspension in VIRGO

For both of the *thermal* noise effects, the internal vibrations of the mirrors as well as the pendulation mode, as well as for the *seismic* disturbances, the sensitivity goal in strain of $h = 2\delta\ell/\ell$ can only be reached if we choose the armlength ℓ long enough. This is where our need for kilometer dimensions comes from. The steep rise at the left-hand side of the sensitivity curve "Earth" in Figure 2 is mainly due to the seismic and vibrational noise.

4.4   Shot noise

Particularly at higher frequencies, the sensitivity is limited by another fundamental source of noise, the so-called shot noise, a fluctuation in the measured interference coming from the "graininess" of the light. These statistical fluctuations fake apparent fluctuations in the optical path difference ΔL that are inversely proportional to the square root of the light power P used in the interferometer. For the very ambitious aims of the "advanced" detectors, about 10 kW of light power, in the visible or the near infrared, would be required. This is not as unrealistic as it may sound; it can be realized by the concept of "power recycling".

The laser interferometers are planned to monitor the (gravitational-wave induced) changes δL of the light path by observing the dark fringe of the interferometer in one output port. The (unused) light going out at the other port of the beam splitter can be fed back, via a mirror M_P, and in correct phase with the incoming light (Figures 3, 4), so that the circulating light power is significantly enhanced. This scheme was proposed by Ron Drever in 1981, at the same time as Roland Schilling saw it come as a natural consequence in the Garching 30 m prototype, where the appropriate feedback had already been implemented for an efficient frequency stabilisation of the laser. The first implementations were done in that prototype: 1987 with only short arms, in 1996 with the full 30 m arm length [21].

To achieve the sensitivity goals of the current generation of gravitational wave detectors, the light power circulating in the interferometer needs to be of the order 10 kW. With lasers of 10 to 100 W and power recycling gains of 100 to 1000, such values are within current technology.

Shot noise is a 'white' noise, but with the response rolling off as $1/f\tau$ at frequencies above the inverse storage time, the apparent strain noise rises with frequency, as shown in the curves 'Space' and 'Earth' in Figure 2.

4.5   Advanced interferometry configurations

An additional "recycling" scheme was later proposed by Brian Meers, and now forms the baseline for the GEO 600 interferometer: 'signal recycling (SR)' [14]. A further mirror, M_S, is added to the interferometer, this one in the output port (Figures 3, 4). The microscopic position of this mirror can be adjusted such that the signal sideband is also resonant in the interferometer, providing an enhancement of the signal, with possibly reduced measuring bandwidth. Schemes like this "signal recycling (SR)", or the related "resonant sideband extraction (RSE)" [15], are expected to be employed in future upgrades also of the other detectors. They can be used to optimize and tune the detector bandwidth independently of the carrier storage time in the arm cavities. Experimental verification of the combination of power and signal recycling, albeit in a table-top setup, was given in 1995 [22].

The curve marked *"Earth"* in Figure 2 indicates the sensitivities that will eventually be reached with the current large interferometers, at least in their advanced versions.

4.6   Next-generation ground-based detectors

Even though the current detectors are not yet in full operation, it is essential to develop a next generation of detectors early on. The study of new technologies to be employed, of new materials, of advanced interferometric configurations has to be pushed forward, so that the necessary new implementations can be undertaken in or around the year 2005.

Three plans for such next-generation detectors have been put forward, which are entered in the lower part of Table 1: Advanced LIGO, LCGT, and EURO. The status of these three future projects will be sketched below.

Advanced LIGO   Among these, the proposed US project is furthest progressed and well documented [23]. Advanced LIGO makes full use of the common efforts in the LIGO Scientific Collaboration, LSC. For locations, Advanced LIGO will rely on the existing facilities at the sites of Hanford and Livingston. The advantage is clear: no cost for new sites, for civil and vacuum engineering. One draw-back is that the incorporation of more "aggressive" approaches (cryogenics, all-refractive optics, Sagnac) is not so easy to realize, and that the option of lower seismic noise of underground sites is forfeited.

The Advanced LIGO groups of LSC have come up with simulations of the expected sensitivity that indicate that an operation limited only by the optics noise (shot noise, radiation pressure noise) appears possible. The suspension would have to be modeled after the GEO 600 triple pendulum concept, mirrors be made from large substrates of sapphire (or YAG), and the schemes of SR or RSE, developed at GEO, have to be used.

As was recently shown by Buonanno and Chen [24], the 'detuned' implementation of SR/RSE can even lead to a (moderate) reduction of what is usually termed the 'Standard Quantum Limit'.

LCGT   The concept of the Japanese project LCGT (Large Cryogenic Gravitational-Wave Telescope) is also rather well defined; it will use supercooled (cryogenic) mirrors. The location of LCGT will be deep inside the mountain that houses the famous neutrino detector Super-Kamiokande. The ground noise is by nearly two orders of magnitude lower than at ground level. The armlength will be 3 km, and an existing tunnel can be used for one arm. Funding is not yet secured.

EURO   Even more ambitious is the concept of the European detector EURO. The four funding agencies (CNRS, MPG, INFN, PPARC) of France, Germany, Italy, and the UK, agreed to pursue the definition of a common European high-sensitivity detector. However, the completion and commissioning of the current projects, GEO 600 and VIRGO, has the highest priority. Thus, the actual beginning of the project may be as late as 2008. A site deep underground (as for LCGT) is preferred, but not yet decided upon.

Simulations using parameter sets from optimistic but not unreasonable assumptions, verified that an operation limited only by the (quantum-)optical noise, i.e. solely by shot noise and radiation pressure noise, seems possible, using classical techniques, but going to the ultimate frontier of current technologies.

4.7   The technological challenge in advanced detectors

The technology required in the proposed future detectors is at the forefront of current technology, and even beyond. Many new developments are being pushed just by this goal of improved gravitational wave detectors, but it is fortunate that some other developments required are also driven by commercial interest.

Optical materials   Of great importance is the effort to develop better optical materials. The high light powers (up to the order of Megawatt!) will require extremely low absorption losses, in the reflective and anti-reflective coatings in reflection, as well as in the bulk material of the substrate in transmission. Three materials have the greatest promise: (1) very pure fused silica, specially prepared to have low OH-content, for low absorption, (2) sapphire single crystals, having a naturally low absorption, but with a high constant of birefringence, (3) YAG (Yttrium aluminum garnet) as used for laser crystals.

For all three materials, the great technological challenge is to produce substrates of the order of half a meter in diameter, and of similar thickness, with a high level of homogeneity. For substrates for the end mirrors, which do not require light transmission, also silicon is a possible option.

High mechanical Q   At the same time, to keep thermal vibrations low, the optical components need to have extremely high mechanical quality Q. Investigation into the properties of the materials (see above) are carried out by several institutions, and Q values of the order $Q \sim 10^7$ at room temperature and 10^9 at cryogenic temperatures have been accomplished [25, 26]. And also much research was done on how to suspend these components in such a way as to preserve their high Q [25, 26].

5   Space interferometer LISA

Only a space mission allows us to investgate the gravitational wave spectrum at very low frequencies. For all ground-based measurements, there is a natural, insurmountable boundary towards lower frequencies. This is given by the (unshieldable) effects due to varying gravity gradients of terrestrial origin: moving objects, meteorological phenomena, as well as motions inside the Earth. To overcome this "brick wall", the only choice is to go far enough away, either into a wide orbit around the Earth, or better yet further out into interplanetary space. The European Space Agency (ESA) and NASA have agreed to collaborate on such a space mission called LISA, "Laser Interferometer Space Antenna" [28].

5.1   The LISA configuration

Once we have left our planet behind and find ourselves in outer space, we have some great benefits for free: to get rid of terrestrial seismic and gravity gradient noise, to have excellent vacuum along the arms, and in particular to be able to choose the arm length large enough to match the frequency of the astrophysical sources we want to observe.

LISA consists of three identical spacecraft, placed at the corners of an equilateral triangle (Figure 7). The sides are to be 5 million km long

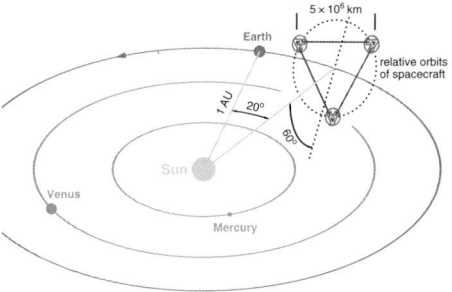
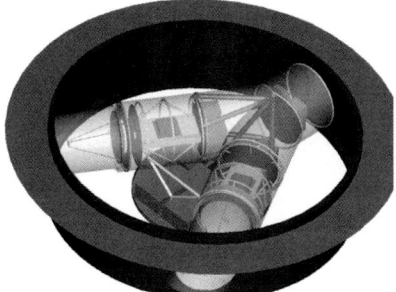

Figure 7: Orbits of the three spacecraft of LISA, trailing the Earth by 20°

Figure 8: View of one LISA spacecraft, housing t wo optical assemblies

$(5 \times 10^9 \, \text{m})$. This triangular constellation is to revolve around the Sun in an Earth-like orbit, about 20° (i. e. roughly 50 million km) behind the Earth. The plane of this equilateral triangle needs to have an inclination of 60° with respect to the ecliptic to make the common rotation of the triangle most uniform. The small orbit correction manoeuvres required can be made with field-effect ion thrusters. The three spacecraft form a total of three, but not independent, Michelson-type interferometers, here of course with 60° between the arms.

The spacecraft at each corner will have t wo optical assemblies that are pointed, subtending an angle of 60°, to the t wo other spacecraft (indicated in Figure 8, with the Y-shaped thermal shields shown semi-transparent). An optical bench, with the test-mass housing in its center, can be seen in the middle of each of the t wo arms, and a telescope of 30 cm diameter at the outer ends. Each of the spacecraft has t wo separate lasers that are phase-locked so as to represent the "beam-splitter" of a Michelson interferometer.

5.2 Gravitational sensors

The distances are measured from test masses housed *drag-free* in the three spacecraft. The three LISA spacecraft each contain t wo test masses, one for each arm forming the link to another LISA spacecraft. The test masses, 4 cm cubes made of an Au/Pt alloy of low magnetic susceptibility, reflect the light coming from the YAG laser and define the reference mirror of the interferometer arm. These test masses are to be freely floating in space.

For this purpose these test masses are also used as inertial references for the drag-free control of the spacecraft that constitutes a shield to external forces. Development of these sensors is done at various institutions. Figure 9 shows a sensor modelled after already space-proven developments at ONERA [29], other configurations are being discussed [30].

These sensors feature a three-axis electrostatic suspension of the test mass with capacitive position and attitude sensing. A resolution of $10^{-9} \, \text{m}/\sqrt{\text{Hz}}$

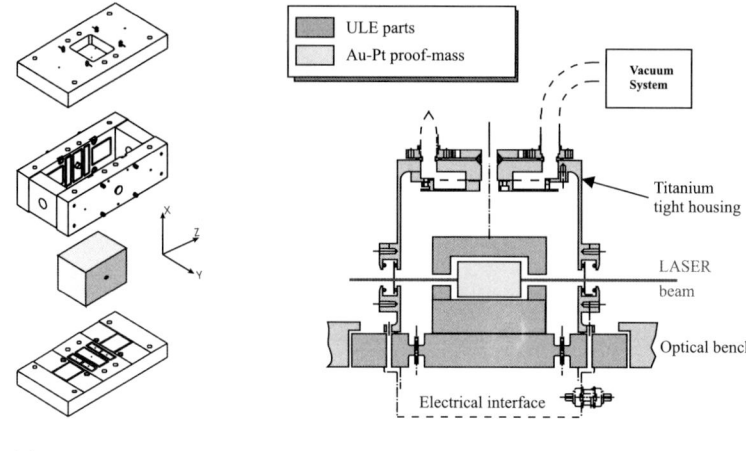

(a) The sensor cage	(b) Sensor configuration

Figure 9: Layout of gravitational sensor: (a) exploded view, (b) with housing.

is needed to limit the disturbances induced by relative motions of the space-craft with respect to the test mass: for instance the disturbances due to the spacecraft self-gravity or to the test-mass charge.

5.3 FEEP thrusters

The very weak forces required to keep up drag-free operation, less than $100\,\mu\mathrm{N}$, are to be supplied by field-effect electrical propulsion (FEEP) devices: a strong electrical field forms the surface of liquid metal (Cs or In) into a cusp from which ions are accelerated to propagate into space with a velocity (of the order $60\,\mathrm{km/s}$) depending on the applied voltage. Such FEEP thrusters have been developed at various European institutions, and their characteristics will be studied in a technology demonstration mission (Section 5.7).

5.4 Noise in LISA

With the $30\,\mathrm{cm}$ optics planned, from $1\,\mathrm{W}$ of infrared laser power transmitted, only some $10^{-10}\,\mathrm{W}$ will be received after 5 million km, and it would be hopeless to have that light reflected back to the central spacecraft. Instead, also the distant spacecraft are equipped with lasers of their own, phase-locked to the incoming laser beam.

Shot noise Due to the low level of light power received, shot noise plays a major role in the total noise budget of spurious displacements. Again, with the response rolling off at frequencies above the inverse round-trip time, this shot noise leads to the frequency-proportional rise in the sensitivity 'Space' in Figure 2.

Acceleration noise At frequencies below $1\,\mathrm{mHz}$, the noise is mainly due to accelerations of the test mass that cannot be shielded even by the drag-free scheme: forces due to gravitating masses on the spacecraft when temperature changes their distances, charging of the test masses due to cosmic radiation, residual gas in the test mass housing. The noise rolls off roughly as $1/f^2$.

Noise total With a myriad of other, smaller, noise contributions the total apparent path noise amounts to something like $\widetilde{\delta L} \approx 20 \times 10^{-12}$ m/$\sqrt{\text{Hz}}$. For signals monitored over a considerable fraction of a year, the sensitivity is about $h \approx 3 \times 10^{-24}$, indicated in Figure 2 by the curve marked "Space".

The prospects Some of the gravitational wave signals are guaranteed to be much larger. Failure to observe them would cast severe doubts on our present understanding of the laws that govern the universe. Successful observation, on the other hand, would give new insight into the origin and development of galaxies, existence and nature of dark matter, and other issues of fundamental physics.

5.5 LISA data analysis

Due to the low frequency band of the LISA detection, the amount of data is rather low. (The very limited transfer rate from spacecraft to Earth would not allow much more than the envisaged 1.5 kbit/sec.)

LISA cannot maintain a sufficient equality of the (three) arm lengths. As a matter of fact, the arms will have annual changes in length of the order 100 000 km. Thus in data analysis clever schemes need to be employed to suppress faked signals resulting from short-term fluctuations in laser frequency. Very promising is using time-delayed combinations of the six individually measured beat signals in the three spacecraft [31, 32]. Nevertheless, these schemes require a *knowledge* of the lengths of the arms to better than 30 m to be able to apply the proper time delays to the various time series.

5.6 Status of LISA

LISA is approved by ESA as a cornerstone mission under Horizons 2000. A System and Technology Study [28] has substantiated that improved technology, lightweighting, and collaboration with NASA can lead to a considerable reduction of cost. Thus, a new, *faster, cheaper, and better* approach, together with NASA, is being pursued, under the auspices of an international LISA Science Team. Launch is foreseen for 2011, not very long after first operation of the next-generation ground based detectors.

LISA has a nominal lifetime of 2 years, but the equipment and thruster supply are chosen to allow even 10 years of operation.

A collection of papers given at the *Third International LISA Symposium, 2000,* is presented in a special issue of Classical and Quantum Gravity [27].

5.7 Technology demonstrator

Some of LISA's essential technologies (gravitational sensor, interferometry, micronewton thrusters) are to be tested in a mission LTP (LISA Technology Package) on board an ESA SMART-2 satellite.

The package will contain, on a common optical bench, two gravitational sensors, similar to the one of Section 5.2 The relative motion between the two freely floating test masses will be monitored with high accuracy by interferometry. The sensitivity in this (scaled-down) experiment will come to within one power of ten to the proposed LISA sensitivity.

This package is to be launched into an orbit relatively far away from Earth in August 2006.

5.8   LISA follow-on

Even as early as now concepts are being discussed for a successor to LISA, on the possible enhancements in sensitivity and/or frequency band. One scheme would try to bridge the frequency gap between ground and space detectors, by reducing the arm lengths, leaving the general configuration unchanged. Another concept is to have a square constellation instead of the triangle, providing independent interferometers. These can be used to detect and measure a stochastic background of gravitational waves, similar to, but reaching much further back than the $3\,\mathrm{K}$ electromagnetic background radiation.

6   Conclusion

The difficulties (and thus the great challenge) of gravitational wave detection stem from the fact that gravitational waves have so little interaction with matter (and space), and thus also with the measuring apparatus. Great scientific and technological efforts, large detectors, and a working international collaboration are required to detect and to measure this elusive type of radiation.

And yet – just on account of their weak interaction – gravitational waves (just as neutrinos) can give us knowledge about cosmic events to which the electromagnetic window will be closed forever. This goes for the processes in the (millisecond) moments of a supernova collapse, as well as of the many mergers of binaries that might be hidden by galactic dust. And it is also true for the distant, but violent, mergers of galaxies and their central (super)massive black holes. A LISA follow-on mission, but also combinations of terrestrial detectors, might probe the GW background from the very beginning of our universe ($10^{-14}\,\mathrm{s}$ or even only $10^{-22}\,\mathrm{s}$ after the big bang) [33].

In this way, gravitational wave detection can be regarded as a new window to the universe, but to open this window we must continue on our way in building and perfecting our antennas. It will only be after these large interferometers are completed (and perhaps even only after the next generation of detectors) that we can reap the fruits of this enormous effort: a sensitivity that will allow us to look far beyond our own galaxy, perhaps to the very limits of the universe.

References

[1] A. Einstein, Sitzungsber. Preuss. Akad. Wiss. (1916), 688–696

[2] A. Einstein, Sitzungsber. Preuss. Akad. Wiss. (1918), 154–167

[3] K. Thorne, *Gravitational Radiation*, in: *300 Years of Gravitation*, ed. S.W. Hawking, W. Israel, Cambridge University Press (1987), 330–458

[4] K. Thorne, *Gravitational Radiation – A New Window Onto the Universe*, in: *Gravitation*, ed. R.E. Schielicke, Rev. Mod. Astron. **10** (1997), 1–28

[5] A. Buonanno, T. Damour, *Effective one-body approach to general two-body dynamics*, Phys. Rev. D **59** (1999), 084006 1–24

[6] B.F. Schutz, *Lighthouses of gravitational wave astronomy – Prospects with LIGO and LISA*, in: *Lighthouses of the Universe*, ESO Astrophys. Symp., Springer (2002), in press

[7] R. Weiss, *Electromagnetically coupled broadband gravitational antenna*, in: *Quarterly Progress Report, Research Laboratory of Electronics*, MIT **105** (1972), 54–76

[8] D. Shoemaker et al., *Noise behavior of the Garching 30 meter prototype gravitational wave detector*, Phys. Rev. D **38** (1988), 423–432

[9] A. Abramovici et al., *LIGO: The Laser Interferometer Gravitational-Wave Observatory*, Science **256** (1992), 325–333

[10] A. Lazzarini, *The Status of LIGO*, in: *Gravitational Wave Detection II*, Universal Academy Press, Tokyo (2000), 1–13

[11] A. Vicere, *The VIRGO experiment*, in: AIP Conf. Proc. **555** (2001), 138–145

[12] A. Rüdiger and K. Danzmann, *The GEO 600 gravitational wave detector – status, research, development*, in: *Gyros, clocks, interferometers: Testing relativistic gravity in space*, Lec. Not. Phys. **562** (2001), 131–140

[13] B. Willke, *The GEO 600 gravitational wave detector*, Class. Quantum Grav. (2002), in press

[14] B.J. Meers, *Recycling in laser-interferometric gravitational-wave detectors*, Phys. Rev. D **38** (1988), 2317–2326

[15] J. Mizuno et al., *Resonant sideband extraction: a new configuration for interferometric gravitational wave detectors*, Phys. Lett. A **175** (1993), 273–276

[16] M.K. Fujimoto, *Overview of the TAMA Project*, in: *Gravitational Wave Detection II*, Universal Academy Press, Tokyo (2000), 41–43

[17] about 1000 hour run: http://tamago.nao.ac.jp/tama/daq/recom/recom3/

[18] D.E. McClelland et al., *Second-generation laser interferometer for gravitational wave detection*, Class. Quantum Grav. **18** (2001), 4121–4126

[19] I. Zawischa et al., *The GEO 600 laser system*, Class. Quantum Grav. (2002), in press

[20] A. Rüdiger et al., *A mode selector to suppress fluctuations in laser beam geometry*, Opt. Acta **28** (1981), 641–658

[21] D. Schnier et al., *Power recycling in the Garching 30-m prototype interferometer for gravitational-wave detection*, Phys. Lett. A **225** (1997), 210–216

[22] G. Heinzel et al., *An experimental demonstration of resonant sideband extraction for laser-interferometric gravitational wave detectors*, Phys. Lett. A **217** (1996), 305–314

[23] *Proposal for continuing LIGO Operations (FY 2002–2006): Research and Development*, LIGO Laboratory document M000352-00-M (Dec 2000), 117–162

[24] A. Buonanno and Y. Chen, *Optical noise correlations and beating the standard quantum limit in advanced gravitational wave detectors*, Class. Quantum Grav. **18** (2001), L95–L101

[25] M.V. Plissi et al., *Aspects of the suspension system for GEO 600*, Rev. Sci. Instrum. **69** (1998), 3055–3061

[26] V.B. Braginsky et al., *Isolation of test masses in the advanced laser interferometric graqvitational-wave antennae*, Rev. Sci. Instrum. **65** (1994), 3771–3774

[27] Proceedings Third International LISA Symposium, Golm/Berlin, July 2000, Class. Quantum Grav. **18** (2001), 3965–4164

[28] LISA: System and Technology Study Report, ESA document ESA-SCI (2000) 11, July 2000

[29] V. Josselin, M. Rodrigues, P. Touboul, *Inertial sensor concept for the gravity wave missions*, Acta Astronautica **49/2** (2001), 95–103

[30] A. Cavalleri et al., *Progress in the development of a position sensor for LISA drag-free control*, Class. Quantum Grav. **18** (2001), 4133–4144

[31] J.W. Armstrong et al., *Sensitivities of alternate LISA configurations*, Class. Quantum Grav. **18** (2001), 4059–4066

[32] M. Tinto et al., *Time-delay interferometry for LISA*, Phys. Rev. D (2002), in press

[33] B. Allen, *The stochastic gravity-wave background: sources and detection*, in: *Relativistic gravitation and gravitational radiation*, Cambridge University Press (1997), 373–417 (p. 381/382)

ASTRONOMISCHE GESELLSCHAFT: Reviews in Modern Astronomy **15**, 113–131 (2002)

Observations of Weak Polarisation Signals from the Sun

Achim Gandorfer

Institute of Astronomy, ETH Zentrum
Scheuchzerstrasse 7, CH-8092 Zurich, Switzerland
gandorfer@astro.phys.ethz.ch, www.astro.phys.ethz.ch

Abstract

New highly sensitive polarimetric instruments and observational techniques allow to observe weak polarisation signals in the visible and near ultraviolet part of the solar spectrum. Many of these signals are caused by scattering processes in the upper photosphere and lower chromosphere and thus reflect the thermodynamics of these layers. Also magnetic fields lead to polarisation via the Zeeman effect or alter scattering polarisation via the Hanle effect. The observation of both effects requires highest polarimetric sensitivity in combination with very high spectral resolution. In the following the instrumental and observational concepts are described. Observations of scattering polarisation and the signatures of Hanle and Zeeman effect are presented to demonstrate the diagnostic potential of highly sensitive solar polarimetry.

1 Introduction

Polarimetric studies of astrophysical situations have become an important diagnostic technique over the last decades.

As polarised light originates in physical situations where the spatial symmetry is broken, investigations of type and degree of polarisation allow to obtain information on the symmetry relations at the point where the light is coming from. In astronomy this is of special importance since the electromagnetic radiation emitted by any astrophysical object is often the only source of information on the object's physical conditions.

Classical spectroscopy provides insights in a variety of scalar quantities like thermodynamics or chemical parameters and – using model assumptions – also in the past and the future evolution of the system, be it a star, a nebula or any other object in the universe. However, scalar spectroscopy will never yield information on vector quantities, or more general, it will never give answers on the question about physical directions, like anisotropies. For

example the magnetic field strength in sunspots may be derived easily by the
Zeeman splitting of Fraunhofer lines (as was indeed first observed by G. E.
Hale in 1908) to be of the order of 0.3 T, but it is impossible by just analysing
the intensity spectrum to answer the question how the magnetic field vector
is oriented.

Apart from magnetic fields, which represent the most prominent sources of
polarised light in astronomy, there are other – nonmagnetic – physical mechanisms that induce polarisation. Scattering on electrons (Thomson scattering),
free atoms and molecules (Rayleigh scattering) and dust (Mie scattering) can
be detected by the degree of linear polarisation induced by the symmetry
breaking.

Solar Polarimetry and the origin of polarized radiation from the Sun

The investigation of solar magnetic fields – which are responsible not only for
the structuring of the solar atmosphere but also play the fundamental role in
all activity phenomena on the Sun – has long been based on the observation of
the Zeeman effect which causes spectral lines to split up and to be polarised in
the presence of a strong directed magnetic field. While the strength of the field
is related to the wavelength shift of the components, their polarisation pattern
is related to the geometry of the field, i.e. its orientation. Field components
parallel to the line of sight introduce circular polarisation whereas the transverse components lead to linear polarisation. Since the introduction of the
magnetograph by Babcock (1953) and Kiepenheuer (1953) in the early fifties
the circular polarisation and thus the line of sight component of the magnetic
field has been regularely measured with ever increasing accuracy and sensitivity. However, the observation of the full field vector requires the observation
of the full polarisation signal including the linear polarisation and has long
been unpopular, since the linear polarisation signals are usually one order of
magnitude weaker than the circular polarisation. With the improvements in
instrumentation our knowledge of the nature of the photospheric magnetic
field has evolved to an ever more detailed picture. The basic concept is that
of a highly intermittent field that manifests itself mostly in strong kG flux
tubes that occupy only around 1 % of the solar surface and are embedded in
an almost field free atmosphere (Stenflo 1973). These flux tube signals are
ideally observed using Zeeman polarimetry since the Zeeman signal is largest
if the magnetic line splitting is comparable to the Doppler width of the line,
which is the case for kG fields in the photosphere. Zeeman polarimetry fails
however to detect weak or turbulent magnetic fields with mixed polarities
within the spatial resolution element, which lead to cancellation effects.

Scattering polarisation and Hanle effect

Scattering can break the spatial symmetry and thus introduces polarisation.
On the Sun different scattering mechanisms contribute to the polarisation

of both the continuous and the line spectrum. The symmetry breaking is due to the anisotropic illumination of a scattering particle. The anisotropy of the radiation field is caused by the temperature stratification of the solar atmosphere, which manifests itself also in the limb darkening. As the radiation field is basically radial, when observing at the center of the solar disk no polarisation will be seen, since the observational situation is rotationally symmetric around the line of sight.

This symmetry is broken when going from disk center to the Sun's limb, and thus the scattering polarisation shows a typical center-to-limb variation, vanishing at disk center and beeing maximum at the extreme solar limb. With growing wavelength the limb darkening decreases and thus the magnitude of the scattering polarisation also reduces as we go up upwards in wavelength (see Fluri & Stenflo 1999). While the continuous spectrum is formed by scattering at free electrons (*Thomson* scattering) or atoms (*Rayleigh* scattering), the line spectrum is due to *resonant* or *fluorescent* scattering. A blueprint example of a resonant scattering event in quantum mechanics is the $J = 0 \to 1 \to 0$ transition, which is the quantum mechanical analogon to the classical dipole antenna that leads to total polarisation of the scattered radiation in the case of $90°$ scattering. For other combinations of the quantum numbers the scattering amplitudes (the *line polarisability*) is generally smaller.

All these considerations are valid for the non magnetic case. Magnetic fields modify the scattering polarisation via the *Hanle effect*. The Hanle effect is the influence of magnetic field on the polarisation caused by coherent scattering in spectral lines. As the Hanle effect is basically a coherence phenomenon it exclusively occurs if there is coherent scattering. It can be described by a relaxation of the quantum mechanical interferences between Zeeman sublevels that are excited via the incident radiation field. If the Zeeman splitting is smaller than the natural width of the line (as given by the inverse lifetime of the upper level of the transition), then the different magnetic sublevels overlap, and an incident radiation creates a coherent superposition of these sublevels. In the classical description an incident wave excites a classical oscillator. The linear oscillation can be decomposed into two circular oscillations that are coherent. In the absence of magnetic fields the scattered radiation is purely linearly polarised. In a magnetic field the atom will undergo a Larmor precession around the magnetic field direction and the two circular oscillation frequencies mix with the Larmor precession frequency. When looking along the magnetic field the classical oscillator describes a rosette motion. As in the classical representation the oscillator radiates energy the oscillation will be damped resulting in a damped rosette motion. If the Larmor precession rate is high as compared with the inverse lifetime of the oscillator, when looking along the magnetic field an observer sees a complete rosette, and no linear polarisation will be seen. If, however, the Larmor precession is slower than the radiative decay time, a more or less damped rosette will be seen. For a rigorous treatment of the Hanle effect we refer to the review by Trujillo-Bueno (2001).

The Hanle effect manifests itself in two distinct ways: As a depolarisation and as a rotation of the plane of polarisation. The important point is that for the Hanle rotation both signs occur dependent on the field orientation, while the sign of the Hanle depolarisation is always "negative", i.e. a magnetic field always *reduces* scattering polarisation. As a consequence Hanle depolarisation can be used as a measure of even turbulent magnetic fields with mixed polarities within each spatial resolution element that would give rise to no net Zeeman polarisation signal due to cancellation of the contributions with different signs. The other crucial point is that the Hanle effect is most sensitive when the Larmor precession is comparable to the lifetime of the upper level. Thus the Hanle effect is most sensitive in a field strength regime, for which the Zeeman splitting is comparable to the *natural* width of the line. The Zeeman effect manifests itself, if the Zeeman spitting gets comparable to the *Doppler* width of the line. While the Zeeman polarisation gets stronger when the magnetic field increases, and there is no polarisation at all in the absence of a magnetic field, the Hanle effect manifests itself in the opposite way: The maximum polarisation is related to the field free scattering signal, and the polarisation decreases in the presence of a (weak) magnetic field.

The Hanle effect has early been used as a diagnostic tool for magnetic fields in solar prominences (e. g. Leroy et al. 1977; Bommier 1980; Querfeld et al. 1985) using the He I D_3 5876 Å line. In this case the depolarisation and the rotation can be studied without taking radiative transfer into account, because the line is optically thin.

However, Hanle diagnostics of photospheric or chromospheric fields on the solar disk need detailed understanding of the physical processes involved in the formation of the scattering signals as well as their propagation through an optically thick plasma, taking into account state of the art of radiative transfer calculations.

So, until recently the Hanle effect on the disk has been only used for estimates in the turbulent field strength regime. Stenflo (1982) constrained the turbulent field to be 10–100 G, a range that was later narrowed to 10–20 G by detailed analysis of the Sr I 4607 Å line by Faurobert-Scholl (1993) and Faurobert-Scholl et al. (1995).

Nowadays observations with unprecedented polarimetric accuracy have opened new windows to see inside the different physical mechanisms contributing to the wealth of scattering polarisation structures. The features of this "second solar spectrum", a name which aptly describes the fact that the scattered radiation has a totally different spectral appearance than the ordinary Fraunhofer spectrum of the solar intensity (Ivanov 1991), are now accessible for detailed theoretical analysis. Together with ever more sophisticated concepts of radiative transfer calculations the mystery of the "second solar spectrum" and the nature of the solar magnetic field may be unveiled in near future.

One might have the impression that an increase of the polarimetric sensitivity to ever smaller values is only a *quantitative* improvement and only a marginal step forward as compared to existing observational techniques, as one could think that an improved polarimetric sensitivity does not lead to fundamental new insights. In the following it will be shown that this is not the case and that high polarimetric sensitivity opens a new world of diagnostic possibilities. High sensitive solar polarimetry is therefore a *qualitative* step forward in the means of analysing solar magnetic fields.

2   Instrumentation
for highly sensitive solar polarimetry

Polarimetry is polarisation sensitive photometry. The big advantage of polarimetry as compared to classical photometry lies in the fact that for most applications only the degree of polarisation, i. e. the relative amount of polarised light, must be known. Polarimetry in this case is a *differential* measurement where two or more photometric images are combined to form polarisation images.

Such measurements suffer from three main noise sources.

1. *Seeing* introduces errors when the two images that are compared are taken sequentially ("single beam polarimetry").

2. *Gain table* or *flat field* noise arises when the two differential images are spatially separated and detected simultaneously with two detectors or two separated areas of the same detector. This is the main error source in beam splitter based polarimeters ("two beam polarimetry").

3. Photometric shot noise reflects the statistic behaviour of light and is thus unavoidable.

While the last noise source can only be suppressed by increasing the photometric statistics (long integration time), the other two errors can be avoided using instrumental concepts.

Fast polarisation modulation

Intensity fluctuations or image motions due to seeing can severely corrupt polarisation measurements. To avoid influences of seeing the differential polarisation images must be obtained quasi-simultaneously, i. e. the polarisation must be modulated at frequencies higher than the seeing fluctuation frequencies. For typical solar telescopes with apertures of 50 cm to 1 m the seeing variations go up to frequencies of 100 Hz. To completely avoid influences of seeing on the polarisation measurement a modulation frequency of at least 1 kHz should be chosen. Fast polarisation modulation can be achieved in different ways. As mechanical modulators like rotating waveplates are not suitable, electro-optical devices must be used. These can be Pockels cells, piezoelastic modulators (PEMs), or ferroelectric liquid crystals (FLCs). A PEM is a slab of fused silica, which is excited via a piezoelectric transducer at its mechanical resonance frequency. The oscillating mechanical stress induces birefringence, and thus the retardance of the PEM varies sinusoidally

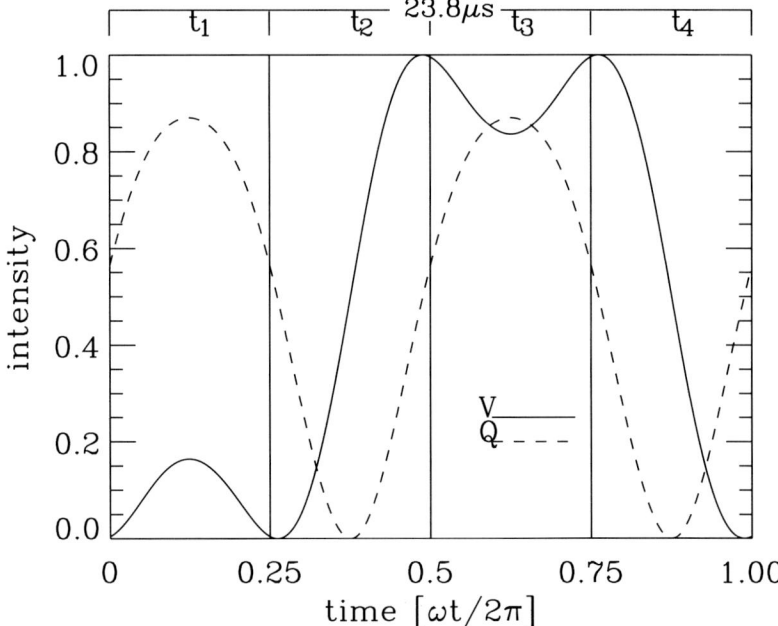

Figure 1: Polarisation modulation with one piezoelastic modulator followed by a linear polariser. While the circular polarisation is modulated at ω, one component of the linear polarisation is modulated at 2ω. $\omega/2\pi$ is the mechanical vibration frequency of the device (in our case 42 kHz). The polarisation is proportional to the amplitude of the modulation and can be extracted by dividing the modulation period in 4 sampling intervals t_1, t_2, t_3, t_4 as indicated by the vertical lines. If the detector integrates the light in each of these four sampling intervals to corresponding charge packages A, B, C, D, the sum of the four images is proportional to the intensity I of the beam, while Q is proportional to the value A−B+C−D and Stokes V is proportional to A+B−C−D. The fractional polarisation Q/I and V/I is therefore independent of the response of the detector, which divides out.

according to $\delta(t) = A\sin(\omega t)$, where A is the maximum retardance at a given wavelength. A can be tuned to any wavelength by varying the drive voltage of the device, $\omega/2\pi$ is typically 40–50 kHz. PEMs have been used with great success for the measurement of small polarisation signals. Their optical stability is very good, and they can be tuned to any wavelength in a wide wavelength range. With one PEM only one component of the linear polarisation (Stokes Q, or U) and the circular polarisation are modulated simultaneously (Gandorfer and Povel 1997). To measure the complete Stokes vector a second PEM must be used, which however normally has a different modulation frequency (Povel 1995). Attempts to use a frequency and phase coupled pair of two PEMs to modulate the complete Stokes vector (Stenflo et al. 1992) have failed until now.

Astronomische Gesellschaft: Reviews in Modern Astronomy 15, 93–112 (2002)

Synchronous demodulation
using charge coupled device image sensors

Standard CCD image sensors are not compatible with the requirement of fast modulation. In the example of Fig. 1 each sampling interval is only around 5 μs long, much less than the typical read-out speed of the detector. Even if it would be possible to read out the frames in synchrony with the fast modulation, almost no photo charge would have been collected, and thus the read-out noise would dominate over the photon shot noise. Therefore Povel et al. (1990) developed the technique of fast on-chip demodulation by creating fast hidden buffers on the CCD sensor. The goal was to allow fast polarisation modulation in combination with imaging sensors, and integration over many modulation cycles without reading out the CCD. In the Zurich *IMaging POLarimeter* ZIMPOL I every second row of the CCD sensor is covered with an opaque mask, so that the screened rows serve as temporary buffer storage zones. In this way two interlaced frames can be handled simultaneously in the CCD, by shifting the photo charges back and forth in synchrony with the modulation. After many thousands of modulation cycles the sensor is read out. The fractional polarisation image is extracted by forming the ratio of the difference of the two images divided by their sum. In this way the resulting fractional polarisation image is absolutely free from any pixel-to-pixel variation of the quantum efficiency (gain table) since the two differential images that are compared have been obtained with the same set of physical pixels. With a ZIMPOL I camera the intensity and one Stokes parameter can be recorded with extremely high sensitivity. For details on ZIMPOL I we refer to Povel (1995, 2002).

In the ZIMPOL principle the two main noise sources are avoided simultaneously, and the noise in the fractional polarisation images is always dominated by the photon shot noise down to noise levels of 5×10^{-6}. In the second generation Zurich *IMaging POLarimeter* ZIMPOL II three out of four pixel rows are covered with an opaque mask, thus allowing to handle four independent image planes simultaneously in the CCD. The principle is sketched in Fig. 2. With appropriate modulation schemes (Gandorfer 1999a) the complete Stokes vector can be determined with unprecedented sensitivity (Gandorfer 1999b).

In a single CCD exposure around 300,000 electrons can be collected in each pixel (full well capacity), which limits the photometric (and thus the polarimetric) noise to a value of the order of $2–3 \times 10^{-3}$. To increase the signal-to-noise ratio several frames must be added, reducing the rms noise level by a factor of $\sqrt{N}$, where N is the number of frames. If one wants to measure a weak polarisation signal with a noise level of 10^{-4} or better, typically 1000 images have to been added. If one observes with high spectral and/or spatial resolution one needs even with large solar telescopes total integration times of 30 min! If the polarimetric noise level is still too high to see tiny physical effects, in principle both spectral and spatial resolution could be reduced. However, as most signals are very localized in the spectral domain, one must observe at very high spectral resolution. This means that weak signals are mostly observed with moderate or low spatial resolution.

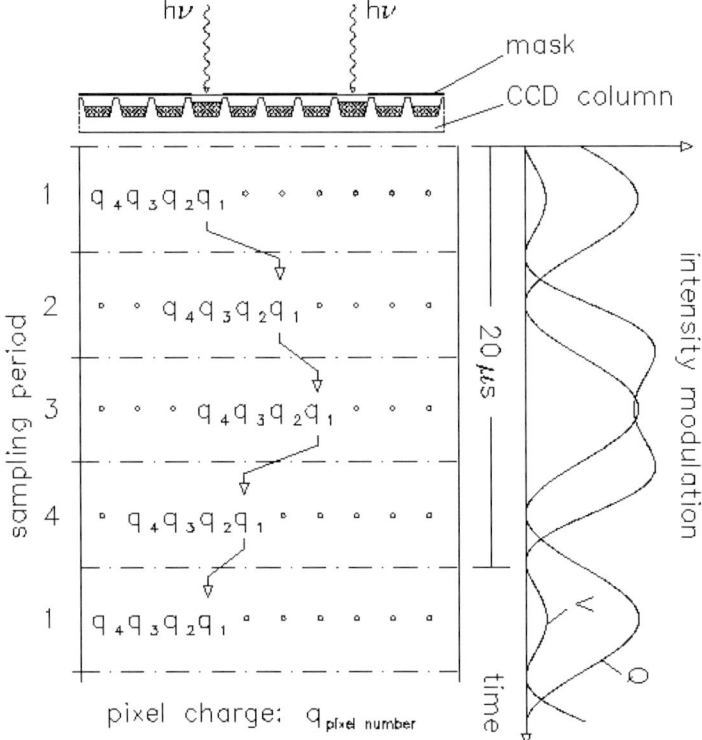

Figure 2: Principle of the ZIMPOL II demodulator. Upper part: Cross section of the CCD in a plane perpendicular to the pixel rows, showing the pixels and the opaque mask. Right part: Intensity modulation for Stokes Q and V. The modulation period is 20 μs (corresponding to 50 kHz). Central part: Position of the charge packages during the four sampling intervals. From Povel (2002). For a detailed description see also Gandorfer and Povel (1997).

3   Observing the second solar spectrum

The linearly polarised spectrum that is produced by coherent scattering has a structural richness that is comparable to that of the intensity spectrum, but its appearance is entirely different. This so called "second solar spectrum" provides us with a new window for diagnostics of the Sun.

As the second solar spectrum is formed by scattered photons the amplitude of the scattering polarisation increases steeply when observing at the Sun's limb. The observational geometry is such that the electric vector of the scattered light is parallel (in rare cases perpendicular) to the solar limb.

The natural first step in observing the second solar spectrum would be to get a systematic overview of all the spectral signatures using spectropolarimetry. Stenflo et al. (1983a, b) obtained a survey of the second solar spectrum in

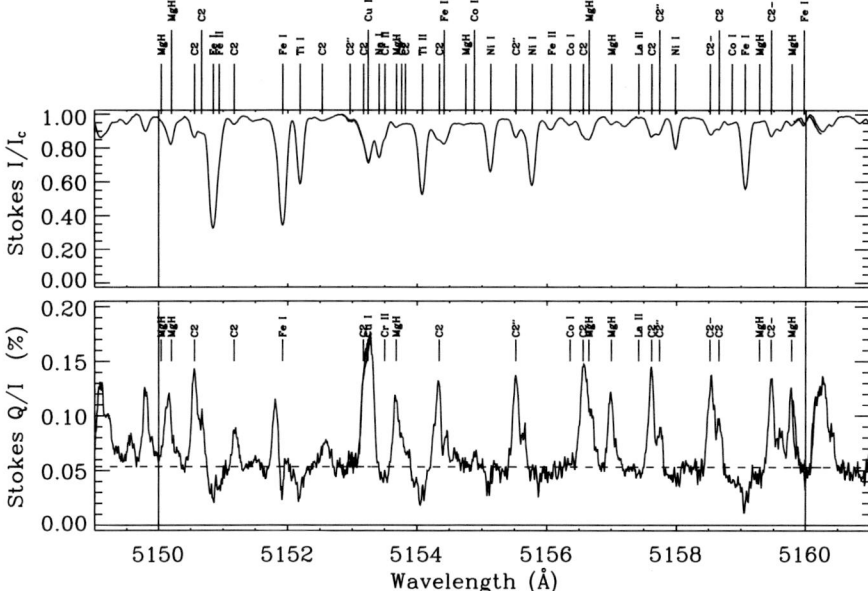

Figure 3: Instructive example of the two faces of the Sun, taken from the atlas of the second solar spectrum by Gandorfer (2000): In the upper panel the Fraunhofer spectrum between 5150 Å and 5160 Å is shown, while the second solar spectrum is displayed in the lower part of the figure. Each page of the atlas covers 10 Å with 1 Å overlap with the preceding and following pages. The scale is always from 0 to 0.2 %, which is sufficient to display most signals in an appropriate way. The continuum has been shifted to the theoretical value given by Fluri and Stenflo (1999) and is marked by the dashed line in the figure. The polarised features are due to scattering at the C_2 molecule.

the whole visible range of the spectrum, but the sensitivity of their observations was not sufficient to unveil the total wealth of polarised features in the spectrum.

As the observation of scattering polarisation requires very long integration times and the spectral field of view is rather limited when observing with very high spectral resolving power of around 250,000, which is necessary to resolve the line profiles, a systematic overview over a broad wavelength interval can only be made at a telescope with basically unlimited access.

During 1999 and 2000 an atlas of the second solar spectrum has been obtained using the ZIMPOL II polarimeter at the 45 cm Gregory-Coudé telescope of the Istituto Ricerche Solari Locarno (IRSOL) which is ideally suited for this type of observations. To reduce the noise level in the observations, which were done placing the spectrograph entrance slit parallel to and 5 arcsec inside the visible solar limb, all spectra have been averaged along the spatial direction; 500 images have been added. The spectra have noise levels around

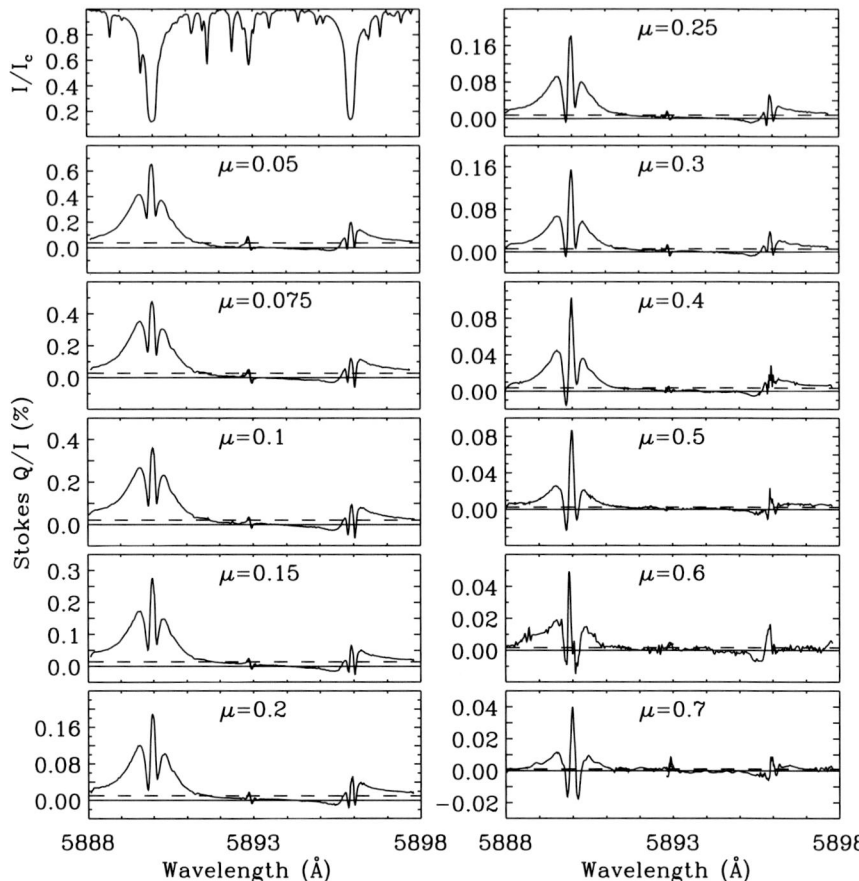

Figure 4: Center to limb variation of the complicated polarisation structure of
Na I D_2–D_1. The intensity spectrum shown in the upper left-hand panel has been
recorded at $\mu = 0.1$. All the other panels display the polarised profiles for varying
distances from the solar limb. μ, the cosine of the heliocentric angle, goes from 0.05
to 0.7. As the amplitude of the polarisation decreases steeply when going away from
the limb, the scale has been adapted. From Stenflo et al. (2000a).

2–3×10^{-5} in combination with a spectral resolution of 300,000. The total
survey, which has been published in the form of a book (Gandorfer 2000) to
serve as a reference guide for future observations and theoretical approaches
to the field, covers the wavelength range 4625 Å to 6995 Å. Currently the
near ultraviolet part of the second solar spectrum is beeing investigated using
a modified ZIMPOL II sensor which is sensitive to wavelengths below 4500 Å.

A very important point is that the second solar spectrum is not invariant,
but varies systematically with limb distance and statistically due to influences
of local magnetic fields via the Hanle effect.

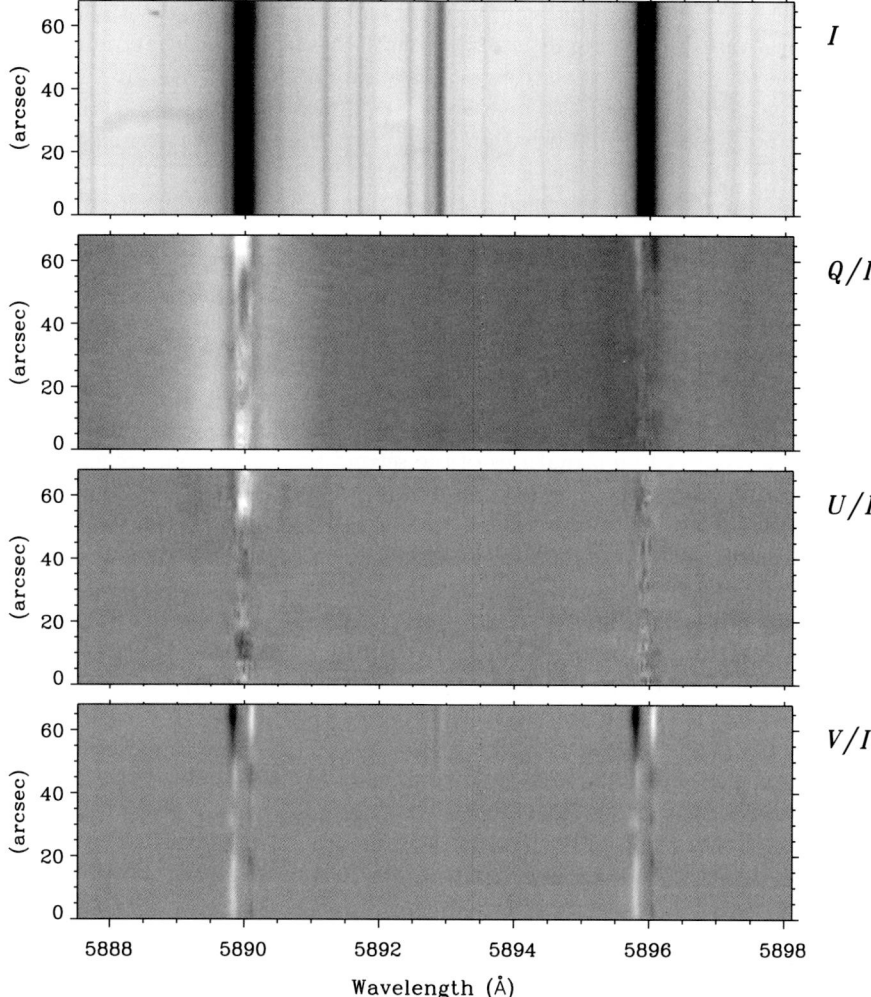

Figure 5: Example of the mixed Hanle Zeeman regime: Stokes spectra of the Na I D_2–D_1 line pair obtained with the spectrograph slit parallel to and 5 arcsec inside the solar limb in a moderately active region. The scattering polarisation structure can easily be seen in the Stokes Q/I image. Stokes Q is defined to be parallel to the solar limb. While the contributions from the longitudinal magnetic field give rise to circular polarisation, the Hanle effect by the weak components of the field lead to fluctuations in the linear polarisation in the line core of the D_2 line. In the fractional Stokes U image contributions from Hanle rotation can be seen. From Stenflo et al. (2001).

To study the center-to-limb variation one-dimensional spectra can be obtained with the spectrograph slit placed at different limb distances. Especially near the limb the exact limb distance must be precisely known, which can be made easier using an image stabilizing system (see Gandorfer 2000 for details). As the amplitude of the polarisation steeply decreases when going away from the limb, only strong signals can be observed at significant limb distances. Among the strongest lines are the Na I D_2–D_1 lines around 5890 Å and 5896 Å. These two lines show a complicated polarisation structure that has not been explainable within the current framework of atomic physics and polarised radiative transfer (Stenflo et al. 2000a, b, 2001). Their precise center-to-limb variation has been obtained using the ZIMPOL system at the 1.5 m McMath-Pierce telecope of the National Solar Observatory (Kitt Peak) by Stenflo et al. (2000a). The center-to-limb curves are shown in Fig. 4. Note the steep decrease of the signal, and the adapted scale, when going from the limb of the Sun to its center. These observations require highest polarimetric sensitivity and accurate control of the instrumental polarisation.

The second factor which influences the amplitude of the scattering polarisation signal are magnetic fields via the Hanle effect and/or the Zeeman effect. The polarisation signatures of the second solar spectrum are expected to have their largest amplitudes in the absence of magnetic fields, since the Hanle effect manifests itself primarily as a depolarisation. To study the Hanle effect different techniques have been used. The first is to observe parts of the second solar spectrum at different position angles around the solar limb and to associate the relative differences in the polarisation amplitudes with different field strength regimes in the areas observed. The interpretation of those data is based either on a differential approach (Stenflo et al. 1998) or on the technique of *Hanle histograms* (Bianda et al. 1998b, 1999). In regions with strong signals, where it is not necessary to sacrifice the spatial resolution by spatial averaging, the influence of magnetic fields may be studied directly from the two dimensional spectra. Such observations are only possible at the largest solar telescopes with a well adapted instrumentation. Using ZIMPOL II at the McMath-Pierce facility the mixed Hanle-Zeeman regime in the Na I D_2–D_1 doublet could be observed with moderate spatial resolution in regions with varying magnetic activity. An example of such an observation, which will be much easier with the upcoming generation of large solar telescopes, is shown in Fig. 5.

A natural way to study the spatial topology of the scattering polarisation is to observe the Sun through a narrowband wavelength filter which can be tuned to the center or the wings of a spectral line. If the full polarisation state is recorded in the wings of the line as well as in its core, the different contributions from the scattering polarisation and from Zeeman effect can be separated. An example, which has been obtained using ZIMPOL II in combination with the Universal Birefringent Filter (UBF) at the R. B. Dunn telescope of the National Solar Observatory (Sacramento Peak) in the center and the wings of the Na I D_2 line at 5890 Å, is shown in Fig. 6. The spatial region is approx. 80 arcsec times 80 arcsec at the limb of the Sun. While the steep increase in the linear polarisation is clearly visible in the fractional Q/I images (Stokes Q is defined to be parallel to the solar limb), the Stokes V/I images show the contributions from the longitudinal Zeeman effect due to intranetwork fields with opposite signs in the blue and the red line wings.

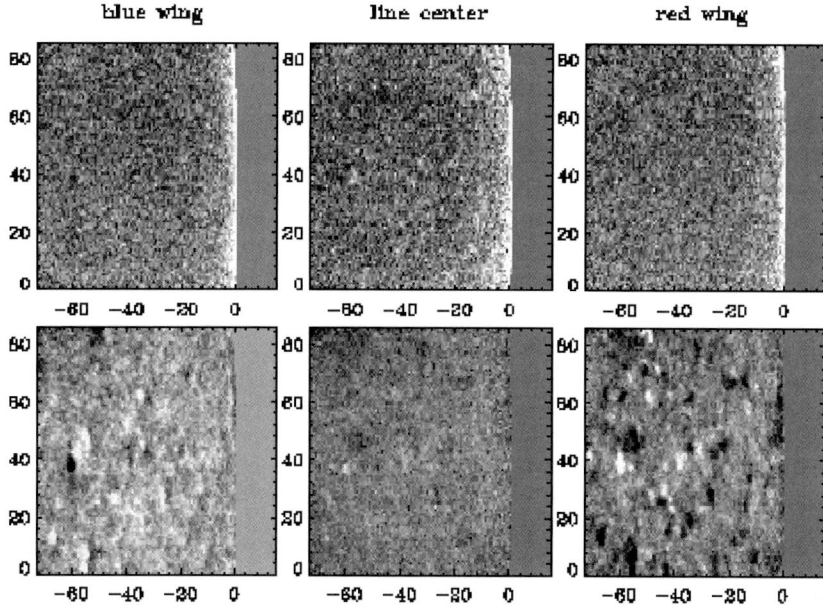

Figure 6: Example of the mixed Hanle Zeeman regime: Stokes images in the core and the wings of the Na I D$_2$ 5890 Å line at the solar limb in a moderately active region. The bandpass of the tunable filter was 240 mÅ. The fractional linear polarisation Q/I is shown in the upper row, while the fractional circular polarisation V/I is shown in the lower row. Positive Stokes Q is defined to be parallel to the solar limb. The scattering polarisation structure can easily be seen in the Stokes Q/I images. Contributions from the longitudinal magnetic field give rise to circular polarisation. The magnetic field does not affect the scattering polarisation in the wings of the spectral line, but reduces the polarisation in the Doppler core, a typical signature of the Hanle effect. The spatial field of view is around 80 arcsec times 80 arcsec, black corresponds to -5×10^{-3}, white to $+5 \times 10^{-3}$ in the degree of polarisation.

The bandpass of the tunable filter was 240 mÅ, the scale in the polarisation images is -5×10^{-3} (extreme black) to $+5 \times 10^{-3}$ (extreme white).

We see that we have different observational techniques to address the various scientific questions related to the second solar spectrum and the physics of small scale polarisation signals.

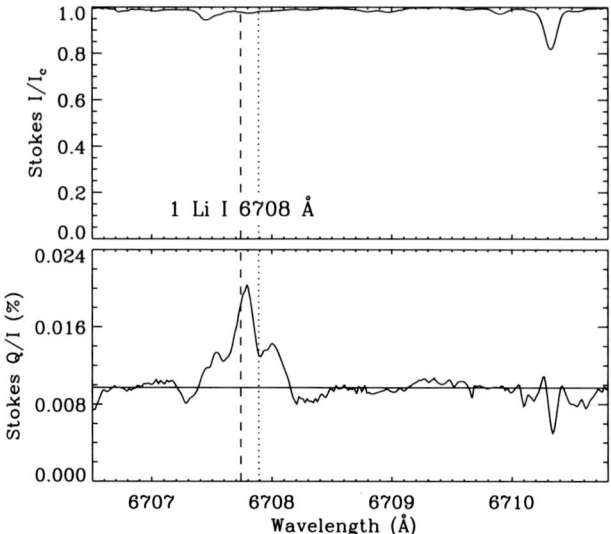

Figure 7: Spectral line diagnostics without detectable absorption lines: The Lithium doublet around 6708 Å stands out in the fractional polarisation spectrum and becomes accessible to spectroscopic investigation. The observation was done using ZIMPOL I at the McMath-Pierce facility of the National Solar Observatory (Kitt Peak) on Sept. 8 1996. 12,000 images have been added to achieve the noise level of 5×10^{-6} rms. From Stenflo, Keller, and Gandorfer (2000b).

4 The second solar spectrum as a diagnostic tool

4.1 Classical Spectroscopy

Most information on the chemical and thermodynamic structure of the solar atmosphere is based on spectroscopic investigation. As the second solar spectrum has a structural richness comparable to that of the Fraunhofer spectrum of the solar intensity, but is totally different in appearance, it is evident that the second solar spectrum contains different, complementary information. This can be used for detailed investigation of element or isotope abundances. Correct values of element and isotope abundances are of crucial importance for models of the solar interior.

As an example the Lithium doublet around 6708 Å is shown. Because of the abnormally low abundance of lithium in the Sun due to mixing of the surface layers to depths, where nuclear burning of lithium can take place, there is no detectable absorption that could be used for classical spectroscopy in these lines. However the resonance structure of the the Li I 6708 Å line of multiplet no. 1 stands out with high contrast in the polarised spectrum, as shown by Fig. 7. For a detailed analysis of the observed triplet structure

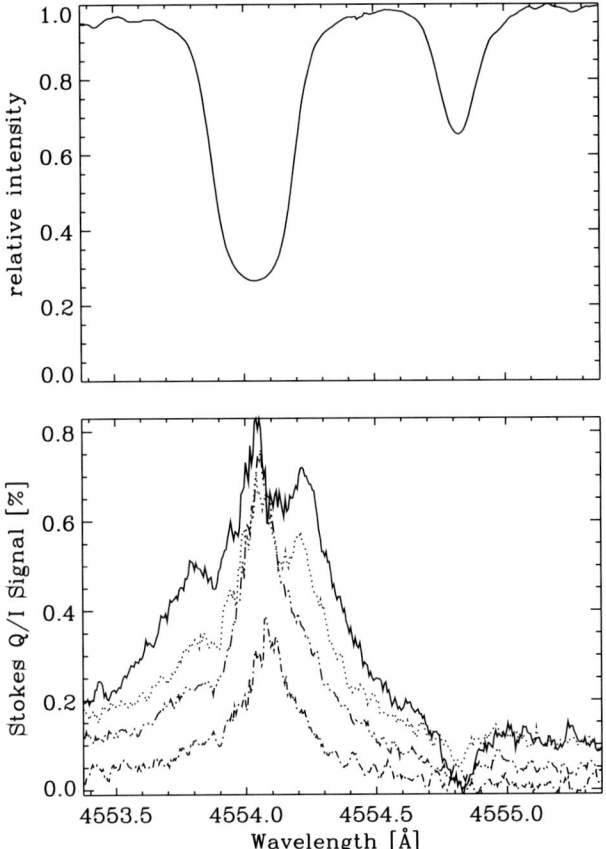

Figure 8: Fraunhofer and second solar spectrum of Ba II 4554 Å. The fractional polarisation profiles reveal the hyperfine structure splitting of the odd isotopes. The
different curves reflect the strong center to limb variation of the scattering polarisation; the strongest polarisation has been obtained for $\mu = 0.05$. The other curves
refer to $\mu = 0.1$, 0.15, and 0.2 μ being the cosine of the heliocentric angle. From
Gandorfer and Povel (1997).

we refer to Stenflo, Keller, and Gandorfer (2000b). Another example can be
seen in Fig. 8. While the Ba II line at 4554 Å is a single line in the intensity
spectrum and cannot be separated into its hyperfine structure components due
to saturation effects, the hyperfine structure splitting of the odd isotopes is
clearly seen in the fractional polarisation spectrum of the line. Other targets
for spectroscopic investigation are the elements of the rare earths that are
difficult to observe in the Fraunhofer spectrum but give numerous large signals
in the second solar spectrum. The reason why the ions of elements like Nd,
Y, Sc, Pr, or Ce are strongly polarising is not yet clarified. In general, before
the second solar spectrum can be used for diagnostic purposes for the Sun,
the nature of scattering polarisation must be understood.

4.2   Atomic physics

The structural richness of the second solar spectrum is related to the fact that different physical mechanisms contribute to the wealth of polarised features. The different scattering mechanisms respond to the incident radiation with a characteristic wavelength dependence. While the continuous spectrum is formed by scattering at free electrons (Thomson scattering) and atoms (Rayleigh scattering), the spectral lines reflect the contributions from resonant or flourescent scattering. The amplitude of the line polarisation depends strongly on the intrinsic polarisability of the line, which is related to the quantum mechanical structure of the transition. The intrinsic polarisability of the line may be derived from quantum mechanical calculations within the framework of atomic or molecular physics. However, for many lines the calculations are not straightforward and the situation is often further complicated due to hyperfine structure or effects of quantum mechanical interference of fine structure components (Stenflo 1997). These effects are normally insignificant for the formation of the absorption line profile, but dominate the polarisation structure. The most prominent example of quantum mechanical interference is the complicated polarisation structure in the Na I D_2–D_1 lines pair, which was already shown in Fig. 4. Complicated theoretical concepts like so called lower level polarisation and optical depopulation pumping are discussed by various authors to explain the observed signatures. Direct measurements of these effects provide a rigorous test for theoretical approaches to these exotic mechanisms (c. f. Landi 1998; Trujillo-Bueno 2001, Manso-Sainz and Trujillo-Bueno 2001).

4.3   Magnetic field diagnostics

Since the Hanle effect is most sensitive to field strengths for which the corresponing Zeeman splitting is comparable to the natural line width, we can address a totally different field strength regime. While the Zeeman effect is ideally suited to observe the directed flux tube fields in the photosphere, which are of the order of 1 kG, Hanle effect based diagnostics gives information on the turbulent, weak background field with intrinsic field strengths in the range 1–100 G.

Hanle effect based magnetic field diagnostics is still in its explorative state, but already, using idealised model assumptions, important properties of the weak background field could be observed.

Here only examples will be presented, as a detailed discussion of the state of the art in Hanle effect diagnostics would be beyond the scope of this paper.

Using a differential approach, Stenflo, Keller, and Gandorfer (1998) showed that the turbulent background field that fills up to 99 % of the photospheric volume has not a unique field strength, but shows large spatial variations in the range of 4–40 G. This is consistent with the observations by Bianda et al. (1998a, b, 1999) who use another, more statistical approach to derive the field strength distribution in the quiet solar photosphere and lower chromosphere.

The question whether or not the second solar spectrum exhibits temporal variations in phase with the solar activity cycle, as has been suspected in Gandorfer (2001), is still open. For a further discussion of the field we refer to the recent reviews by Stenflo (2001) and Trujillo-Bueno (2001).

5 Outlook

During the last five years the number of observations dealing with the problem of scattering polarisation and Hanle effect has dramatically increased. While direct observations using new improved polarimeters and new observational techniques are reported from various groups (especially from the new THEMIS observatory, c. f. Trujillo-Bueno et al. 2001; Arnaud et al. 2001; at IRSOL: Bianda et al. 1998a, b, 1999, 2001; with the german telescopes on Tenerife: c. f. Dittmann et al. 2001), new theoretical concepts in polarised radiative transfer and the modelling of the second solar spectrum (c. f. Fluri et al. 2001) allow ever more detailed interpretation of the various observations.

We will enter a new era with the advent of new large solar telescopes (ATST, Keil et al. 2001; GREGOR, von der Lühe 2000), which will offer the possibility to observe weak polarisation signals with unprecedented spatial resolution.

Personally I have no doubt that the observation of scattering polarisation offers a unique potential for our attempts to analyse and understand the nature of the Sun's magnetic field.

Acknowledgements

The ZIMPOL instruments have been invented, developed and built at the Institute of Astronomy of ETH Zurich by H. P. Povel, U. Egger, Stefan Hagenbuch, Peter Steiner and Frieder Aebersold. M. Bianda provided observing time at the IRSOL facility. D. Gisler helped during the observations at the Sacramento Peak Observatory. I would like to thank J. O. Stenflo for his support and for numerous discussions. This work has been partly supported by the Swiss National Science Foundation, which is gratefully acknowledged.

References

Arnaud, J., Vigneau, J., Faurobert, M., Paletou, F. 2001, in: "Advanced Solar Polarimetry", Sigwarth, M. (ed.), ASP conference series 236, 151

Babcock, H.W. 1953, ApJ 118, 387

Bianda, M., Stenflo, J.O. 2001, in: "Advanced Solar Polarimetry", Sigwarth, M. (ed.), ASP conference series 236, 117

Bianda, M., Solanki, S.K., Stenflo, J.O. 1998a, A&A 331, 760

Bianda, M., Stenflo, J.O., Solanki, S.K. 1998b, A&A 337, 565

Bianda, M., Stenflo, J.O., Solanki, S.K. 1999, in "Solar Polarisation", Proc. 2nd Solar Polarisation Workshop, Nagendra, K.N., Stenflo, J.O. (eds.), ASSL 243, 31

Bommier, V. 1980, A&A 87, 109

Dittmann, O., Trujillo-Bueno, J., Semel, M., Lopez-Ariste, A. 2001, in: "Advanced Solar Polarimetry", Sigwarth, M. (ed.), ASP conference series 236, 125

Faurobert-Scholl, M. 1993, A&A 268, 765

Faurobert-Scholl, M., Feautrier, N., Machefert, F., Petrovay, K., Spielfiedel, A. 1995, A&A 298, 289

Fluri, D., Stenflo, J.O. 1999, A&A 341, 902

Fluri, D.M., Nagendra, K.N., Frisch H. 2001, A&A, in press

Gandorfer, A. 1999a, Opt. Engineering 38(8), 1402

Gandorfer, A. 1999b, in "Solar Polarisation", Proc. 2nd Solar Polarisation Workshop, Nagendra, K.N., Stenflo, J.O. (eds.), ASSL 243, 297

Gandorfer, A. 2000, "The Second Solar Spectrum. A high spectral resolution polarimetric survey of scattering polarization at the solar limb in graphical representation", Vol. I 4625 Å to 6995 Å. Zurich: VdF. ISBN: 3 7281 2764 7

Gandorfer, A. 2001, in: "Advanced Solar Polarimetry", Sigwarth, M. (ed.), ASP conference series 236, 109

Gandorfer, A., Povel, H.P. 1997, A&A 328, 381

Ivanov, V.V. 1991, in: "Stellar Atmospheres – Beyond Classical Models", Crivellari, L., Hubeny, I., Hummer, D.G., (eds.), Proc. NATO, 81

Keil, S., Rimmele, T.R., Keller, C.U., ATST team 2001, in: "Advanced Solar Polarimetry", Sigwarth, M. (ed.), ASP conference series 236, 597

Keller, C.U. 1996, Sol. Phys. 164, 243

Kiepenheuer, K.O. 1953, ApJ 117, 447

Landi degl'Innocenti, E. 1998, Nature 392, 256

Leroy, J.L., Ratier, G., Bommier, V. 1977, A&A 54, 811

Manso-Sainz, R., Trujillo-Bueno, J. 2001, in "Advanced Solar Polarimetry", Sigwarth, M. (ed.), ASP conference series 236, 213

Povel, H.P. 1995, Opt. Engineering 34, 1870

Povel, H.P. 2002, in: "Magnetic fields across the Hertzsprung Russel diagram", Mathys, G., Solanki, S.K. (eds.), ASP conference series, in press

Povel, H.P., Aebersold, H., Stenflo, J.O. 1990, Applied Optics 29, 319

Querfeld, C.W., House, L.L., Smartt, R.N., Bommier, V., degl'Innocenti, E.L. 1985, Sol. Phys. 96, 277

Stenflo, J.O. 1973, Sol. Phys. 32, 41

Stenflo, J.O. 1997, A&A 324, 344

Stenflo, J.O. 2001, in: "Advanced Solar Polarimetry", Sigwarth, M. (ed.), ASP conference series 236, 97

Stenflo, J.O., Keller, C.U. 1996, Nature 382, 588

Stenflo, J.O., Keller, C.U. 1997, A&A 321, 927

Stenflo, J.O., Keller, C.U., Povel, H.P. 1992, LEST Foundation Technical Report 54

Stenflo, J.O., Keller, C.U., Gandorfer, A. 1998, A&A 329, 319

Stenflo, J.O., Gandorfer, A., Keller, C.U. 2000, A&A 355,

Stenflo, J.O., Keller, C.U., Gandorfer, A. 2000, A&A 355, 319

Stenflo, J.O., Twerenbold, D., Harvey, J.W. 1983a, A&AS 52, 161

Stenflo, J.O., Gandorfer, A., Wenzler, T., Keller, C.U. 2001, A&A

Stenflo, J.O., Twerenbold, D., Harvey, J.W., Brault, J.W. 1983b, A&AS 54, 505

Trujillo-Bueno, J. 2001, in "Advanced Solar Polarimetry", Sigwarth, M. (ed.), ASP conference series 236, 161

Trujillo-Bueno, J., Collados, M., Paletou, F., Molodji, G. 2001, in "Advanced Solar Polarimetry", Sigwarth, M. (ed.), ASP conference series 236, 161

von der Lühe, O., Schmidt, W., Soltau, D., Kneer, F., Staude, J. 2000, in: "The solar cycle and terrestrial climate", Wilson, A. (ed.), ESA SP-463

Wiehr, E. 1975, A&A 38, 303

Wiehr, E. 1978, A&A 67, 257, 303

Astronomische Gesellschaft: Reviews in Modern Astronomy **15**, 133–149 (2002)

A Statistical Analysis of the Extrasolar Planets and the Low-Mass Secondaries

Tsevi Mazeh and Shay Zucker

School of Physics and Astronomy, Raymond and Beverly Sackler
Faculty of Exact Sciences, Tel Aviv University, Tel Aviv, Israel
e-mail: `mazeh, shay@wise.tau.ac.il`

Abstract

We show that the astrometric Hipparcos data of the stars hosting planet candidates are not accurate enough to yield statistically significant orbits. Therefore, the recent suggestion, based on the analysis of the Hipparcos data, that the orbits of the sample of planet candidates are not randomly oriented in space, is not supported by the data. Assuming random orientation, we derive the mass distribution of the planet candidates and shows that it is flat in log M, up to about 10 M_{Jup}. Furthermore, the mass distribution of the planet candidates is well separated from the mass distribution of the low-mass companions by the 'brown-dwarf desert'. This indicates that we have here two distinct populations, one which we identify as the giant planets and the other as stellar secondaries. We compare the period and eccentricity distributions of the two populations and find them surprisingly similar. The period distributions between 10 and 1650 days are flat in log period, indicating a scale-free formation mechanism in both populations. We further show that the eccentricity distributions are similar – both have a density distribution peak at about 0.2–0.4, with some small differences on both ends of the eccentricity range. We present a toy model to mimic both distributions. The toy model is composed of Gaussian radial and tangential velocity scatters added to a sample of circular Keplerian companions. A scatter of a dissipative nature can mimic the distribution of the eccentricity of the planets, while scatter of a more chaotic nature could mimic the secondary eccentricity distribution. We found a significant paucity of massive giant planets with short orbital periods. The low-frequency of planets is noticeable for masses larger than about 1 M_{Jup} and periods shorter than 30 days. We point out how, in principle, one can account for this paucity.

1   Introduction

More than fifty candidates for extrasolar planets have been announced over
the past six years (e. g., Schneider 2001). In each case, precise stellar radial-
velocity measurements indicated the presence of a low-mass unseen companion, with a minimum mass between 1 and about 10 Jupiter masses (M_{Jup}).
The identification of these unseen companions as planets relied on their masses
being in the planetary range.

However, the actual masses of the planet candidates are not known. The
radial-velocity data yield only $M_2 \sin i$, where M_2 is the secondary mass and
i is the inclination angle of its orbital plane, which cannot be derived from
the spectroscopic data. Nevertheless, the astronomical community considered
the planet-candidate masses as being close to their derived minimum masses
– $M_2 \sin i$. This is so because at random orientation the most probable inclination is $90°$, and the expected value of $\sin i$ is close to unity.

Very recently some doubt has been cast about the validity of the random
orientation assumption. Gatewood, Han, & Black (2001) and Han, Black, &
Gatewood (2001) analysed the Hipparcos astrometric data of the stars hosting
planet candidates *together* with the stellar precise radial-velocity measurements and derived in some cases very low inclination angles for the orbital
planes. Han, Black, & Gatewood (2001) found eight out of 30 systems with an
inclination smaller or equal to $0.5°$, four of which they categorized as highly
significant. The probability of finding such small inclinations in a sample of
orbits that are *isotropically* oriented in space is extremely small, indicating
either a problematic derivation of the astrometric orbit, or, as suggested by
Han, Black, & Gatewood (2001), some serious orientation bias in the inclination distribution of the sample of detected planet candidates.

However, the analysis of the Hipparcos data can be misleading. As has
been shown by Halbwachs et al. (2000), one can derive a small *false* orbit with
the size of the typical positional error of Hipparcos, about 1 milli-arc-second
(= mas), caused by the scatter of the individual measurements. Therefore,
one should carefully evaluate the statistical significance of any astrometric
orbit of that size derived from the Hipparcos data. In Section 2 we summarize our work (Zucker & Mazeh 2001a) that evaluates the significance of
the astrometric orbits by applying a permutation test to the Hipparcos data.
Similarly to the results of Pourbaix (2001) and Pourbaix & Arenou (2001),
we also find that the significance of all the Hipparcos astrometric orbits of the
planet candidates are less than 99 %, including ϱ CrB that attracted much
attention after the publication of Gatewood, Han, & Black (2001) suggestion.
We therefore conclude that the Hipparcos data does not prove the anisotropy
of the orientations of the orbital planes of the planet candidates.

After showing that the random orientation in space is still a reasonable
assumption, not confronted by any available measurement, we present in Section 3 our work (Zucker & Mazeh 2001b) that uses this assumption to derive
the mass distribution of the planet candidates. This is done with MAXLIMA,
a MAXimum LIkelihood MAss algorithm which we constructed to derive the

mass distribution. Similar to the results of Jorrisen, Mayor, & Udry (2001), we show that the mass distribution of the planet candidates is separated from the one of the secondary masses by the so-called 'brown-dwarf desert' (e. g., Marcy & Butler 2000). This indicates that we are dealing with two different classes of objects. One is the giant planets, with masses not far from the planetary mass range, while the other is the low-mass secondaries, with stellar mass range.

One could speculate that the separation between the two different mass distributions indicates different formation processes. The commonly accepted paradigm is that planets were probably formed by coagulation of smaller, possibly rocky, bodies, whereas stars were probably formed by some kind of fragmentation of larger bodies. In other words, planets were formed by small bodies that grew larger, whereas stars, binary included, were formed by fragmentation of large bodies into smaller objects (e. g., Lissauer 1993; Black 1995). This could imply, for example, that the distribution of orbital eccentricities of giant planets and low-mass binaries would be substantially different. All the solar planets have nearly circular orbits, whereas binaries have eccentric orbits (e. g., Mazeh, Mayor, & Latham 1996). We could also expect the periods of planets to be longer than 10 years, like the giant planets in the solar system. Many studies of the newly discovered planets showed that this is not the case (e. g., Marcy, Cochran, & Mayor 2000). Moreover, following Heacox (1999) who based his analysis upon only 15 binaries and a handful of planet candidates, we show in Section 4 that within some reasonable restrictions, the eccentricity and period distributions of the two samples are surprisingly similar. Similar results have been obtained by Stepinski & Black (2001a, b, c). In Section 5 we consider a toy model that can generate the eccentricity distribution of both populations.

2 The significance of the astrometric orbits

In this section we present our work (Zucker & Mazeh 2001a) where we evaluate for each of the extrasolar planets the statistical significance of its astrometric orbit, derived from the Hipparcos data *together* with its radial-velocity measurements. We first derived the best-fit orbit by assuming that the spectroscopic and astrometric solutions have in common the following elements: the period, P, the time of periastron passage, the eccentricity, e, and the longitude of the periastron. In addition, the spectroscopic elements include the radial-velocity amplitude, K, and the center-of-mass radial velocity. We have three additional astrometric elements – the angular semi-major axis of the photocenter, a_0, the inclination, i, and the longitude of the nodes. In addition, the astrometric solution includes the regular astrometric parameters – the parallax, the position and the proper motion.

In most cases the elements are not all independent. From the spectroscopic elements we can derive the projected semi-major axis of the primary orbit. This element, together with the inclination i and the parallax, yields

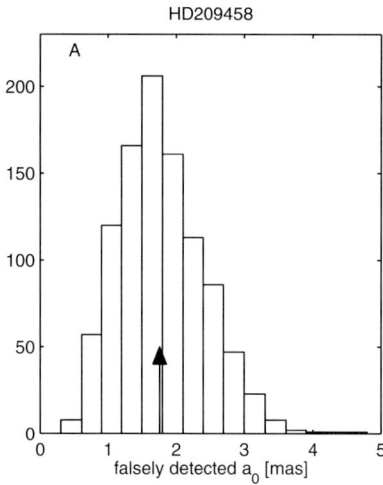 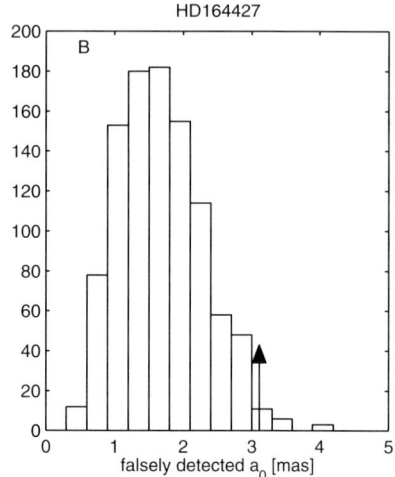

Figure 1: Histograms of the sizes of the falsely derived semi-major axes in the simulated permuted data of HD 209458 (A) and HD 164427 (B). The size of the actually detected axes are marked by an arrow.

the angular semi-major axis of the primary, a_1. Assuming the secondary contribution to the total light of the system is negligible, this is equal to the observed a_0.

To find the statistical significance of the derived astrometric orbit in each case we applied a permutation test (e. g., Good 1994) to the Hipparcos data. For each star we generated simulated permuted astrometric data and analyzed them either together with the actual individual radial velocities of that star, or by imposing the published spectroscopic elements. Details of the analysis are given by Zucker & Mazeh (2001a).

The distribution of the falsely detected semi-major axes indicated the range of possible false detections. For example, a_{99} – the 99th percentile, denotes the semi-major axis size for which 99 % of the simulations yielded smaller values. Consequently, an astrometric orbit is detected with a significance of 99 % if and only if the actually derived semi-major axis, a_{derived}, is larger than a_{99}.

As an illustration, Figure 1 A shows the histogram of the semi-major axis values derived by random permutations of the Hipparcos data of HD 209458. This star's inclination is known to be close to 90° through the combination of radial velocity and transit measurements (Charbonneau et al. 2000; Mazeh et al. 2000; Henry et al. 2000; Brown et al. 2001). The Hipparcos derived semi-major axis, a_{derived}, is 1.76 mas, which is marked in the figure by an arrow. One can clearly see that many random permutations led to larger semi-major axes, a fact that renders this derived value insignificant. The derived value is obviously false since the known inclination implies a value of less than a micro-arc-second.

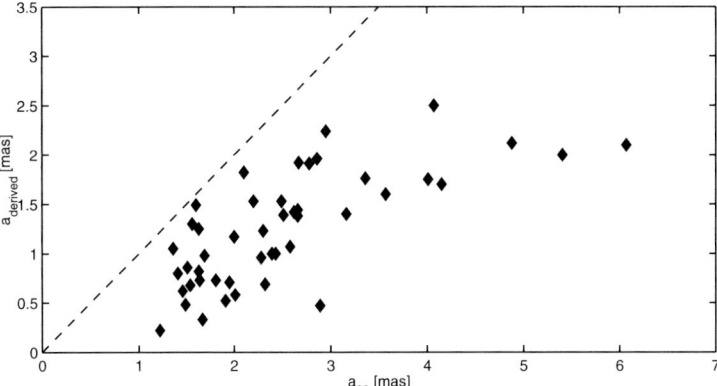

Figure 2: The derived semi-major axes of the planet candidates as a function of the 99th percentile of the falsely derived semi-major axes. The dashed line represents the line $a_{\text{derived}} = a_{99}$.

In Figure 1 B we show an opposite case, HD 164427, where the derived astrometric orbit is quite significant. Note that a_{derived} is relatively large – 3.11 mas, which made the significant detection possible. However, this is not a planet-candidate case. The minimum mass suggests this secondary is a brown-dwarf candidate, whereas the astrometric orbit shows the secondary mass is in the stellar regime.

As of March 2001, the Encyclopedia of extrasolar planets included 49 planet candidates with minimum masses smaller than 13 M_{Jup}. We (Zucker & Mazeh 2001a) analyzed all but t wo of the planet candidates. One star had no Hipparcos data, and the other star was known to have t wo companions. Figure 2 presents our results by depicting a_{derived} versus a_{99}. The figure indeed shows that all points fall to the right of the line $a_{\text{derived}} = a_{99}$. This means that all our derived astrometric motions are not significant in the 99 % level. This includes the planets of v And and HD 10697 whose derived orbits were previously published by us (Mazeh et al. 1999; Zucker & Mazeh 2000), but the new analysis renders their orbits less significant.

Note, however, that this does not mean that the orbits derived are all false. Figure 2 shows that some of the systems are close to the border line, indicating that the orbits of these systems were detected with significance close to 99 %. The systems with significance higher than 90 % are listed in Table 1. Here we list the Hipparcos number and the stellar name, the confidence level of the derived astrometric orbit, the derived semi-major axis, its uncertaint y and the derived inclination; the derived secondary mass together with its 1 σ range.

To summarize, the combination of the Hipparcos data together with the radial-velocity measurements did not yield any astrometric orbit with signif-icance higher than 99 %. Apparently, the Hipparcos precision is not good enough to detect a 1 mas orbit, even with the combination of the radial-velocity measurements. The analysis shows that the data are consistent with no astrometric detection at all, although one or t wo true astrometric orbits, which imply low inclinations, are still possible. However, such a finding would *not* prove that the orbits of the sample of planet candidates are not randomly oriented in space.

Table 1:
Derived planet-candidate orbits with confidence level higher than 90 %

HIP number	Name	Signif- icance	$a_{derived}$ (mas)	σ_a (mas)	$i_{derived}$ (deg)	$M_{derived}$ ($M_\odot$)	Mass Range (1σ)
5054	HD 6434	0.96	1.34	0.67	−0.08	0.45	(0.20, 0.77)
43177	HD 75289	0.90	1.05	0.52	0.03	1.13	(0.45, 2.19)
78459	ρ CrB	0.98	1.49	0.46	0.54	0.12	(0.086, 0.17)
90485	HD 169830	0.92	1.25	0.64	2.1	0.081	(0.039, 0.124)
94645	HD 179949	0.90	1.92	0.68	0.034	3.4	(1.57, 6.49)
98714	HD 190228	0.95	1.82	0.77	4.5	0.064	(0.037, 0.093)
100970	HD 195019	0.92	2.24	0.78	0.32	0.92	(0.51, 1.47)

3 The mass distribution of the extrasolar planets

Assuming the orbits of the detected planet candidates are randomly oriented in space we can now proceed to derive their mass distribution. To do that we have to account for the unknown orbital inclination and for the fact that stars with too small radial-velocity amplitudes could not have been detected as radial-velocity variables. Therefore, planets with masses too small, orbital periods too large, or inclination angles too small were not detected.

Numerous studies accounted for the effect of the unknown inclination of spectroscopic binaries (e. g., Mazeh & Goldberg 1992; Heacox 1995; Goldberg 2000), assuming random orientation in space. Heacox (1995) calculated first the minimum-mass distribution and then used its relation to the actual mass distribution to derive the latter. This calculation amplified the noise in the observed data, and necessitated the use of quite heavy smoothing of the observed data. Mazeh & Goldberg (1992) introduced an iterative algorithm whose solution depended, in principle, on the initial guess.

Very recently Jorissen, Mayor, & Udry (2001) studied the planet distribution by considering only the effect of the unknown inclination. Like Heacox (1995), Jorrisen, Mayor, & Udry derived first the distribution of the minimum masses and then applied two alternative algorithms to invert it to the distribution of planet masses. One algorithm was a formal solution of an Abel integral equation and the other was the Richardson-Lucy algorithm (e. g., Heacox 1995). The first algorithm necessitated some degree of data smoothing and the second one required a series of iterations. The results of the first algorithm depended on the degree of smoothing applied, and those of the second one on the number of iterations performed. In addition, Jorissen, Mayor,

& Udry (2001) did not apply any correction to the observational selection effect.

We (Zucker & Mazeh 2001b) followed Tokovinin (1991, 1992) and constructed a maximum likelihood algorithm – MAXimum LIkelihood MAss, to derive an histogram of the mass distribution of the extrasolar planets. MAXLIMA derives the histogram directly by solving a set of numerically stable linear equations. It does not require any smoothing of the data, except for the bin size of the histogram, nor any iterative procedure. MAXLIMA also offers a natural way to correct for the undetected planets. This is done by considering each of the detected systems as representing more than one system with the same $M_2 \sin i$, depending mainly on the period distribution. The details of the algorithm are given in Zucker & Mazeh (2001b).

To apply MAXLIMA to the current known sample of extrasolar planets we (Zucker & Mazeh 2001b) considered all known planets and brown dwarfs orbiting G- or K-star primaries as of April 2001. To acquire some degree of completeness to our sample we have decided to exclude planets with periods longer than 1500 days and with radial-velocity amplitudes smaller than 40 m/s. The values of these two parameters determine the correction of MAXLIMA for the selection effect, for which we assumed a period distribution which is flat in $\log P$. This choice of parameters also implies that our analysis applies only to planets with periods shorter than 1500 days. We further assumed that the primary mass is $1 M_\odot$ for all systems.

The results of MAXLIMA are presented in the lower panel of Figure 3 on a logarithmic mass scale. The value of each bin is proportional to the estimated number of planets found in the corresponding range of masses in the known sample of planet candidates, after correcting for the undetected systems. To estimate the uncertainty of each bin we ran 5000 Monte Carlo simulations and found the r.m.s. of the derived values of each bin. Therefore, the errors plotted in the figure represent only the statistical noise of the sample. Obviously, any deviation from the assumptions of our model for the selection effect induces further errors into the histogram, the assumed period distribution in particular. This is specially true for the first bin, where the actual number of systems is small and the correction factor large.

To compare the mass distribution of the planet candidates with that of the stellar secondaries we (Zucker & Mazeh 2001b) plot the latter on the same scale in an adjacent panel of Figure 3. We plot here only two bins, with masses between 100 and 1000 $M_{\rm Jup}$, using subsamples of binaries found by the Center for Astrophysics (= CfA) radial-velocity search for spectroscopic binaries (Latham 1985) in the Carney & Latham (1987) sample of the high-proper-motion stars (Latham et al. 2001; Goldberg et al. 2001).

Note that the upper panel does *not* have any estimate of the values of the bins with masses smaller than 100 $M_{\rm Jup}$. This is so because the CfA search does not have the sensitivity to detect secondaries in that range. On the other hand, the lower panel does include information on the bins below 100 $M_{\rm Jup}$. This panel presents the results of the high-precision radial-velocity searches, and these searches could easily detect stars with secondaries in the range of,

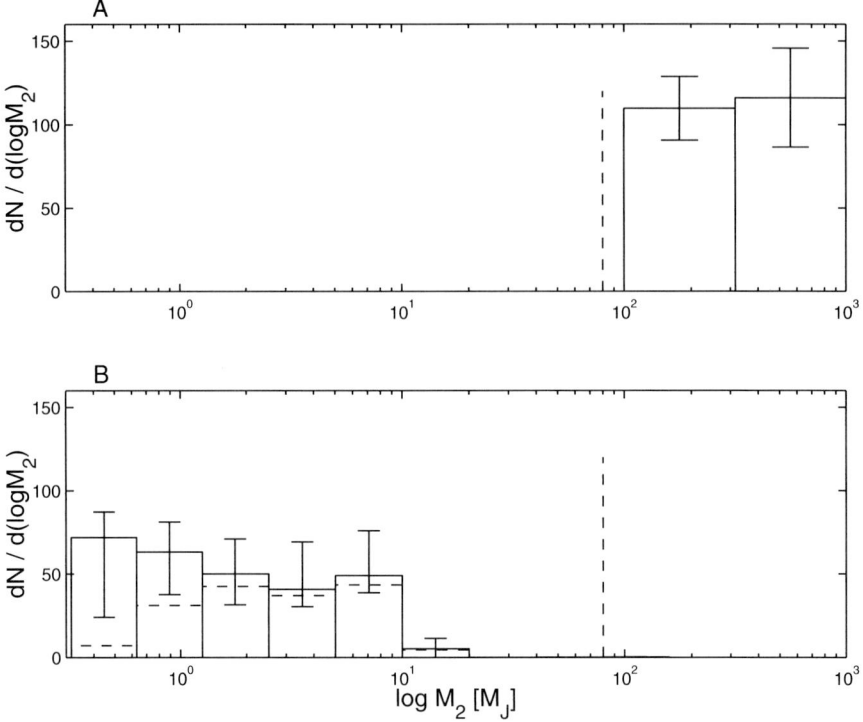

Figure 3: The mass distributions of the planets (lower panel) and the stellar companions (upper panel). The horizontal dashed lines represent the mass distribution without the correction for the selection effect. The vertical dashed line marks the stellar–sub-stellar border line.

say, 20–100 M_{Jup}. The lower panel shows that the frequency of secondaries in this range of masses is close to zero.

The relative scaling of the planets and the stellar companions is not well known (see Zucker & Mazeh 2001b for a detailed discussion). Nevertheless the comparison is illuminating. It suggests that we have here two distinct populations, separated by a 'gap' of about one decade of masses, in the range between 10 and 100 M_{Jup}. We will assume that the two populations are the giant planets, at the low-mass side of Figure 3, and the stellar companions at the high-mass end of the figure. The present analysis is not able to tell whether the gap extends up to 60, 80 or 100 M_{Jup}.

The gap between the two populations was already noticed by many previous studies (Basri & Marcy 1997; Mayor, Queloz, & Udry 1998; Mayor, Udry, & Queloz 1998; Marcy & Butler 1998). Those papers binned the mass distribution linearly. Here we follow our previous work (Mazeh, Goldberg, & Latham 1998; Mazeh 1999a, b; Mazeh & Zucker 2001) and use a logarithmic scale to study the mass distribution, because of the large range of masses, 0.5–1000 M_{Jup}, involved. The gap or the brown-dwarf desert is consistent also with the finding of Halbwachs et al. (2000), who used Hipparcos data and found that many of the known brown-dwarf candidates are actually stellar companions.

The distribution we derived in Figure 3 suggests that the planet mass distribution is almost flat in $\log M$ over five bins – from 0.3 to 10 M_{Jup}. Actually, the figure suggests a possible slight rise of the distribution toward smaller masses. At the high-mass end of the planet distribution the mass distribution dramatically drops off at 10 M_{Jup}, with a small high-end tail in the next bin. Although the results are still consistent with zero, we feel that the small value beyond 10 M_{Jup} might be real. The dramatic drop at 10 M_{Jup} and the small high-mass tail agree with the findings of Jorissen, Mayor, & Udry (2001), despite the differences in the algorithm used to derive the distribution, and the logarithmic scale we use for the distribution.

4   Eccentricity and period distribution of the two populations

Having established the difference between the mass distribution of the giant planets and that of the low-mass secondaries in spectroscopic binaries, we turn now to compare the period and eccentricity distributions of the two populations. For the latter we (Mazeh & Zucker 2001) use the results of a very large radial-velocity study of the Carney & Latham (1987) high-proper-motion sample, which yielded about 200 spectroscopic binaries (Latham et al. 2001; Goldberg et al. 2001). Goldberg (2000) separated statistically between the binaries of the Galactic halo and those coming from the disk. We consider in this section only the 59 single-lined spectroscopic binaries (= SB1s) of the Galactic disk. For the giant planet sample we use again the sample of 66 planet candidates listed in Schneider (2001) as of April 2001.

Figure 4 A shows the cumulative period distribution of the two samples. The figure suggests similar general trends, except in the two ends of the distributions. We therefore plotted in Figure 4 B the two distributions only in the range between 10 and 1650 days. The similarity is astounding, since the two distributions are identical. Both are consistent with a straight line, which implies a flat distribution in $\log P$.

We speculate that at the short period range, below 10 days, some dynamical interaction changed the distribution of either one or both distributions. Such an interaction could also change the eccentricity distribution of the orbits. In order not to be distracted by this possible interaction when we consider the eccentricity distribution, we choose to consider only the eccentricities of the orbits with periods between 10 and 1650 days. The cumulative distributions are plotted in Figure 5 A. We again see a similar trend in both distributions, except in both ends of the range [0,1]. To illuminate the difference we plotted the density distribution in Figure 5 B. We derived the distribution by convolving the actual data points with a Gaussian kernel with a width of 0.08. It is clear that both distributions peak at about 0.2–0.4. However, the

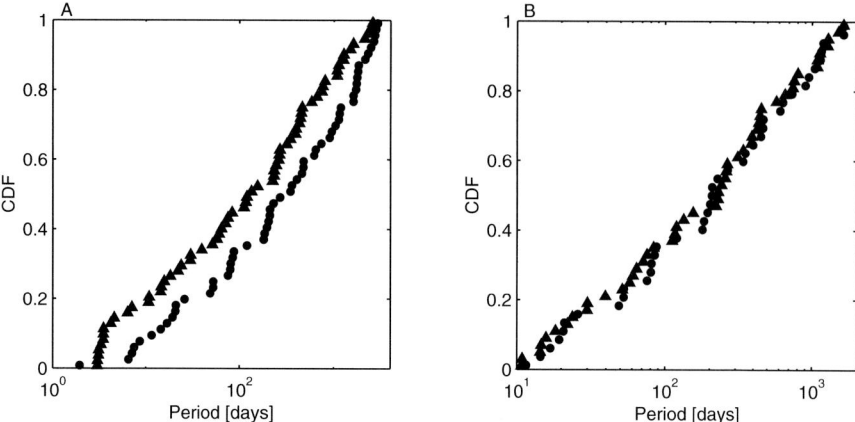

Figure 4: The period cumulative distribution for the planet candidates (triangles) and the Galactic disk SB1s (circles). A. All the stars in the samples. B. After restricting the samples to $10 < P < 1560$ d.

distribution of the spectroscopic binaries drops sharply toward zero, whereas the planet distribution does not. The eccentricity distribution of the binaries displays a tentative 'shoulder' at the large eccentricities, whereas that of the planets displays such a possible shoulder at the small eccentricities.

Any paradigm that assumes the two populations were formed differently has to explain why their eccentricity as well as period distributions are so much alike. Although we do not try to explain any of the two similarities, we suggest in the next section a toy model that can generate the two eccentricity distributions.

5   A toy model to generate the eccentricity distributions of the two samples

Consider a sample of low-mass companions that orbit their parent stars in circular Keplerian orbits. For simplicity let us choose the units such that the orbital radii of all orbits are of length unity, and so are their orbital tangential velocities. Now let us introduce a Gaussian scatter to the velocities of the companions of the sample, with two independent components. One component is tangential and the other is radial. The tangential component changes the moduli of the velocities, while the radial one changes mainly their directions.

The new scatter determined the new velocity *distribution*. Denote the center of the distribution by v_0 and its r.m.s. by σ_v. Suppose that the velocity angles are distributed around $90°$, with r.m.s. of σ_θ. Note that the distribution has three parameters, v_0, σ_v and σ_θ.

We can now calculate the eccentricity distribution of the sample, and see if such a simple-minded toy model can mimic the observed distributions of

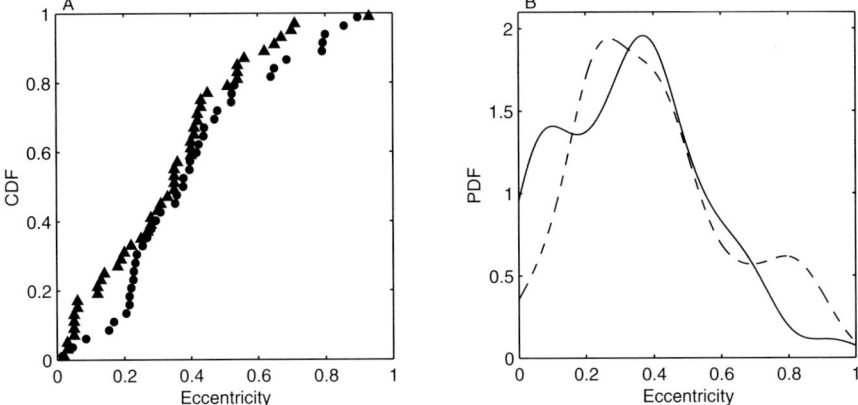

Figure 5: A. The eccentricity cumulative distribution for the planet candidates (triangles) and the Galactic disk SB1s (circles), restricted to $10 < P < 1560$ d. B. Estimated probability density function of the same samples, using a 0.08-wide kernel. The continuous line represents the planets and the dashed line represents the SB1s.

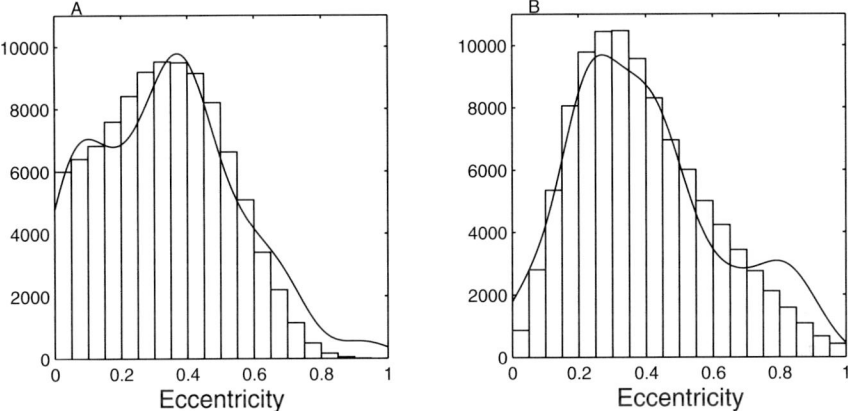

Figure 6: The simulated eccentricities histograms, together with the empirically estimated distributions of the planet candidates (A) and the stellar companions (B).

the giant planets and the low-mass companions. Figure 6 compares the two. We found that we can approximate the giant planet distribution with $v_0 = 0.82$ and $\sigma_v = 0.13$, while $\sigma_\theta = 0$, whereas the low-mass stellar companions necessitated $\sigma_\theta = 25°$, $\sigma_v = 0.05$ and $v_0 = 1.08$.

The fact that we succeeded to mimic the two actual distributions is not surprising. As the old statistical saying goes: "You can fit an elephant with any model with two parameters, and you can make him dance with three". However, the specific values of the parameters found are somewhat intriguing. Suppose that both populations started with Keplerian *circular* orbits, and two mechanisms introduced the scatter into the two populations. Suppose the nature of the mechanism that operated on the planet population was dissipative, like the dissipation generated by an interaction of a planet with a swarm of small particles in a disk. Such a mechanism could *decrease* the velocity without changing its direction. This would result with a null σ_θ and v_0 less than unity, the difference being of the same order of σ_v. On the other hand, the spectroscopic binaries could be subject to a more chaotic, eruptive disturbing mechanism, like the gravitational interaction with a few large bodies. In such a process one could expect a spread of the velocity directions and moduli, without significantly changing v_0. This simple-minded picture is consistent with our findings.

We should emphasize that the aforementioned discussion is not meant to explain how the eccentricities were formed, nor why the two distributions are similar with some definite small differences. The model might only serve as a starting point for any theoretical study to account for the observed distributions.

6   The paucity of short-period massive planets

In Section 3 & 4 we have discussed the distributions of masses, periods and eccentricities of the extrasolar planets. In this section we move to examine one aspect of the inter dependence of these variables. To explore this possible dependence we performed a Principal Component Analysis (e. g., Kendall & Stuart 1958), which immediately pointed out to the significant correlation between the (minimum) masses and periods of the extrasolar planets. This is depicted in Figure 7, in which we plotted the period as a function of the (minimum) masses of the known planets, as of April 2001. We choose to plot the two axes with logarithmic scales, because the frequency of planets is flat in $\log M$ and $\log P$, as has been shown in previous sections.

Most of the correlation between the periods and masses of the extrasolar planets could be accounted for by a selection effect, that prevents planets which are not massive enough from being discovered if their periods are too long. Such systems have radial-velocity amplitude, K, which is too small to be detected by the present planet-search projects. This is easily seen in the small-mass–long-period corner of the diagram, bounded by the $K = 25\,\text{ms}^{-1}$ line. There are only four planets above this line. However, a close examination of Figure 7 reveals an additional feature – a significant paucity of planets at the opposite, large-mass–short-period corner of the diagram. Only three planets appear at that corner, all marked by a circle. This is certainly not a selection effect, because planets at that part of the diagram have the largest radial-velocity amplitude, and therefore are the easiest to detect.

It is not clear yet what is the shape of the area in which we find low frequency of planets. That corner might have a rectangular shape bordered by $\log(M_2 \sin i) = 0.2$ and $\log P = 1.5$, or could be of a wedge shape, bordered by the line that goes from $(\log(M_2 \sin i), \log P) = (0, 0.5)$ to $(1, 1.5)$.

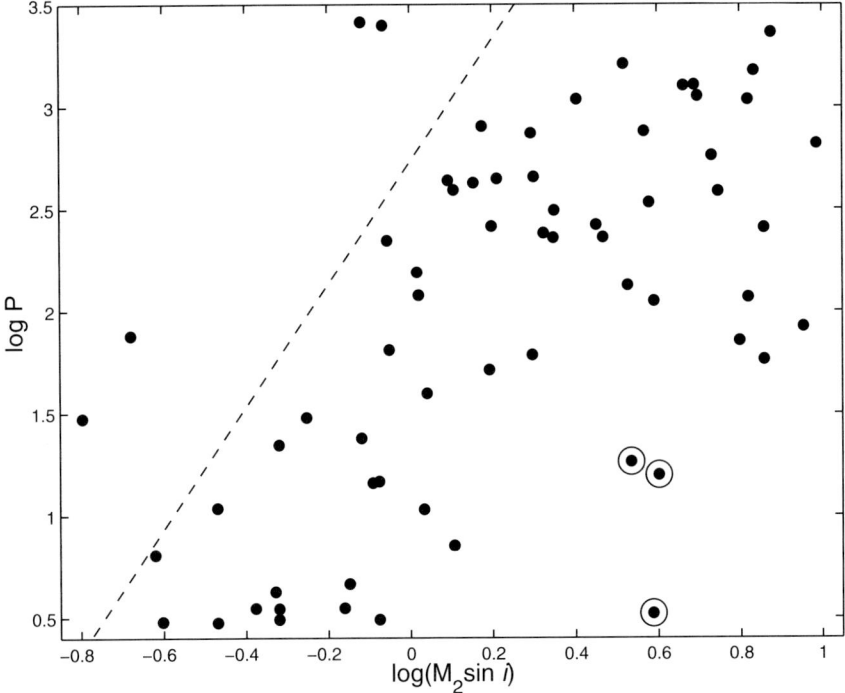

Figure 7: The logarithm of the period vs. the logarithm of the mass of the planet candidates. The dashed line represents a detection limit of 25 mþs^{-1} radial velocity amplitude. The three circled points correspond to the stars HD 195019, Gls 86 and τ Boo (see text).

The three planets that we find in the small-mass–long-period corner are Gls 86, HD 195019 and τ Boo. Interestingly enough, all three systems are wide binaries. Els et al. (2001) discovered very recently that the star Gls 86 has a brown-dwarf companion at about 20 AU projected separation. Pourbaix & Arenou (2001) pointed out that HD 195019 is a known visual binary with a companion fainter by about 3 mag., observed at a separation of 3.5 arcsec in 1988 (Mason et al. 2001). The angular separation of HD 195019 (= WDS 20283+1846) translates to 130 AU projected separation for a parallax of 27 mas (ESA 1997). The third star, τ Boo, is also a known visual binary (WDS 13473+1727), with an M2 companion. Apparently, the period is about t wo thousand years (Hale 1994) and the orbit is very eccentric. The separation between the t wo stars has been measured in 1991 to be 3.4 arcsec (Mason et al. 2001), which translates to about 50 AU projected distance for a parallax of 64 mas (ESA 1997). Planets in binary systems might go through different orbital evolution, and therefore might be considered as special cases. Thus, the low frequency of planets with large masses and short periods seems to be even more real than is seen from the figure.

Statistical assessment of the significance of the low frequency found in this part of the parameter space is under way. Very simple-minded calculations that ignore both the observational selection effects and the binarity of the three stars indicate a significance at the 2–3σ level. Taking into account the selection effect and the binarity of the three stars makes the significance of the low frequency even higher.

The paucity of large-mass planets with short periods and consequently small orbits might be another clue to the formation and orbital evolution of the extrasolar planets. There are now two different scenarios that account for the existence of giant planets in close-in orbits. One of them, accepted by most of the astronomical community, assumes the planets were formed out of a disc of gas and dust at a distance of 5 AU or larger, and have migrated through interaction with the disc to their present position (e. g., Lin, Bodenheimer, & Richardson 1996). The other one is that the planets were formed by some *in situ* disc instability (Boss 1997). In principle, our findings can be accounted for by both scenarios.

From the migration point of view, our findings might indicate that most large-mass planets halted their migration at orbital radius of the order of 0.2 AU. Obviously, the more massive the planet is, the more angular momentum and energy have to be removed from its orbital motion to enable the migration. Angular momentum and energy could be absorbed by the disc of gas and dust through generation of density waves (e. g., Goldreich & Tremaine 1980; Ward 1997) or by a planetesimal disc through gravitational interaction with the planet (e. g., Murray et al. 1998; Del Popolo, Gambera, & Ercan 2001). A too massive planet might move in until the local inner disc cannot absorb its angular momentum and energy. Such a consideration might account for a continuous dependence of the final orbital period on the planetary mass.

Interestingly enough, some studies suggested different migration scenarios for planets with small and large masses (Ward 1997). Massive planets open a gap in the disc, and subsequently go through slow, type II, migration, while small planets do not open a gap in the disc and therefore go through a relatively fast, type I, migration. The apparent paucity of short-period massive planets is consistent with such an evolutionary separation between large and small planets, if we can assume that the separation between the two types of migration occurs at a mass of about 1 M_{Jup}, and that type II migration could halt at about 0.2 AU (e. g., Lin et al. 2000).

According to the instability scenario, the mass of the formed planet depends on the available mass in the disc at the region of instability (e. g., Boss 2000). At small distances the available mass might be smaller, a fact that could result in low frequency of massive planets with short periods.

The fact that all three planets with relatively large masses and short periods are found in binary systems is intriguing. The interaction of the secondary with the protoplanetary disc could modify the structure and evolution of the disc, and therefore the formation and evolution of the planet. We obviously need more data to see whether this feature is statistically significant.

In all the aforementioned scenarios, the paucity of massive planets with short-period orbits is a natural consequence of the formation and evolutionary mechanism. However, detailed theoretical models have to be worked out so we can compare the theory with the observations. If confirmed by the discovery of more planets, the interesting input of the present analysis is the actual boundaries of the low-frequency part of the diagram. A borderline at about 1.5 M_{Jup} and at about 30 days can help us quantitatively understand the formation and evolutionary process of extrasolar planets.

7   Summary

The logarithmic mass distribution derived here shows that the planet candidates are indeed a separate population, probably formed in a different way than the secondaries in spectroscopic binaries. Surprisingly, the eccentricity and period distributions, with some restrictions, are very much the same.

Furthermore, the two period distributions follow strictly a straight line. This indicates flat density distributions on a logarithmic scale, inconsistent with the Duquennoy & Mayor (1991) log-normal distribution. Interestingly, flat logarithmic distribution is the only scale-free distribution, and could be argued to be the most simple distribution. Maybe the two populations were formed by two different mechanisms that still have this scale-free feature in common (Heacox 1999).

The eccentricity distribution of the sample of giant planets and that of stellar companions are similar (Stepinski & Black 2001c). In spite of the similarity, they are not identical, especially if compared to the remarkable similarity between the two period distributions. The eccentricity distributions can be attained by Keplerian orbits whose velocities are normally disturbed in the tangential and the radial directions.

We found a significant paucity of large planets with short orbital periods, and point out how, in principle, one can account for this paucity.

Acknowledgments

We acknowledge support from the Israeli Science Foundation through grant no. 40/00. This research has made use of the SIMBAD database, operated at CDS, Strasbourg, France, and the Washington Double Star Catalog maintained at the U.S. Naval Observatory.

References

Basri, G. & Marcy, G. W. 1997, in AIP Conf. Proc 393, Star Formation, Near and Far, eds. S. Holt & L. G. Mundy (New York: AIP), 228

Black, D. C. 1995, ARA&A 33, 359

Boss, A. P. 1997, Science 276, 1836

Boss, A. P. 2000, ApJL 536, L101

Brown, T. M., Charbonneau, D., Gilliland, R. L., Noyes, R. W., & Burrows, A. 2001, ApJ 552, 699

Carney, B. W. & Latham, D. W. 1987, AJ 92, 116

Charbonneau, D., Brown, T. M., Latham, D. W., & Mayor, M. 2000, ApJ 529, L45

Del Popolo, A., Gambera, M., & Ercan, N. 2001, MNRAS 325, 1402

Duquennoy, A. & Mayor, M. 1991, A&A 248, 485

Els, S. G., Sterzik, M. F., Marchis, F., Pantin, E., Endl, M., & Kürster, M. 2001, A&A 370, L1

ESA 1997, The Hipparcos and Tycho Catalogues, ESA SP-1200

Gatewood, G., Han, I., & Black, D. 2001, ApJ 548, L61

Goldberg, D. 2000, Ph.D. thesis, Tel Aviv University

Goldberg, D., Mazeh, T., Latham, D. W., Stefanik, R. P., Carney, B. W., & Laird, J. B. 2001, submitted to A&A

Goldreich, P. & Tremaine, S. 1980, ApJ 241, 425

Good, P. 1994, Permutation Tests – A Practical Guide to Resampling Methods for Testing Hypotheses, (New York: Springer-Verlag)

Halbwachs, J.-L., Arenou, F., Mayor, M., Udry, S., & Queloz, D. 2000, A&A 355, 581

Hale, A. 1994, AJ 107, 306

Han, I., Black, D., & Gatewood, G. 2001, ApJ 548, L57

Heacox, W. D. 1995, AJ 109, 2670

Heacox, W. D. 1999, ApJ 526, 928

Henry, G. W., Marcy, G. W., Butler, R. P., & Vogt, S. S. 2000, ApJ 529, L41

Jorissen, A., Mayor, M., & Udry, S. 2001, A&A 379, 992

Kendall, M. G. & Stuart, A. 1966, The Advanced Theory of Statistics, vol. 3, (London: Griffin)

Latham, D. W. 1985, in IAU Colloq. 88, Stellar Radial Velocities, eds. A. G. D. Philip & D. W. Latham (Schenectady, L. Davis Press) 21

Latham, D. W., Stefanik, R. P., Torres, G., Davis, R. J., Mazeh, T., Carney, B. W., Laird, J. B., & Morse, J. A. 2001, submitted to A&A

Lin, D. N. C., Bodenheimer, P., & Richardson, D. C. 1996, Nature 380, 606

Lin, D. N. C., Papaloizou, J. C. B., Terquem, C., Bryden, G., & Ida, S. 2000, in Protostars and Planets IV eds. V. Mannings, A. P. Boss, S. S. Russell (Tucson: University of Arizona Press), 1111

Lissauer, J. J. 1993, ARA&A 31, 129

Marcy, G. W. & Butler, R. P. 1998, ARA&A 36, 57

Marcy, G. W. & Butler, R. P. 2000, PASP 112, 137

Marcy, G. W., Cochran, W. D., & Mayor, M. 2000, in Protostars and Planets IV eds. V. Mannings, A. P. Boss, S. S. Russell (Tucson: University of Arizona Press), 1285

Mason, B. D., Wycoff, G. L., Hartkopg, W. I., Douglass, G. G., & Worley, C. E. 2001, Washington Double Star Catalog 2001.0, U.S. Naval Observatory, Washington

Mayor, M., Queloz, D., & Udry, S. 1998, in Brown Dwarfs and Extrasolar Planets, eds. R. Rebolo, E. L. Martin, & M. R. Zapatero-Osorio (San Francisco: ASP), 140

Mayor, T., Udry, S., & Queloz, D. 1998, in ASP Conf. Ser. 154, Tenth Cambridge Workshop on Cool Stars, Stellar Systems, and the Sun, eds. R. Donahue & J. Bookbinder (San Francisco: ASP), 77

Mazeh, T. 1999a, Physics Reports 311, 317

Mazeh, T. 1999b, in ASP Conf. Ser. 185, IAU Coll. 170, Precise Stellar Radial Velocities, eds. J. B. Hearnshaw & C. D. Scarfe, (San Francisco: ASP), 131

Mazeh, T. et al. 2000, ApJ 532, L55

Mazeh, T. & Goldberg, D. 1992, ApJ 394, 592

Mazeh, T., Goldberg, D., & Latham, D. W. 1998, ApJ 501, L199

Mazeh, T., Mayor, M., & Latham D. W. 1996, ApJ 478, 367

Mazeh, T., & Zucker, S. 2001, in IAU Symp. 200, Birth and Evolution of Binary Stars, eds. B. Reipurth and H. Zinnecker (San Francisco: ASP), 519

Mazeh, T., Zucker, S., Dalla Torre, A., & van Leeuwen, F. 1999, ApJ 522, L149

Murray, N., Hansen, B., Holman, M., & Tremaine, S. 1998, Science 279, 69

Pourbaix, D. 2001, A&A 369, L22

Pourbaix, D. & Arenou, F. 2001, A&A 372, 935

Schneider, J. 2001, in Extrasolar Planets Encyclopaedia
http://www.obspm.fr/planets

Stepinski, T. F. & Black, D. C. 2001a, in IAU Symp. 200, Birth and Evolution of Binary Stars, ed. B. Reipurth & H. Zinnecker (San Francisco: ASP) 167

Stepinski, T. F. & Black, D. C. 2001b, A&A 356, 903

Stepinski, T. F. & Black, D. C. 2001c, A&A 371, 250

Tokovinin, A. A. 1991, Sov. Astron. Lett. 17, 345

Tokovinin, A. A. 1992, A&A 256, 121

Ward, W. R. 1997, Icarus 126, 261

Zucker, S. & Mazeh, T. 2000, ApJ 531, L67

Zucker, S. & Mazeh, T. 2001a, ApJ 562, 549

Zucker, S. & Mazeh, T. 2001b, ApJ 562, 1038

ASTRONOMISCHE GESELLSCHAFT: Reviews in Modern Astronomy **15**, 151–164 (2002)

Radiative Transfer
in Turbulent Molecular Clouds

Michael Hegmann

Institut für Theoretische Physik
Johann-Wolfgang-Goethe-Universität Frankfurt
Robert-Mayer-Straße 10, 60054 Frankfurt am Main, Germany
Hegmann@astro.uni-frankfurt.de

Abstract

We present a radiative transfer model which takes stochastic density and velocity fluctuations into account. The variation of both, the density and the velocity component along the line of sight, is assumed to be governed by a Markov process. In order to demonstrate the effects of an inhomogeneous, stochastic density distribution, we present and discuss results obtained for two different physical problems: the transport of continuum photons into an absorbing dusty layer and the formation of CO rotational lines in the case of NLTE.

1   Introduction

Radiation emitted at different wavelengths from the dense and cold molecular medium gives information on the emitting and absorbing material along the line of sight. In principle, the hydrogen density, molecular abundances, the temperature of the gas phase or the dust, and the structure of a nonthermal velocity field may be determined. Unfortunately, it is not always easy to extract all this information from the observations. A quantitative analysis of continuum or line radiation has in general to be based on extensive radiative transfer calculations. That is, solutions of the radiative transfer equation

$$\frac{dI_\nu}{ds} = -\kappa_\nu I_\nu + \varepsilon_\nu \qquad (1)$$

have to be found. Here, I_ν is the specific intensity at frequency ν, κ_ν the absorption coefficient, ε_ν the emissivity and s the distance along the line of sight. These radiative transfer calculations require a detailed physical model specifying the geometry of the cloud and the spatial variation of the physical parameters.

For this the following problem arises: Observations with high spectral and angular resolution indicate that interstellar clouds have a tendency to be extremely inhomogeneous down to the smallest accessible scales, and beyond. A deterministic description of the physical conditions along the line of sight, accounting for all these small fluctuations, seems to be impossible. Therefore, we use an approach which describes distinct physical parameters of the cloud in a statistical sense only. In its original form, this theory has been developed by Gail and collaborators (Gail et al. 1974, Gail & Sedlmayr 1974, Gail et al. 1975), in order to describe the formation of spectral lines in stellar atmospheres.

However, their approach has many applications. We focused our efforts on the problem of radiative transfer in a medium with an inhomogeneous stochastic density and velocity distribution. In order to demonstrate the basic effects of stochastic velocity and density fluctuations on the radiative transfer, we will concentrate in the following on two specific physical problems: the transport of continuum photons through an absorbing dusty layer and the formation of CO rotational lines in the case of NLTE.

2 A model for the radiative transfer in a turbulent and clumpy medium

2.1 A stochastic description of the velocity and density distribution

In this section, we give a short overview about our model and the basic equations used in the following. For a more detailed description, see Hegmann & Kegel (2000) and references therein. We consider radiative transfer in an inhomogeneous medium. The spatial variation of the velocity component v along the line of sight and of the density n is described in a statistical sense only. This means that we have to specify the functional form of the multipoint probability distributions. In principal, the complete hierarchy of multipoint distributions has to be given

$$W_1(n_1, v_1, s_1)$$
$$W_2(n_2, v_2, s_2; n_1, v_1, s_1)$$
$$W_3(n_3, v_3, s_3; n_2, v_2, s_2; n_1, v_1, s_1)$$
$$\cdots \tag{2}$$

for every point s_i along the line of sight. For the sake of simplification, we assume a Markovian structure for the velocity and density field along each line of sight. In this approximation, only two point correlations are accounted for. Further, to guarantee a positive density, we transform the density to a logarithmic scale:

$$\tilde{n}(s) \stackrel{\text{def}}{=} \ln\left(\frac{n(s)}{n_{\text{ref}}}\right), \tag{3}$$

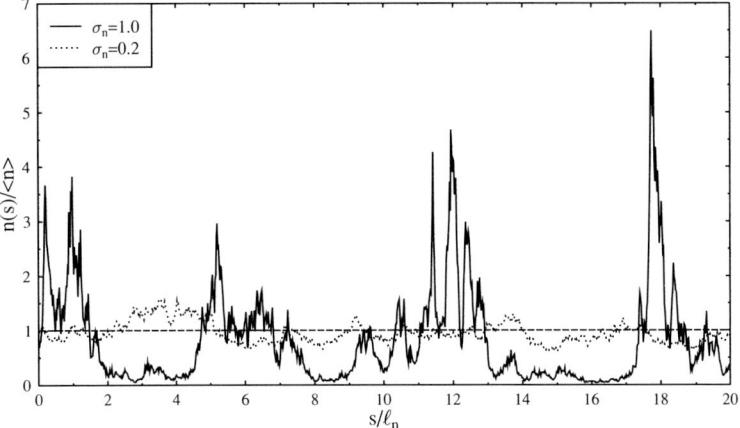

Figure 1: Two concrete realizations of the stochastic density distribution.

with n_{ref} being the reference density and s the geometric coordinate along the line of sight. The one-point distributions of v (the turbulent velocity component along the line of sight) and $\tilde{n}$ are then considered to be Gaussian

$$W_{\text{v}}(v) = \frac{1}{\sqrt{2\pi}\sigma_{\text{v}}} \exp\left(\frac{-v^2}{2\sigma_{\text{v}}^2}\right) \tag{4}$$

$$W_{\text{n}}(\tilde{n}) = \frac{1}{\sqrt{2\pi}\sigma_{\text{n}}} \exp\left(\frac{-\tilde{n}^2}{2\sigma_{\text{n}}^2}\right) \tag{5}$$

and the two-point correlation functions to be exponential

$$f_{\text{v}}(\Delta s) = \frac{\langle v(s)v(s+\Delta s)\rangle}{\sigma_{\text{v}}^2} = \exp\left(-\frac{|\Delta s|}{\ell_{\text{v}}}\right) \tag{6}$$

$$f_{\text{n}}(\Delta s) = \frac{\langle \tilde{n}(s)\tilde{n}(s+\Delta s)\rangle}{\sigma_{\text{n}}^2} = \exp\left(-\frac{|\Delta s|}{\ell_{\text{n}}}\right), \tag{7}$$

where ℓ_{v} and ℓ_{n} are the correlation lengths of the velocity and density field, respectively. In writing equations (4)–(7), we neglect for our present study any correlations between the density and the velocity component v. This is an admissible lower order approximation, since only one velocity component, that along the line of sight, enters the transfer equation.

2.2 On the structure of the density field

The two free parameters ℓ_{n} and σ_{n} describing the stochastic density distribution have a simple physical meaning: The first quantity is a measure to which extent the cloud's mass is concentrated in dense regions, whereas the latter is a length scale for the size and the mean distance of these dense regions.

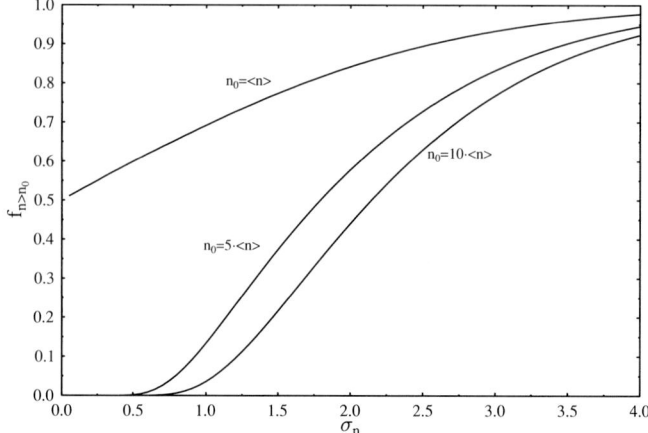

Figure 2: Fraction of the mass contained in clumps denser than n_0.

Although we are essentially interested in the mean values of the relevant physical quantities in the following, it is very instructive to view single realizations of the stochastic processes described by equations (3) to (7). Fig. 1 shows two concrete realizations of the density field along one line of sight. One can clearly see that for $\sigma_n = 1.0$ a clump like structure is obtained, whereas for $\sigma_n = 0.2$ the density distribution is rather smooth. That is, for a high Gaussian width of the logarithmic density distribution most of the cloud's mass is concentrated in dense regions, which one may call clumps, whereas most of the spatial volume is filled by a low density gas. In Fig. 2, the fraction of the mass contained in clumps denser than a given density n_0 is shown as function of σ_n. We note in passing that the reference and the mean density are different. They are related to each other by the following equation:

$$\langle n \rangle = n_{\text{ref}} \cdot \exp\left(\frac{\sigma_n^2}{2}\right). \tag{8}$$

2.3   The generalized radiative transfer equation

Due to the stochastic nature of the underlying density and velocity field the intensity becomes a stochastic variable, too. Within the stochastic approach discussed above, a partial differential equation can be set up which describes the variation of the expectation value of the intensity along each line of sight (see Gail et al. 1974):

$$\frac{\partial q_\nu}{\partial s} = \frac{1}{\ell_n}\left(-\tilde{n}\frac{\partial q_\nu}{\partial \tilde{n}} + \sigma_n^2\frac{\partial^2 q_\nu}{\partial \tilde{n}^2}\right)$$

$$+ \frac{1}{\ell_v}\frac{\partial}{\partial v}\left(-v\frac{\partial q_\nu}{\partial v} + \sigma_v^2\frac{\partial^2 q_\nu}{\partial v^2}\right)$$

$$- \kappa_\nu\left(q_\nu - S_\nu\right). \tag{9}$$

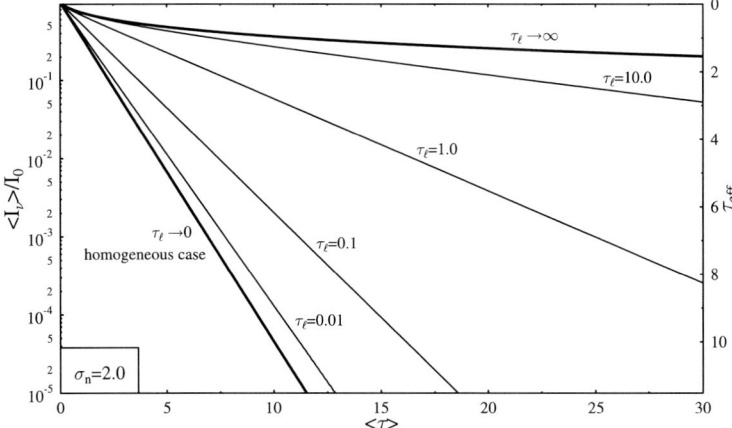

Figure 3: Excpectation value of the intensity for $\sigma_n = 2.0$ and different values for τ_ℓ.

Here, $q_\nu(v, \tilde{n}, s)$ is the conditional intensity, that is the expectation value of the intensity for a given velocity v and a given logarithmic density $\tilde{n}$ at point s. The expectation value of the intensity is obtained from $q_\nu(v, \tilde{n}, s)$ by a simple quadrature:

$$\langle I_\nu(s) \rangle = \int_{-\infty}^{+\infty} q_\nu(\tilde{n}, v, s) W_v(v) W_n(v) \, d\tilde{n} dv \; . \tag{10}$$

3 The absorption of stellar radiation by dust and the infrared emission

3.1 The case of pure absorption

In the following we want to examine the basic effects of stochastic density fluctuations on the transport of continuum photons into an absorbing layer of dust grains. It is convenient to do a study for a simple geometry first: we assume an, on average, homogeneous plane-parallel slab which is illuminated from one side by an isotropic radiation field.

In Figs. 3 and 4, we give the mean intensity $\langle I_\nu \rangle$ along the line of sight perpendicular to the slab as function of the dimensionless spatial coordinate

$$\langle \tau \rangle = \sigma_{\text{abs}} \langle n_{\text{dust}} \rangle s \; , \tag{11}$$

where σ_{abs} is the absorption coefficient per dust grain. Since the mean free path of the photons is defined as

$$l_a = \frac{1}{\sigma_{\text{abs}} \langle n_{\text{dust}} \rangle} \; , \tag{12}$$

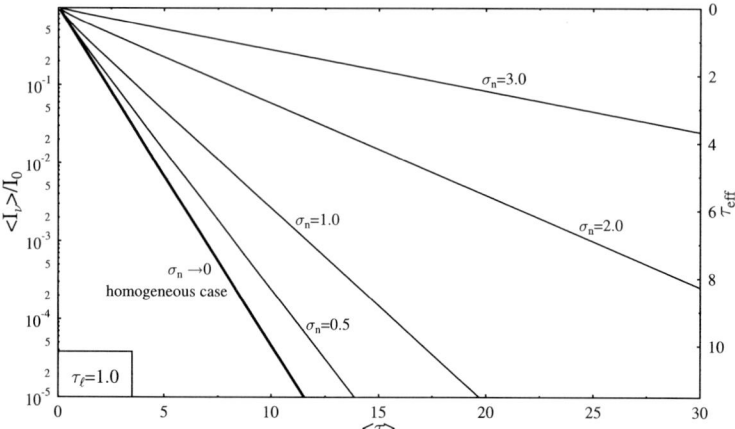

Figure 4: Excpectation value of the intensity for $\tau_\ell = 1.0$ and different values for σ_n.

this new coordinate may be interpreted as the penetration depth along the line of sight in units of l_a. Correspondingly, ℓ_n is substituted by

$$\tau_\ell = \sigma_{abs} \langle n_{dust} \rangle \ell_n = \frac{\ell_n}{l_a} \, . \tag{13}$$

That is, τ_ℓ is the ratio of the correlation length of the density field and the mean free path of the photons. It has to be noted that the quantity $\langle \tau \rangle$ is distinct from the effective optical depth τ_{eff}. For the case of pure absorption, we define τ_{eff} by

$$\tau_{eff} = -\ln \left(\frac{\langle I_\nu \rangle}{I_0} \right) \, . \tag{14}$$

In Fig. 3, σ_n is kept constant and τ_ℓ is varied, whereas in Fig. 4, τ_ℓ is set to a fixed value and σ_n varies. It can be clearly seen, that the effective optical depth of a clumpy system is smaller than that of the equivalent homogeneous system. For fixed $\langle \tau \rangle$, the mean intensity of the penetrating radiation increases with both, an increasing correlation length and an increasing standard deviation of the logarithmic density. These findings are plausible: with σ_n becoming larger, the amount of dust which can be found in dense but spatially small regions increases strongly. The extension of the dense regions and their mean distance scale with the correlation length. Due to this dilution effect the mean intensity of the penetrating radiation field can be substantially larger than would be expected from calculations based on a homogeneous density distribution.

3.2   The infrared emission of dust

As a next example, we study the infrared emission of a dust layer. If the emitting region is in thermal equilibrium, the effects of stochastic density

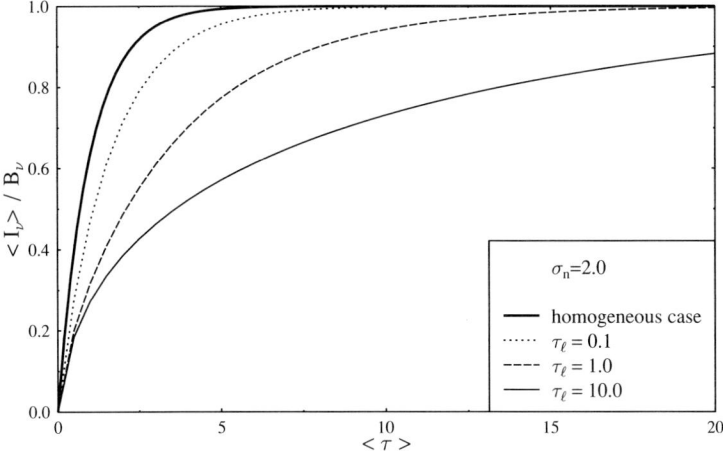

Figure 5: Emission of a layer with density fluctuations, but constant temperature.

fluctuations can be easily foreseen. As we have just seen, the effective optical depth decreases with both, an increasing correlation length and an increasing Gaussian width of the one-point density distribution. Hence, we expect the intensity to decrease with increasing σ_n and ℓ_n. The maximum intensity should be obtained as τ_ℓ approaches zero. For this special case, the expectation value of the intensity is simply given by

$$\lim_{\tau_\ell \to 0} \langle I_\nu \rangle = B_\nu \left(1 - e^{-\langle \tau \rangle}\right) . \tag{15}$$

And indeed, our calculations show the expected behavior. In Fig. 5, we give the mean intensity as function of the width of the layer for different correlation lengths. The intensity is normalized to the Planck function corresponding to the temperature of the dust. Hence, all graphs approach one, as $\langle \tau \rangle$ becomes infinite. Again, the dependence of the mean intensity on the correlation length is remarkably strong.

Let us now consider a somewhat more challenging physical model: we take the case of a slab of dust grains, which is heated from one side by an intense UV radiation field. For such an intense radiation field, the most important way energy is transferred to the dust is by absorption of photons. The grains readjust their temperature by re-emission in the IR regime. For grains, which are not too small, an equilibrium temperature can be easily calculated by equating the absorbed and re-emitted energy

$$\int_0^\infty F(\nu)\sigma_{\mathrm{abs}}(\nu)\,\mathrm{d}\nu = \int_0^\infty \sigma_{\mathrm{abs}}(\nu)B_\nu(T_{\mathrm{grain}})\,\mathrm{d}\nu , \tag{16}$$

in which $F(\nu)$ is the energy flux of the UV radiation field. To do this, one has, of course, to know the absorption coefficient of the dust grains as function of

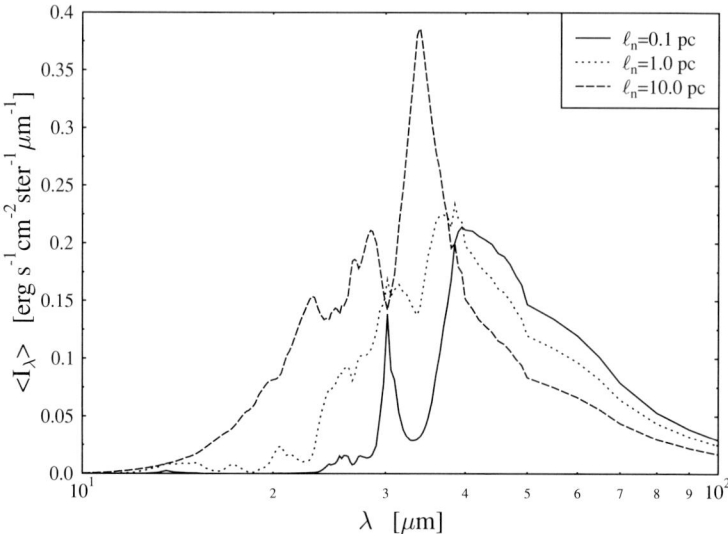

Figure 6: IR spectrum of a plane-parallel layer of Olivin particles, illuminated by an UV field $G_0 = 10^5$ times more intense than the ambient interstellar radiation field. The cloud parameters are: $L = 5$ pc, $\sigma_n = 2.0$, and $n_{H_2} = 10^4$ cm^{-3}.

the frequency. Based on optical constants from Huffman (1975), we use values that have been calculated for spherical Olivin particles with a diameter of 100 nm.

In Fig. 6, we give an emerging IR spectrum, which has been calculated for an incident UV radiation field being 10^5 times more intense than the ambient interstellar radiation field. The width of the slab is taken to be $L = 5$ pc, and the mean hydrogen number density is $n_{H_2} = 10^4$ cm^{-3}. For the dust to gas ratio we take a value of 10^{-11}. This time, the effects of the density fluctuations on the emerging IR radiation cannot be so easily predicted as in the case of a homogeneous temperature. It depends on the respective wavelength λ, whether an increase of the correlation length leads to an increasing or decreasing intensity. Furthermore, even for a fixed λ the intensity may become a non-monotonous function of the correlation length. Two different mechanism have to be taken into account: on the one hand side, with an increasing clumpiness of the medium the UV radiation is able to penetrate much deeper into the slab. Dust grains deep within the cloud are much more heated than would be expected for a homogeneous layer. This effect should lead to a higher intensity in the IR. On the other hand side, for strong density fluctuations the layer becomes more transparent. That is, the effective optical depth in the IR becomes substantially smaller. This should lead to lower IR intensity.

Both effects can be clearly seen in Fig. 7 a. Here, an IRAS t wo-color diagram is shown. Different points in this diagram correspond to different

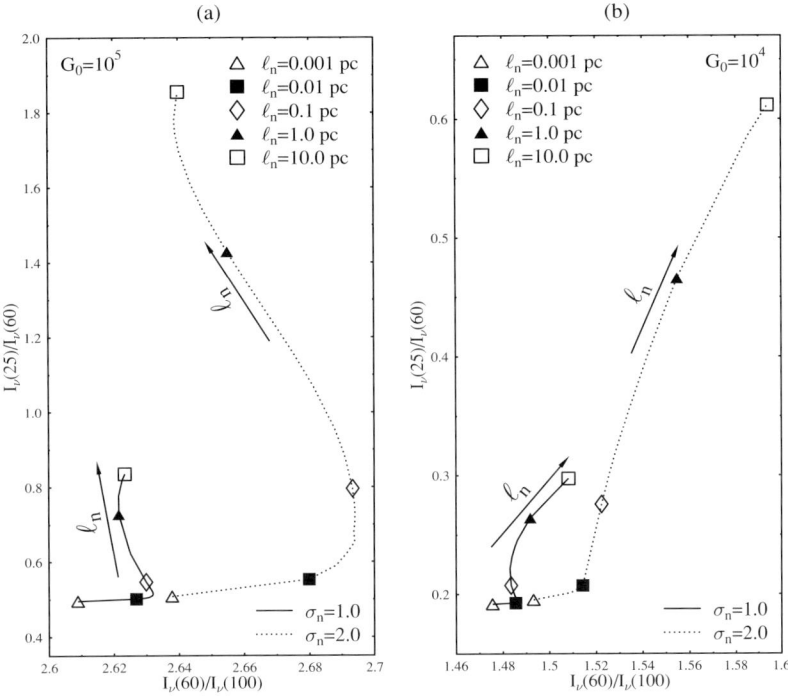

Figure 7:   IRAS two-color diagram for $\langle n_{H_2} \rangle = 10^4$ cm^{-3}, $L = 1$ pc, and $n_{dust}/n_{gas} = 10^{-11}$.

values of ℓ_n and σ_n, respectively. Except for the width of the slab, which is taken to be $L = 1$ pc, the same parameters as in Fig. 6 are chosen. At first, the ratio of the intensities in the 60 μm and the 100 μm band grows with an increasing correlation length. But for $\ell_n \gtrsim 0.1$ pc the graph changes its direction and the ratio now decreases with increasing ℓ_n. However, the absolute change of the ratio is not very impressive, since the slab is almost optically thin for 60 μm and 100 μm, respectively. On the other hand, the 25 μm to 60 μm color varies substantially for different correlation lengths. In Fig. 7b, we give the IRAS colors for a slab which is illuminated by an UV radiation field only 10^4 times more intense than the ambient interstellar radiation field. The values of the other physical parameters are not changed. The slab is much cooler now, that is the IRAS colors are noticeably decreased. But again, it is the 25 μm to 60 μm ratio, which varies almost by a factor of four.

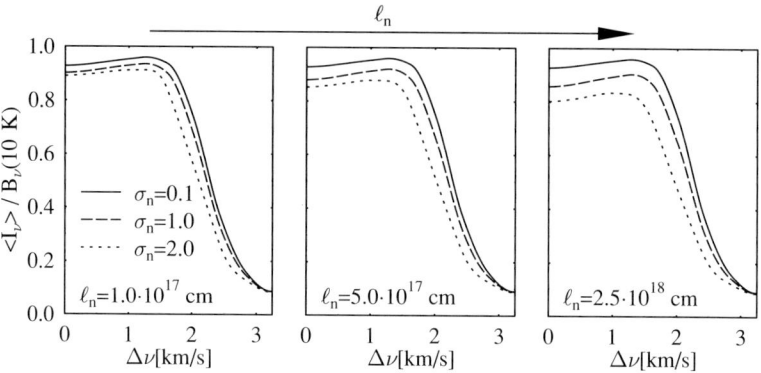

Figure 8: Line profiles for the $J = 1 \rightarrow 0$ transition of CO for $T_{\text{gas}} = 10\,\text{K}$, $L = 3.0\,10^{19}$ cm, $\langle n_{\text{H}_2} \rangle = 10^3$ cm^{-3}, $\langle N_{\text{CO}} \rangle = 2.4\,10^{18}$ cm^{-2}, $\ell_v = 5.0\,10^{17}$ cm, and $\sigma_v/v_{\text{th}} = 10$.

4 Non-LTE line formation

We will now focus on a completely different physical process: the formation
of CO rotational lines. We do not assume local thermodynamic equilibrium.
Hence, the occupation numbers for the different levels of angular momentum
of the molecule, that is different energy levels, are not given by a Boltzmann
distribution. They have to be determined selfconsistently with the line intensities from the rate equations. In our model, the system of rate equations has
to be solved simultaneously with the generalized radiative transfer equation
(9) for every point in physical, velocity and density space. All results presented in the following have been obtained for a 6-level CO molecule and an,
on average, homogeneous plane-parallel slab. The relative CO abundance is
considered to be constant throughout the cloud. We adopted the value given
by Black & Willner (1984), that is $n_{\text{CO}}/n_{\text{H}_2} = 8.0 \cdot 10^{-5}$. For a detailed
description of the physical model and the resulting equations, see Hegmann
& Kegel (2000). Although we account also for a stochastic, turbulent velocity
field, we will again concentrate on the effects of stochastic density fluctuations. The effects of a turbulent velocity field with finite correlation length
on the process of line formation were examined by Albrecht & Kegel (1987),
Kegel et al. (1993), and Piehler & Kegel in (1994) in great detail.

Fig. 8 gives line profiles for the $J = 1 \rightarrow 0$ transition of CO. The adopted
average column density is $\langle N_{\text{CO}} \rangle = 2.4 \cdot 10^{18}$ cm^{-2}, the mean hydrogen number
density is $n_{\text{H}_2} = 10^3$ cm^{-3} and the gas temperature is taken as $T_{\text{gas}} = 10$ K.
The different profiles within one frame correspond to different standard de-
viations of the logarithmic density, whereas from one frame to the other the
correlation length of the density field is altered. It can be clearly seen, that
the line intensities are decreasing for increasing σ_n and ℓ_n. This is not that
surprising: as we have seen in the last two sections, the effective optical depth
of the layer becomes smaller with these two parameters increasing.

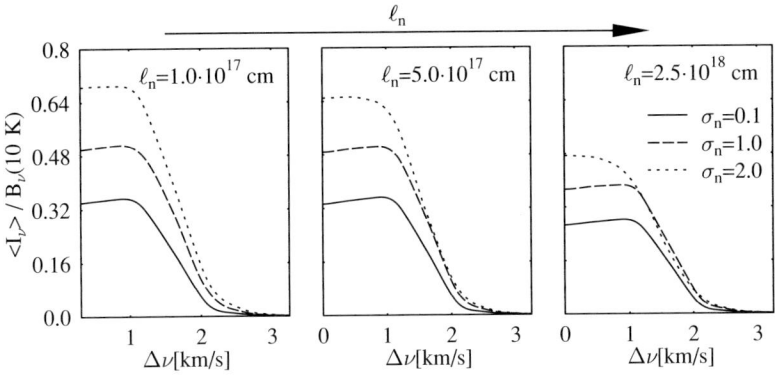

Figure 9: Line profiles for the $J = 4 \rightarrow 3$ transition of CO for $T_{\mathrm{gas}} = 10\,\mathrm{K}$, $L = 3.0 \cdot 10^{19}$ cm, $\langle n_{\mathrm{H}_2} \rangle = 10^3$ cm^{-3}, $\langle N_{\mathrm{CO}} \rangle = 2.4 \cdot 10^{18}$ cm^{-2}, $\ell_{\mathrm{v}} = 5.0 \cdot 10^{17}$ cm, and $\sigma_{\mathrm{v}}/v_{\mathrm{th}} = 10$.

In Fig. 9, the profiles for the $J = 4 \rightarrow 3$ transition are shown. The physical parameters are the same as before. However, we get a completely different picture. The intensities are now substantially raised for an increased Gaussian width of the logarithmic density distribution. If these lines were formed under the conditions of local thermal equilibrium, we would get the same result as before, the effective optical depth would be strongly reduced for a large σ_{n}. But in the case of NLTE both, the mean emission coefficient ε_ν and the mean absorption coefficient are increased. Most of the radiation is emitted from

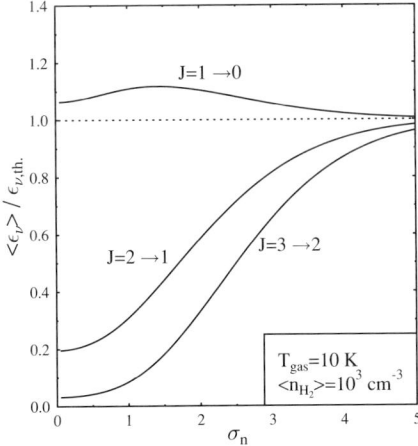

Figure 10: Mean emission coefficients for rotational transitions $J = J + 1 \rightarrow J$ of CO. All values are given for an optically thin line and a 3K-background. The emission coefficients are normalized to the corresponding thermal values.

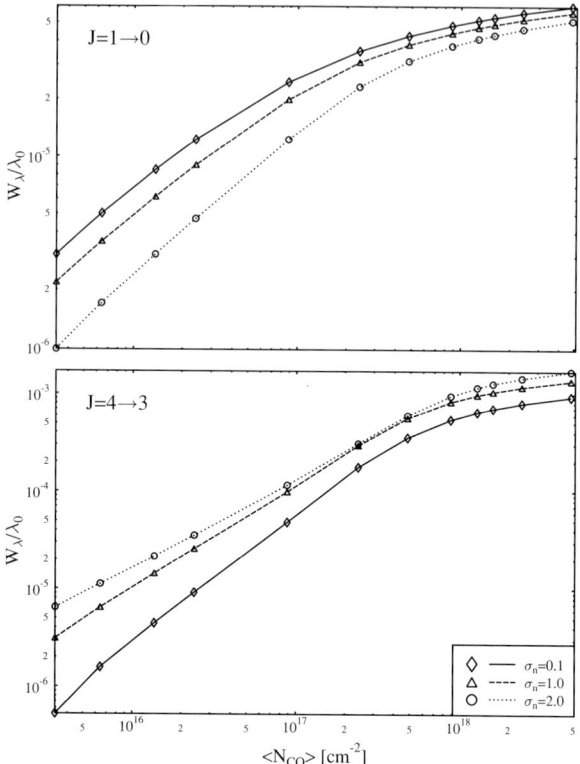

Figure 11: Curves of growth for the $J = 1 \to 0$ and $J = 4 \to 3$ transitions of CO for $T_{\text{gas}} = 10\,\text{K}$, $\langle n_{\text{H}_2} \rangle = 10^3\,\text{cm}^{-3}$, $\ell_{\text{v}} = 5.0 \cdot 10^{17}$, $\ell_{\text{n}} = 5.0 \cdot 10^{17}$, and $\sigma_{\text{v}}/v_{\text{th}} = 10$. The equivalent width is normalized to the wavelength in the line center.

dense clumps, where the high rotational levels can be populated effectively by collision of CO with H_2. This effect can be seen also in Fig. 10. Here, we give the expectation value of the emission coefficient $\langle \varepsilon_\nu \rangle$ for the three lowest rotational transitions of CO as function of σ_{n}. In order to exclude radiative transfer effects, an optically thin layer was assumed this time. It can be clearly seen that for large values of σ_{n} the emission coefficient approaches the corresponding LTE value.

In Fig. 11, we give the curves of growth for the $J = 1 \to 0$ and $J = 4 \to 3$ transitions of CO. That is, the equivalent width

$$\langle W_\lambda \rangle = \int_{\text{line}} \frac{\langle I_\lambda \rangle - B_\lambda(2.7\,\text{K})}{B_\lambda(2.7\,\text{K})} \mathrm{d}\lambda \qquad (17)$$

is plotted as function of the mean CO column density. It can be clearly seen, that the integrated intensities for the high rotational transitions becomes larger as σ_{n} increases, whereas the integrated intensities for the $J = 1 \to 0$

transition are reduced for a large Gaussian width of the logarithmic density distribution. As a result, the $J = 4 \to 3$ lines are much stronger than would be expected from the CO $J = 1 \to 0$ lines.

5   Conclusions

We presented a method of treating the radiative transfer in a medium with an inhomogeneous stochastic density and velocity distribution. To describe the turbulent velocity field and the spatial variation of the density, we applied a stochastic approach based on the assumption of a continuous Markov process. On the basis of three examples, we showed that stochastic density fluctuations have indeed a profound impact on radiative transfer calculations.

First of all, we examined the penetration of photons into a dusty layer. Our results clearly show that for a given mean mass density a clumpy structure of the absorbing medium leads to a substantial reduction of the effective optical depth. That is, photons are able to penetrate much deeper into the layer than would be expected from a uniform density model. In particular, the effective optical depth strongly decreases with an increasing correlation length of the stochastic density distribution. This result is especially important for the chemistry in molecular clouds. An increased UV radiation field, will not only effect the formation and destruction of molecules, but will also determine the gas and dust temperature, respectively.

Next, we studied the IR emission of a dusty layer, which is heated by an intense UV radiation field. We could see, that stochastic density fluctuations have great impact on the emerging IR radiation field. With ℓ_n and σ_n becoming large, a dramatic increase of the intensity for wavelengths smaller than approximately 40 μm was found. On the other hand side, the intensity is decreased for $\lambda > 40$ μm.

As a last example, we studied the formation of CO rotational lines in the case of NLTE. In all calculations we adopted the same CO column density and the same gas temperature. Our findings can be summarized as follows: Lines of low rotational transitions become in most cases weaker with an increasing correlation length and an increasing value of the standard deviation of the logarithmic density distribution. In contrast, high rotational lines tend to be substantially stronger than in the case of a uniform density model. Consequently, the line intensity ratio $R(4 \to 3/1 \to 0)$ is strongly increased for large ℓ_n and σ_n.

Acknowledgments

I would like to thank my collaborators Jörg Wittorf, Christian Hengel, and Markus Röllig for their encouragement and support. In particular, I wish to thank Prof. Dr. W. H. Kegel, who contributed much to the results presented here. This work was supported in part by the German Bundesministerium für Bildung, Wissenschaft, Forschung und Technologie (BMBF) in the frame of the project No. 053FM13A.

References

Albrecht, M. A., Kegel, W. H. 1987, A&A 176, 317

Black, J. H., Willner, S. P. 1984, ApJ 279, 673

Boissé, P. 1990, A&A 228, 483

Gail, H. P., Hundt, E., Kegel, W. H., et al. 1974, A&A 32, 65

Gail, H. P., Sedlmayr, E. 1974, A&A 36, 17

Gail, H. P., Sedlmayr, E., Traving, G. 1975, A&A 44, 421

Gail, H. P., Sedlmayr, E., Traving, G. 1980, J. Quant. Spectr. Radiat. Transfer 23, 267

Hegmann, M., Kegel, W. H. 1996, MNRAS 283, 16

Hegmann, M., Kegel, W. H. 2000, A&A 359, 405

Hobson, M. P., Padman, R. 1993, MNRAS 264, 161

Hobson, M. P., Scheuer, P. A. G. 1993, MNRAS 264, 145

Huffman, D. R. 1975, ASS 34, 175

Juvela, M. 1996, A&A 322, 943

Kegel, W. H., Piehler, G., Albrecht, M. A. 1993, A&A 270, 407

Piehler, G., Kegel, W. H. 1994, A&A 297, 841

Traving, G. 1980, A&A 85, 281

Városi, F., Dwek, E. 1999, ApJ 523, 265

Witt, A. N., Gordon, K. D. 2000, ApJ 528, 799

ASTRONOMISCHE GESELLSCHAFT: Reviews in Modern Astronomy **15**, 165–178 (2002)

Seeing the Light through the Dark

João F. Alves

European Southern Observatory
Karl-Schwarzschild-Straße 2
D-85748 Garching bei München, Germany
jalves@eso.org

Abstract

A new powerful method to study large and small-scale structure of dark molecular clouds is shedding new light on how nature proceeds to form stars. In this talk I will review the method, describe the results obtained so far, and point to the wonderful opportunities offered by future facilities to attack this fundamental problem.

1 The Search for the Initial Conditions to Star Formation

Stars and planets form within dark molecular clouds. However, despite 30 years of study little is understood about the internal structure of these clouds and consequently the initial conditions that give rise to star and planet formation (see Figure 1). This is largely due to the fact that molecular clouds are primarily composed of molecular hydrogen, which is virtually inaccessible to direct observation. Because of its symmetric structure the hydrogen molecule possesses no dipole moment and cannot produce a readily detectable signal under the conditions that characterize cold, dark clouds. The traditional methods used to derive the basic physical properties of such molecular clouds therefore make use of observations of trace H_2 surrogates, namely those rare molecules with sufficient dipole moments to be easily detected by radio spectroscopic techniques (Lada 1996), and interstellar dust, whose thermal emission can be detected by radio continuum techniques (e. g., André et al. 2000). However, the interpretation of results derived from these methods is not always straightforward (e. g., Alves, Lada, & Lada 1999; Chandler & Richer 2000). Several poorly constrained effects inherent in these techniques (e. g., deviations from local thermodynamic equilibrium, opacity variations, chemical evolution, small-scale structure, depletion of molecules, unknown emissivity properties of the dust, unknown dust temperature) make the construction

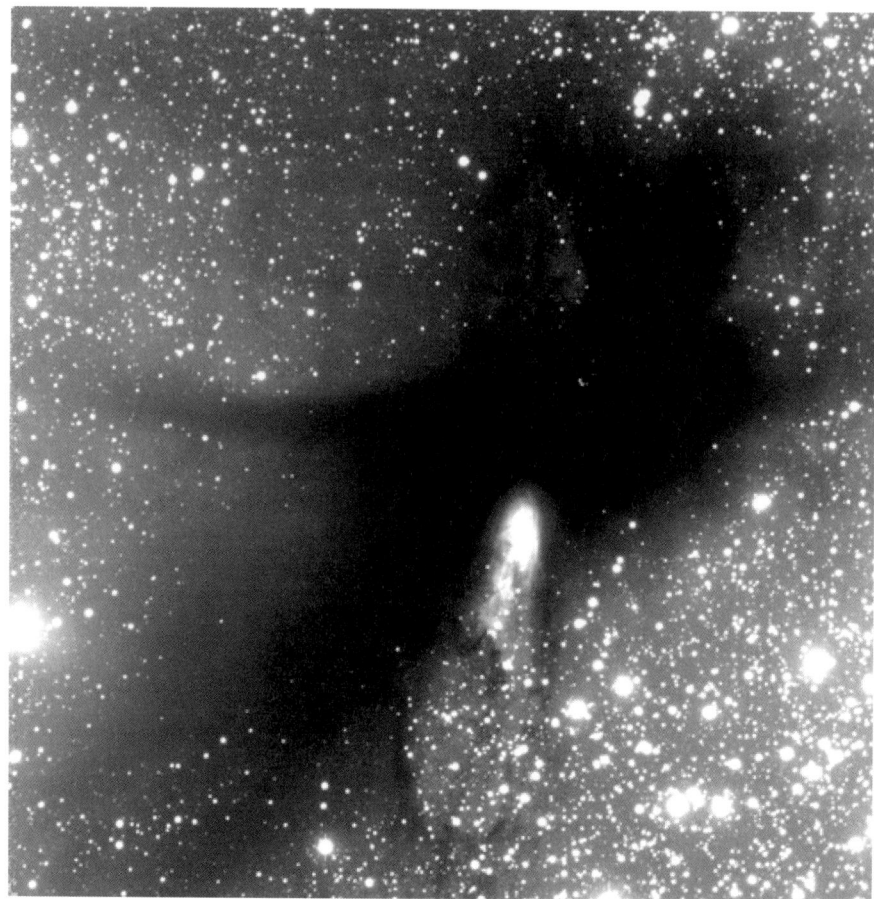

Figure 1: Bok Globule BHR 71 caught in the act of forming a stellar binary (VLT optical image). Bok globules are the simplest configurations of the ISM capable of forming stars and represent our best chance to understand how molecular clouds form and collapse to originate stars and planets.

of an unambiguous picture of the physical structure of these objects—a very difficult task. There is then a need for a less complicate and more robust tracer of H_2 to access not only the physical structure of these objects but also to accurately calibrate molecular abundances and dust emissivity inside these clouds. The deployment of sensitive, large format infrared array cameras on large telescopes however, has fullfilled this need by enabling the direct measurement of the dust extinction toward thousands of individual background stars observed through a molecular cloud. Such measurements are free from the complications that plague molecular-line or dust emission data and enable detailed maps of cloud structure to be constructed.

2   The Method and Results So Far

The most straightforward and reliable way to measure molecular cloud structure is to measure dust extinction of background starlight. We have developed a new powerful technique for measuring and mapping the distribution of dust through a molecular cloud using data obtained in large-scale, multi-wavelength, infrared imaging surveys. This method combines measurements of near-infrared color excess to directly measure extinctions and map the dust column density distribution through a cloud. *It is the most straightforward and unambiguous way of determining the density structure in dark molecular clouds.* Moreover, the measurements can be made at significantly higher angular resolutions and substantially greater optical depths than previously thought possible. We have conclusively demonstrated the efficacy of this technique with our study of the dark cloud complex IC 5146 (Lada et al. 1994, Lada et al. 1999), L 977 (Alves et al. 1998), and Barnard 68 (Alves, Lada, & Lada 2001), where we detected nearly 7000 infrared sources background to these clouds and produced detailed maps of the extinction across the cloud to optical depths and spatial resolution an order of magnitude higher than previously possible ($A_V \sim 40$ magnitudes, spatial resolution ~ 10 arcsec).

We have used our extinction observations to measure the masses, density structure, extinction laws, and distances to these objects. We found the radial density profiles of filamentary clouds (IC 5146 and L 977) to be well behaved and smoothly falling with a power-law index of $\alpha = -2$, significantly shallower than predicted by early theoretical calculations of Ostriker (1964) ($\alpha = -4$). Moreover because we are using pencil beam measurements of dust column density along the line of sight to background stars we were able to demonstrate that the small-scale structure of the clouds is surprisingly smooth with random density fluctuations ($\delta A_V / A_V$) present at very small levels ($< 3\,\%!$). This result is in very good agreement with optical studies of the small scale structure of the diffuse Interstellar Medium (ISM) (Thoraval, Boissé, & Duvert 1999).

When convolved to the appropriate spatial resolution our maps showed structure in the dust distribution which was strikingly well correlated with millimeter-wave CO and CS emission maps of the cloud (see Figure 2), although showing crucial differences at high optical depths where these other tracers of column density become unreliable (see Figure 3). These comparisons enabled us to, for the first time, directly derive CO, CS, and N_2H^+ abundances, and variations of, over an extinction range of 1–30 magnitudes, a range nearly an order of magnitude greater than achieved previously with optical star counting techniques. In a recent experiment we were able to make a direct measurement of molecular depletion in a cold cloud core (Kramer et al. 1999). Finally, a comparison between our extinction data and millimeter continuum emission data, allowed a most accurate measurement of the ratio of dust absorption coefficients at millimeter and near-infrared wavelengths (Kramer et al. 1998).

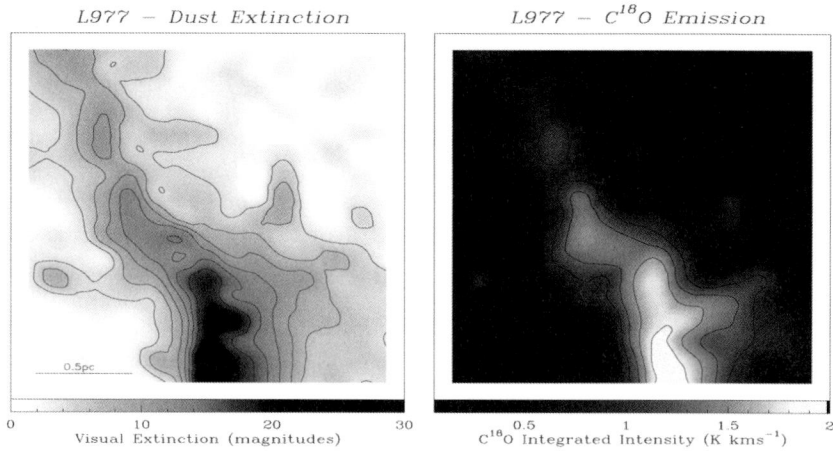

Figure 2: Dust extinction map and $C^{18}O$ molecular line map of L 977. The structure in the dust distribution correlates strikingly well with millimeter-wave $C^{18}O$ emission map of the cloud, although showing crucial differences at high optical depths (see next Figure) where the CO tracer becomes unreliable (from Alves et al. 1999).

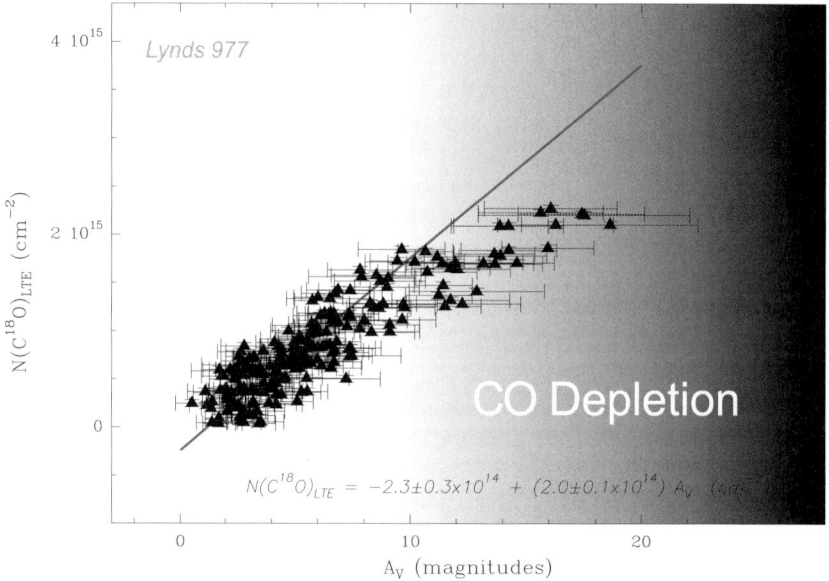

Figure 3: Relation between $N(C^{18}O)_{LTE}$ and visual extinction A_V for molecular cloud L 977. The solid straight red line represents the result of a linear least-squares fit, with errors in both coordinates, over the entire data set. There is a clear deviation from the linear relation at extinctions ≥ 10 magnitudes above which $C^{18}O$ becomes a very poor tracer of H_2 (from Alves et al. 1999). Follow-up molecular line study of this cloud (Tafalla et al. 2001) suggests that, as in the IC 5146 cloud, depletion of CO might be occuring at high optical depths.

3   Barnard 68 as a Stellar Seed

Recently we have been concentrating efforts on mapping the densest regions of the ISM that are likely places of future star formation. We performed very sensitive near-infrared imaging observations to map the structure of a type of dark cloud known as a Bok globule (Bok & Reilly 1947), one of the least complicated configurations of molecular gas known to form stars (Figure 4). The target cloud for our study, Barnard 68 (Figure 5), is itself one of the finest examples of a Bok globule, and was selected because it is a nearby, relatively isolated and morphologically simple molecular cloud with distinct boundaries, a known dis-

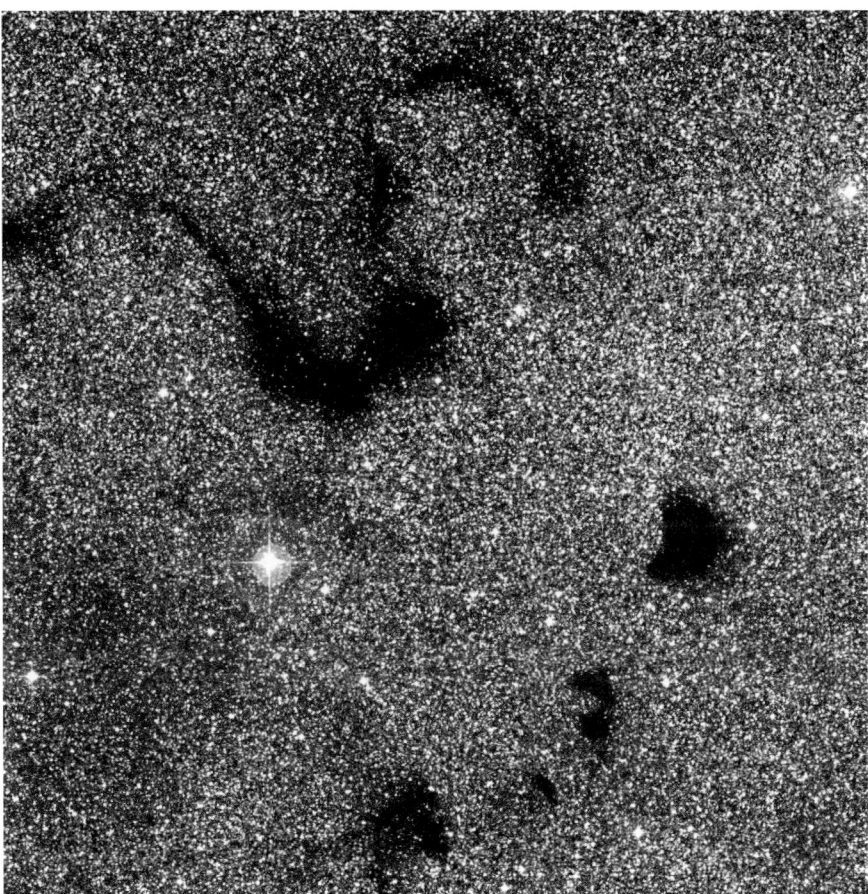

Figure 4: Barnard 72 (the S-nebula) and Barnard 68 (right of center) (DSS plate). At visual wavelengths these clouds are completely opaque owing to extinction of background starlight caused by small (∼ 0.1 μm) interstellar dust particles that permeate the clouds. Complexes of globules like the ones in this image may be the precursors of small young stellar groups, like the well known TW Hya.

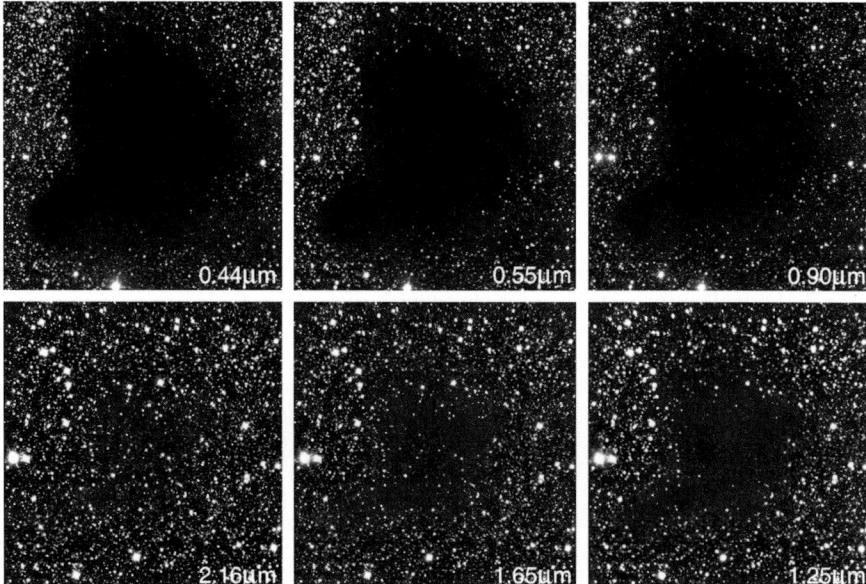

Figure 5: Deep BVIJHK-band (0.44 μm, 0.55 μm, 0.9 μm, 1.25 μm, 1.65 μm, 2.16 μm) images of dark molecular cloud Barnard 68 taken with ESO's Very Large Telescope (VLT) and New Technology Telescope (NTT). The wavelength dependence of interstellar dust extinction in Barnard 68 is clearly depicted in these images. The analysis of the near-infrared colors of the stars seen through the dark cloud allow the construction of a high resolution mass map as traced by dust extinction and the most finely sampled and higher S/N density profile ever obtained for a cold dark cloud (from Alves et al. 2002).

tance (125 pc; Launhardt & Henning 1997), and temperature (16 K; Bourke et al. 1995). It was first discovered by E. E. Barnard (Barnard 1919) and was the target of several optical dust extinction studies by Bart Bok and co-workers (Bok & Reilly 1974; Bok 1977). Although a very dense cloud, Barnard 68 does not present any of the signatures of ongoing star formation, such as IRAS sources, outflows, or mm continuum sources (Avery et al. 1987; Reipurth, Nyman, Chini 1996).

Barnard 68 lies in the direction of the center of the Galaxy but above the galactic plane where it is projected against the rich star field of the galactic bulge. This makes Barnard 68 a particularly ideal candidate for an infrared extinction study for the following reasons. First, the background bulge stars are primarily late-type (giant) stars whose intrinsic infrared colors span a narrow range and can be accurately determined from observations of nearby control fields. Second, the background star field is sufficiently rich to permit a detailed sampling of the extinction across the entire extent of the cloud. Third, the cloud is sufficiently nearby that foreground star contamination is negligible.

We used the SOFI (Moorwood, Cuby, Lidman 1998) near-infrared camera on the European Southern Observatory's (ESO) New Technology Telescope (NTT) to obtain deep infrared J band (1.25 μm), H band (1.65 μm) and Ks band (2.16 μm) images of the cloud over two nights in March 1999. Complementary optical data was obtained with ESO's Very Large telescope (VLT) on Cerro Paranal, fitted with FORS1 CCD camera (Appenzeller et al. 1998), during one night of March 1999. The results of the optical and near-infrared imaging are displayed in Figure 5. At optical wavelengths obscuring dust within the cloud renders it opaque and completely void of stars. However, due to the wavelength dependence of dust extinction (i. e., opacity), the cloud is essentially transparent at infrared wavelengths enabling otherwise invisible stars behind the cloud to be images. We detected 3708 stars simultaneously in the deep H and K band images out of which $\sim$ 1000 stars, lying behind the cloud, are not visible at optical wavelengths. Because dust opacity decreases sharply with wavelength, the colors of stars that are detected through a dust screen appear reddened compared to their intrinsic colors.

We have accurately sampled the dust extinction and column density distribution through the Barnard 68 cloud at more than a thousand positions with extraordinary (pencil beam) angular resolution. Although the individual measurements are characterized by high angular resolution, our mapping of the dust column density in the cloud is highly undersampled. Consequently, we smoothed these data to construct the first 10 arsec resolution map of dust extinction of a cold dark cloud (Figure 6) and an azimuthally averaged radial extinction (dust column density) profile of the cloud (Figure 7). This is the most finely sampled and highest signal-to-noise radial column density profile ever obtained for a dense and cold molecular cloud. For the first time, the internal structure of a dark cloud has been specified with a detail only exceeded by that characterizing a stellar interior.

3.1 Bonnor-Ebert Spheres

The extinction profile in Figure 7 is the projection of the cloud volume density profile function, and therefore provides an exquisite view of the internal structure of this dense dark cloud. As early as 1948 Bart Bok pointed out that roughly spherical homogenous looking clouds, such as Barnard 68, resemble single dynamical units much like the polytropic models of Lane and Emden used to describe stellar structure (Lane 1870; Emden 1907). Can Barnard 68 be described as a self-gravitating, polytropic sphere of molecular gas? To investigate the physical structure of the cloud we begin with the assumptions of an isothermal equation of state and spherical symmetry.

The fluid equation that describes a self-gravitating, isothermal sphere in hydrostatic equilibrium is the following well-known variant of the Lane-Emden equation:

$$\frac{1}{\xi^2} \frac{d}{d\xi} \left(\xi^2 \frac{d\psi}{d\xi} \right) = e^{-\psi} \tag{1}$$

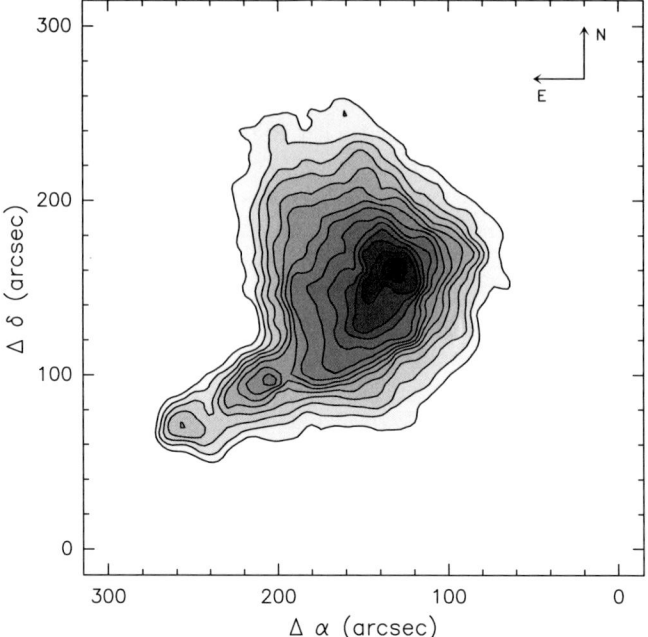

Figure 6: Map of dust extinction obtained from the convolution of the approximately 3000 line-of-sight extinction measurements to all detected stars. Around a 1000 of these are detected through the denser regions of the cloud. The peak extinction measured through the center of the cloud is 33 magnitudes of visual extinction. The contours start at $A_V = 4$ mags (a 4 σ measurement) and increase in steps of 2 mags. The integration of this dust map allows a precise mass determination for Barnard 68, MB68 $= 2.1 \pm 0.1$ $(d/125 \text{ pc})^2$ M$_\odot$, where d is the true distance to the cloud (from Alves et al. 2002).

where ξ is the non-dimensional radius,

$$\xi = \frac{r}{a} \, (4\pi \, G \, \rho_c)^{\frac{1}{2}} \tag{2}$$

while $\psi(\xi)$ equals,

$$\psi(\xi) = -ln(\rho/\rho_c) \tag{3}$$

for which r is the distance from the center of the sphere, a is the isothermal sound speed inside the gas cloud ($a = (KT/m)^{\frac{1}{2}}$), ρ is the density, and ρ_c is the central density. For an isothermal sphere bounded by a fixed external pressure there is a family of solutions characterized by a single parameter (Ebert 1955, Bonnor 1956):

$$\xi_{max} = \frac{R}{a} \, (4\pi \, G \, \rho_c)^{\frac{1}{2}} \tag{4}$$

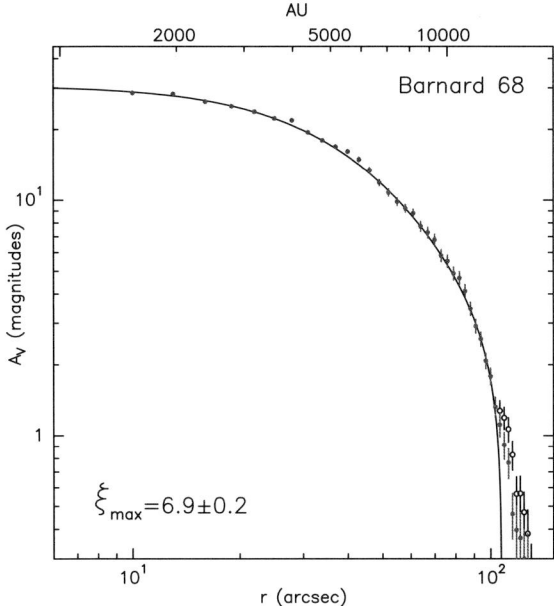

Figure 7: Azimuthally averaged radial density profile of Barnard 68. The filled circles show the data points for the averaged profile of a sub-sample of the data that do not include the cloud's South-East prominence, seen in Figure 6. The open circles, on the other hand, include this prominence. The error bars were computed as the rms dispersion of the extinction measurements in each averaging annulus and are smaller than the data points for the central regions of the cloud. This is the most finely sampled and highest S/N radial density profile ever obtained for a dense molecular (from Alves et al. 2002).

Here ξ_{max} is the value of ξ at the outer boundary, R. Each of these solutions corresponds to a unique cloud mass density profile. Bonnor demonstrated that for $\xi_{max} > 6.5$ such a gaseous configuration would be unstable to gravitational collapse (Bonnor 1956). The high quality of our extinction data permits a detailed comparison with the Bonnor-Ebert predictions and we find that there is a particular solution, $\xi_{max} = 6.9 \pm 0.2$, that fits the data extraordinarily well as seen in Figure 7. For the known distance (125 pc), and temperature (16 K), Barnard 68 has a physical radius of 12500 AU, a mass of 2.1 solar masses, and a pressure at its boundary of $P = 2.5 \times 10^{-12}$ Pa. This surface pressure is an order of magnitude higher than that of the general ISM (McKee 1999) but it is in rough agreement with the pressure inferred for the Loop I superbubble, where Barnard 68 is embedded, derived from X-ray observations with the ROSAT satellite (Breitschwerdt et al. 2000).

The close correspondence of the observed extinction profile with that predicted for a Bonnor-Ebert sphere strongly suggests that Barnard 68 is indeed an isothermal, pressure confined, and self-gravitating cloud. It is also likely to be in a state near hydrostatic equilibrium with thermal pressure primarily supporting the cloud against gravitational contraction. For Barnard 68, ξ_{max} is very near and slightly in excess of the critical radial parameter and the cloud may be only marginally stable and on the verge of collapse. If this is the case we should expect molecular radio-spectroscopy of this cloud to reveal a quiet, non-turbulent cloud with narrow molecular emission lines. Indeed, preliminary results from our radio-spectroscopy observing campaign of Barnard 68 (with the IRAM 30 m Radio Telescope at Pico Veleta, Granada) reveal that the linewidth of the $C^{18}O$ line in this cloud is $\Delta v \sim 0.18$ kms^{-1}, one of the narrowest lines ever observed in molecular clouds, in perfect agreement with the Bonnor-Ebert sphere nature of Barnard 68.

3.2 Reverse Engineering: Distance and Gas to Dust Ratio to Few Percent

The exact physical state of the Barnard 68 cloud is further constrained by the fact that ξ_{max}, which is solely derived from the shape of the observed column density distribution, uniquely specifies the combination of central density, sound speed and physical size that characterizes the cloud (i. e., Equation 4). Independent knowledge of any two of these parameters directly determines the value of the third. For example, in Barnard 68 we independently measure the dust extinction (which is directly related to the mass column density, via the gas-to-dust ratio in the cloud) and the angular size of the cloud (which is directly related to its physical size via the cloud's distance). Thus, if we know the temperature of the cloud, our measurement of its extinction and angular size (combined with the constraint that $\xi_{max} = 6.9 \pm 0.2$) independently gives the distance to the cloud, provided the gas-to-dust ratio is assumed. The temperature of the molecular gas in Barnard 68 has been previously measured using observations of emission from the (1,1) and (2,2) metastable transitions of the ammonia molecule and found to be 16 ± 1.5 K (Bourke et al. 1995). For a canonical gas-to-dust ratio (1.9×10^{21} protons/magnitude; Bohlin, Savage, & Drake 1978), we derive a distance to Barnard 68 of 112 ± 3 pc. Here the quoted uncertainty (3 %) arises solely from the uncertainty in ξ_{max}. Accounting for the uncertainty in the temperature measurement the overall uncertainty increases only to 8 %, or ± 9 pc. From its association with the Ophiuchus complex the distance to the cloud has been estimated to be 125 ± 25 pc (de Geus et al. 1989), which within the uncertainties agrees with our derivation. This, in turn, implies that the gas-to-dust ratio in this dense cloud must be close to the canonical interstellar value. Indeed, if we independently know the distance to the cloud, our modeling directly yields the gas-to-dust ratio in the cloud. Assuming a distance of 125 pc our measurements allow a high precision determination (2 %) for the gas-to-dust ratio in this cloud of $1.73 \pm 0.04 \times 10^{21}$ protons/magnitude. If we also account for the uncertainties in distance and temperature the overall uncertainty (accuracy) of our determination increases to $\pm 0.4 \times 10^{21}$ protons/magnitude, or 23 %. Within the overall error this ratio is the same as the long accepted value characterizing low-density interstellar gas and is the first independent and relatively accurate determination of this important astrophysical parameter in a dense molecular cloud core.

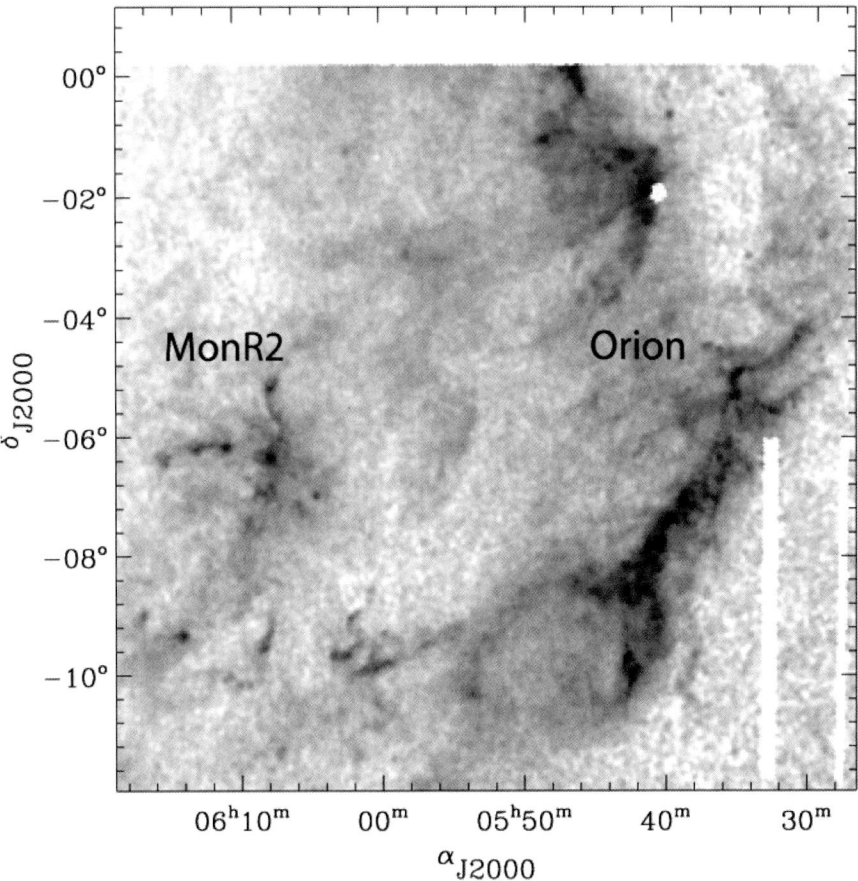

Figure 8: A 12×12 deg^2 extinction map of the Orion and Mon R2 giant molecular cloud complexes. This huge extinction map is an example of what the 2MASS database will be able to offer regarding a global and robust description of the structure of molecular clouds, from the large to small scales (from Lombardi & Alves 2001).

3.3 On the Origin of Small Stellar Groups

Recently there has been special attention for isolated and sparsely populated association of young low mass stars similar to the recently identified TW Hydra association (Rucinsky & Krautter 1983). The TW Hydra association is a stellar group near the solar system consisting of a handful of young low mass, sunlike stars, which has been a primary target of recent substellar objects search (e. g. Neuhäuser et al. 2000a, Neuhäuser et al. 2000b). The existence of such a young stellar group presents an interesting problem to astronomers because its origin is difficult to explain given its youth and relatively large distance from known sites of star formation. Bok globules such as those in the

Barnard 68 group are thought to be remnant dense cores produced as a result of the interaction of massive O stars and molecular clouds (Reipurth 1983). Over their short lifetimes such massive stars through ionization, stellar winds and ultimately supernova explosions, very effectively disrupt the molecular clouds from which they formed. In the process large shells of expanding gas

are created. When surrounding clouds are disrupted by the passage of these shells, a few of their most resilient dense cores will be left behind, embedded within the shell's hot interior. Remnant cores with just the right mass can establish pressure equilibrium with the hot gas within the shell and survive to become Bok globules. Eventually, as a result of processes described above, these clouds will evolve to form low mass, sunlike stars which are relatively isolated and far from the original birthplace of the O stars. We suggest that Barnard 68, and its neighboring globules B69, B70 and B72 (Figure 4) may be the precursors of a small stellar group, like TW Hya. Up to 35 % of all Bok globules contain newly formed stars (Launhardt & Henning 1997) and thus it is likely that our observations of the starless Barnard 68 cloud provide the first detailed description of the initial conditions prior to the collapse of dark globules and the formation of isolated, low mass stars.

4   The Future

Significant progress in extinction mapping studies will result when the 2MASS all sky near-infrared imaging surveys are completed and released (see Figure 8). This surveys will be sufficiently sensitive to produce moderate depth extinction maps (i. e., $A_V \leq 25$ mags) of many nearby dark clouds in those directions of the Galaxy where field stars suffer little extraneous extinction. In the immediate future large aperture telescopes, such as the VLT outfitted with ISAAC and NAOS/CONICA, will provide the additional capability to perform deeper surveys of the higher extinction regions ($25 \leq A_V \leq 60$ mags) in these clouds. Finally, space-based infrared telescopes, such as the NGST, should enable the regions of deepest extinction ($A_V > 60$ mags) to be probed. Together, such observations promise to render a very complete understanding of the physical and chemical structure of molecular clouds and of the general initial conditions to the star formation process.

Acknowledgements

The author acknowledges Charles Lada and Marco Lombardi for fruitful discussions and Monika Petr-Gotzen for helpful suggestions during the preparation of the VLT observations.

References

Alves, J., Lada, C.J., Lada, E.A., Kenyon, S., & Phelps, R. 1998, ApJ 506, 292

Alves, J., Lada, C.J., & Lada, E.A. 1999, ApJ 515, 265

Alves, J., Lada, C.J., & Lada, E.A. 2001, Nature 409, 159

Alves, J., Lada, C.J., Lada, E.A., & Scanapiecco, E. 2002, ApJ, submitted

Appenzeller, I., Fricke, K., Fürtig, W., Gässler, W., Häfner, R., Harke, R., Hess, H., Hummel, W., Jürgens, P., Kudritzki, R., Mantel, K., Meisl, W., Szeifert, T., & Tarantik, K. 1998, *Successful Commissioning of FORS1 – the First Optical Instrument on the VLT*, The ESO Messenger 94, 1–6

André, P., Ward-Thompson, D., & Barsony, M. 2000, *Protostars and Planets IV*, (Tucson: University of Arizona Press), eds. Mannings, V., Boss, A.P., Russell, S.S.

Bok, B. & Reilly, E. 1947, ApJ 105, 255

Bok, B. 1948, Centennial Symposia (Harvard Observatory Monographs No. 7); Harvard-College Observatory, Cambridge

Bonnor, W.B. 1956, MNRAS 116, 351

Bourke, T., Hyland, A., Robinson, G., James, S., & Wright, C. 1995, MNRAS 276, 1067

Bourke, T., Garay, G., Lehtinen, K., Koehnenkamp, I., Launhardt, R., Nyman, L., May, J., Robinson, G., & Hyland, A. 1997, ApJ 476, 781

Breitschwerdt, D., Freyberg, M., & Egger, R. 2000, A&A 361, 303

Chandler, C. & Richer, J. 2000, ApJ 530, 851

de Geus, E., de Zeeuw, P., & Lub, J. 1989, A&A 216, 44

Ebert, R. 1955, Z. Ap. 217

Emden, R. 1907, Gaskugeln (Leipzig:Teubner)

Kramer, C., Alves, J., Lada, C., Lada, E., Sievers, A., Ungerechts, H., & Walmsley, M. 1997, A&A 329, 33

Kramer, C., Alves, J., Lada, C.J., Lada, E.A., Sievers, A., Ungerechts, H., & Walmsley, C.M. 1999, A&A 342, 257

Lada, C.J., Lada, E.A., Clemens, D.P., & Bally, J. 1994, ApJ 429, 694

Lada, C.J. 1996, Proceedings of IAU Symposium *CO: Twenty-Five Years of Millimeter-Wave Spectroscopy*, eds. W.B. Latter et al. (Kluwer-Dordrecth), 387

Lada, C.J., Alves, J., & Lada, E.A. 1999, ApJ 512, 250

Lane, J. 1870, American Journal of Science and Arts, Series 2, 4, 57

Launhardt, R. & Henning, T. 1997, A&A 326, 329

Lombardi, M. & Alves, J. 2001, A&A 377, 1023

Mckee, C. 1999, in *The Origin of Stars and Planetary Systems*, Eds. C. Lada & N. Kylafis (Dordrecht: Kluwer)

Moorwood, A., Cuby, J., & Lidman, C. 1998, *SOFI Sees First Light at the NTT*, The ESO Messenger 91, 9–13

Neuhäuser, R., Brandner, W., Eckart, A., Guenther, E., Alves, J., Ott, T., Huélamo, N., & Fernández, M. 2000a, A&A 354, 9

Neuhäuser, R., Guenther, E., Petr, M., Brandner, W., Huélamo, N., & Alves, J. 2000b, A&A 360, 39

Ostriker, J. 1964, ApJ 140, 1056

Reipurth, B. 1983, A&A 117, 183

Reipurth, B., Nyman, L., & Chini, R. 1996, A&A 314, 258

Rucinski, S. & Krautter, J. 1983, A&A 121, 217

Tafalla, M., Alves, J., & Lada, C. 2002, in prep.

Thoraval, S., Boissé, P., & Duvert, G. 1999, A&A 351, 1051

Astronomische Gesellschaft: Reviews in Modern Astronomy **15**, 179–197 (2002)

Obscured Active Galactic Nuclei

Roberto Maiolino

Osservatorio Astrofisico di Arcetri
Largo E. Fermi 5, 50125 Firenze, Italy
maiolino@arcetri.astro.it

Abstract

I discuss some of the issues related to the obscuration in Active Galactic Nuclei (AGNs) and propose some models which could account for some of the properties observed in this class of objects.

I discuss the obscuration of AGNs at high luminosities and show that in QSOs the covering factor of the obscuring dusty medium is reduced with respect to Seyfert nuclei. This effect may be the result of the interaction of dust grains, accelerated by the strong radiation pressure, with the gas and with other grains which is likely to cause the disruption of dust.

At higher redshift the fraction of obscured AGNs is likely to be higher, as a consequence of the higher rate of interactions and distorted morphologies which drive gas into the nuclear region. To this regard, I discuss some evidences for a connection between non-axisymmetric potentials in the host galaxy and nuclear obscuration in local Sy2s.

I discuss the case of "elusive" AGNs hosted in starburst galaxies which are not identified by means of the classical optical tracers. These are probably AGNs which are completely buried, i. e. obscured in all directions. I show that high angular resolution mid-IR observations may be a powerful tool to identify this class of objects.

Finally, I tackle the hotly debated issue of the relative role of AGN and starburst in the budget of the bolometric luminosity. By combining sub-mm and mid-IR images I show that some nearby Sy2s are clearly dominated by the circumnuclear star formation. Some candidate type 2 QSOs have properties similar to starburst-dominated Sy2s, suggesting that also the former are mostly powered by starburst activity.

1 Introduction

Active galactic nuclei (AGNs) are characterized by a powerful compact source of UV and X-ray radiation surrounded by ionized gas in rapid motion (~ 5000 km/s) confined in a relatively small region (~ 0.01–0.1 pc). The latter is named Broad Line Region (BLR), since this ionized gas emits strong broad

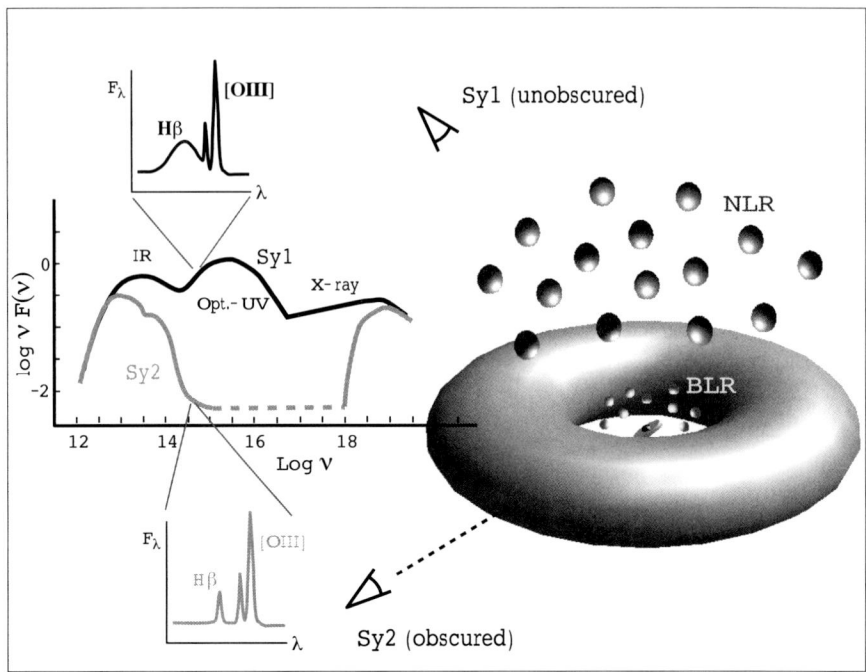

Figure 1: Simple scheme illustrating the basic idea of the (strict) unified model and the implications for the observed properties.

lines, historically identified in the optical but detected also in the UV and in the IR. Ionized gas is also present on much larger scales (10–100 pc) and moving more slowly ($\sim$ 500 km/s) with respect to the BLR clouds. This region is named Narrow Line Region (NLR). In un-obscured AGNs all of these components are observed and the nucleus is classified as "type 1"; in the case of low luminosity AGNs these are named Seyfert 1 (Sy1). A simple sketch illustrating the various components in AGNs is shown in Fig. 1, along with their typical spectral components.

In type 2 AGNs a dusty-gaseous medium intercepts our line of sight to the nuclear UV source and to the BLR, therefore absorbing the nuclear optical-UV radiation (and part of the X-ray radiation) and the broad lines (Fig. 1). Instead, the NLR is not significantly obscured because more extended than the obscuring medium. The latter is generally assumed to have a toroidal geometry, but the actual geometry is not relevant to the discussion in this paper. While in the low-luminosity range a large number of Seyfert 2s have been identified (a factor of 4 more numerous than Sy1; Maiolino & Rieke 1995), at higher luminosities evidence for a significant population of type 2 QSOs (the obscured counterpart of "classical" QSOs) is still sparse. This issue is discussed in detail in the next section.

2 The shortage of type 2 QSOs

During the past few years, various studies have presented claims for obscured QSOs in various systems (e. g. Veilleux et al. 1999, Hines et al. 1999, Norman et al. 2002, Crawford et al. 2001, Stern et al. 2002). However, at least in the local Universe, there is an upper limit to the number of obscured QSOs which is given by the number of ultraluminous infrared galaxies (ULIRGs). The latter is a class of dusty systems with infrared luminosities in excess of $10^{12} L_\odot$, i. e. in the same range as the bolometric luminosity of "classical" QSOs. Since in any obscured QSO the absorbed radiation must be reprocessed into the IR by the circumnuclear dust, the conservation of energy requires that type 2 QSO can only be hosted in ULIRGs. The ratio between ULIRG and (type 1) QSO gives an upper limit on the fraction of obscured QSOs, i. e.

$$\frac{QSO2}{QSO1} < \frac{ULIRG}{QSO1} \simeq 1 \tag{1}$$

where the ratio ULIRG/QSO is taken from Kim & Sanders (1998). Since a significant fraction of ULIRGs are powered by starburst activity and not by AGNs (Genzel et al. 1998, Lutz et al. 1998) the upper limit given above is actually very conservative. This upper limit must be compared with the ratio between Sy2 and Sy1:

$$\frac{Sy2}{Sy1} \simeq 4 \tag{2}$$

(Maiolino & Rieke 1995). The comparison of Eqs. 1 and 2 clearly indicates that the fraction of obscured AGNs decreases at higher luminosities.

An alternative, independent way to illustrate the tendency for the dust obscuration to decrease at higher luminosities is the comparison of the ratio between optical and infrared luminosities in Sy1 and in optically selected QSOs. Indeed, if most of the IR bump is due to dust reprocessing of the "blue-bump" (Fig. 1), then the ratio L_{IR}/L_{opt-UV} gives the covering factor of the dusty medium surrounding the nuclear source. We have derived the ratio L_{IR}/L_{opt-UV} for a sample of Seyfert galaxies and QSOs. The total luminosity of the "blue-bump" was derived from the B luminosity by assuming the SED given in Elvis et al. (1994). The total infrared luminosity of Seyfert galaxies was derived from the work of Perez-Garcia & Rodriguez-Espinosa (2001) based on ISO data. Out of their fit to the infrared SED we only took the component associated to the AGN and excluded the component associated to the host galaxy or to any starburst component. Instead, in the case of QSOs we used the total infrared emission (from Haas et al. 2000), which is a conservative approach since the contribution from any galactic/starburst component would (artificially) increase the apparent covering factor. In Fig. 2 we show the the covering factor of the circumnuclear dusty medium derived from the ratio L_{IR}/L_{opt-UV} as a function of the IR luminosity. At low, Sy-like luminosities there is a large spread and the covering factor reaches values of $\sim 80\%$.

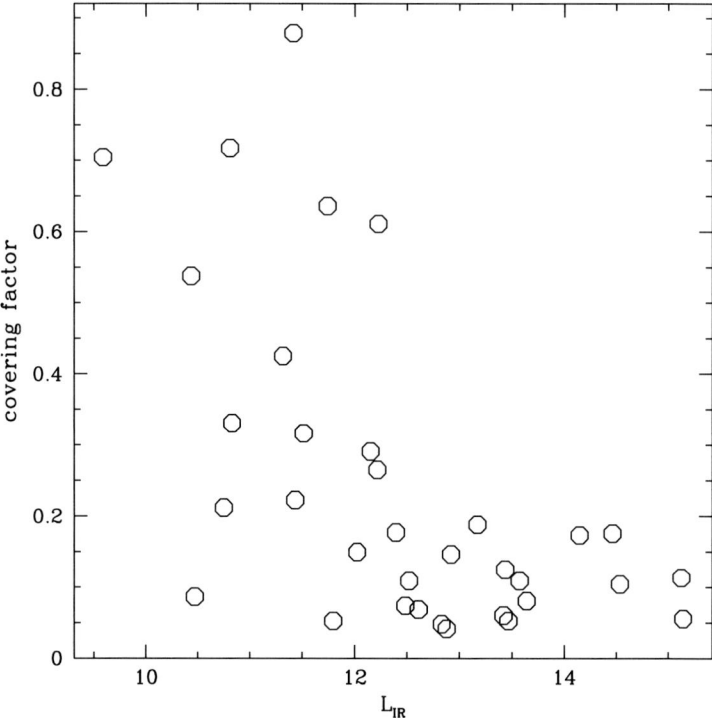

Figure 2: Covering factor of the dust in type 1 AGNs, inferred by the ratio between optical-UV and infrared luminosity, as a function of the IR luminosity. Note that at high, QSO-like luminosities the covering factor is, on average, significantly lower than in Sy1s.

At high, QSO-like luminosities the covering factor is on average significantly lower and never exceeds $\sim 20\,\%$. This result shows that the dust obscuration decreases at higher luminosities, thus confirming the finding obtained above based on the fraction of ULIRGs with respect to QSOs.

In the next section I discuss a possible physical scenario which might explain the effect of reduced dust obscuration at higher luminosities which was discussed in this section.

3   The behavior of dust in the circumnuclear region of AGNs

The presence of dust exposed to the radiation pressure in the circumnuclear region of AGNs has important effects on the overall dynamics of the gas. Indeed, the ratio between gravitational force and the outward force due to the radiation pressure is much higher for the dust grains than for the gas. More specifically, for a t ypical Galactic dust composition, the Eddington luminosity for the dust is lower by a factor of about 10^{-5} than the Eddington luminosity of the gas. Therefore, even if the AGN is radiating at a sub-Eddington rate (typically 0.1 L_{Edd}) this luminosity is highly super-Eddington for the dust which is, therefore, accelerated outwards. This important feature of the circumnuclear medium of AGNs has generally been overlooked, with the exception of a few studies (e. g. Laor & Draine 1993, Scoville & Norman 1995, SN95 hereafter). Here I discuss some of the expected properties of the dust in the circumnuclear region which are probably relevant to the issues discussed in the previous section. A more detailed discussion will be given in a forthcoming paper (Maiolino & Woltjer, in prep.).

3.1 Dust in the ionized gas

The powerful radiation from the AGN strongly charges the dust grains in the circumnuclear region due to photo-ejection of electrons by UV photons. The charge of dust grains is given by the equilibrium between photoelectric effect and electron recombination. Following the analysis in SN95, the equilibrium charge is roughly given by

$$Z_{eq} \simeq 10^3 a_{0.1\mu} \left(\frac{L_{46}}{n_8 r_{pc}^2} \right)^{2/9} T_4^{1/9} \qquad (3)$$

where $a_{0.1\mu}$ is the grain size in units of 0.1 μm, L_{46} is the bolometric luminosity of the AGN in units of 10^{46} erg s^{-1} (note that this is different from $L_{\gamma 13}$ given in SN95), r_{pc} is the distance from the nucleus in pc, while n_8 and $T_4^{1/9}$ are the electron density and temperature in units of 10^8 cm^{-3} and 10^4 K, respectively. From Eq. 3 it is clear that for the typical conditions of the ionized gas in the circumnuclear region of AGNs (outside the sublimation radius) the grains are highly ionized ($Z \sim 10^3$). These grains are accelerated outwards due to the radiation pressure, but the strong Coulomb drag due to the interaction between the strongly charged grains and the free electrons prevents the velocity of the grains (with respect to the gas) to grow significantly. More specifically, at the equilibrium between radiation pressure and Coulomb drag, the relative velocity between gas and dust is about $v_{eq} \simeq 10^{-2}$ km s^{-1}, i. e. dust and gas are strongly coupled. As a result, radiation pressure on dust causes an overall outflow of the whole gaseous medium, with terminal velocities up to 10^4 km s^{-1}, and may be responsible for the outflowing winds in Broad Absorption Line (BAL) QSOs, as well as in Seyfert galaxies (Chang et al. 1987, SN95, Laor & Brandt 2002).

3.2 Dust in the neutral medium

The broad line clouds are inside the dust sublimation radius and, therefore, dust-free (Laor & Draine 1993). Since they are characterized by columns of gas in excess of 10^{23} cm^{-2} they absorb all of the ionizing radiation up to a few keV, but the non-ionizing radiation red-ward of the Ly-edge passes through these clouds. When this filtered, non-ionizing radiation reaches the dusty medium outside the sublimation radius, then the flux impinging on the dust grains is still high (between 0.3 and 0.5 of the bolometric luminosity), but the radiation is not capable of ionizing the gas nor to charge the grains (since the threshold energy for the photoelectric effect of grains, 4–10 eV, is close to the Ly-edge). In these circumnuclear regions, partially shaded by the BLR clouds, the dust is still accelerated outward due to the radiation pressure, but the coupling with the gas is much lower. Indeed, in this case Coulomb friction is negligible and the drag force on the outflowing gas is dominated by the neutral collisions with the gas. At the equilibrium between radiation pressure and gas drag force the relative velocity of dust grains with respect to the gas is given by

$$v_{eq} \simeq 23 \ \frac{L_{46}}{T_4^{1/2} n_8 r_{pc}^2} \ \mathrm{km \, s^{-1}} \tag{4}$$

implying a much lower coupling between dust and gas. As discussed by various authors (Netzer 1987, Collin-Souffrin & Dumont 1990, Maiolino et al. 2001c, and references therein), the BLR probably is not distributed in a spherically isotropic geometry, but flattened. Therefore, according to what discussed above, the equatorial dusty regions are partially shielded by the BLR and here the coupling between dust and gas is low; instead, outside the solid angle covered by the BLR the dusty medium is ionized, the grains are highly charged and coupled to the gas. This scenario is illustrated in Fig. 3.

In the neutral phase the fate of dust grains depends on their velocity relative to the gas. At velocities larger than about 20 km s^{-1} sputtering by atoms and molecules returns most of the mass of the dust grains in the interstellar gas. Another important phenomenon are grain-grain collisions. Since the dusty medium has generally opacity $\tau_{dust} > 1$ dust grains hide each other. Grains that are in the shade suffer a much lower outward acceleration with respect to grains directly exposed to the nuclear radiation; this causes a differential velocity between grains and, therefore, collisions. If the collision velocity is higher than ~ 10 km s^{-1} then the impact is likely to vaporize at least the smaller of the two grains (Tielens et al. 1994). For collision velocities higher than ~ 1 km s^{-1} shattering causes fragmentation of the grains (Jones et al. 1996). Instead, for collision velocities lower than $\sim 10^{-2}$ km s^{-1} smaller grains will stick on larger ones and coagulate (Dominik & Tielens 1997). These regimes must be compared with the expected v_{eq} given by Eq. 4 as a function of the luminosity and of the distance from the nuclear source. Yet, density and temperature of the gas, which enter in Eq. 4, are poorly known, and in the circumnuclear region various gas phases probably cover a wide range of densities and temperatures. In the circmunuclear 10–100 pc of AGNs millimetric observations have measured densities of about 10^6 cm^{-3} (e. g. Tacconi et al. 1996), while on sub-pc scales the detection of mega-maser disks implies densities of the order of 10^8 cm^{-3}. Here I will assume a gas density which decreases linearly with the radius and, more specifically, $n_8 = 0.1 \ r_{pc}$; other

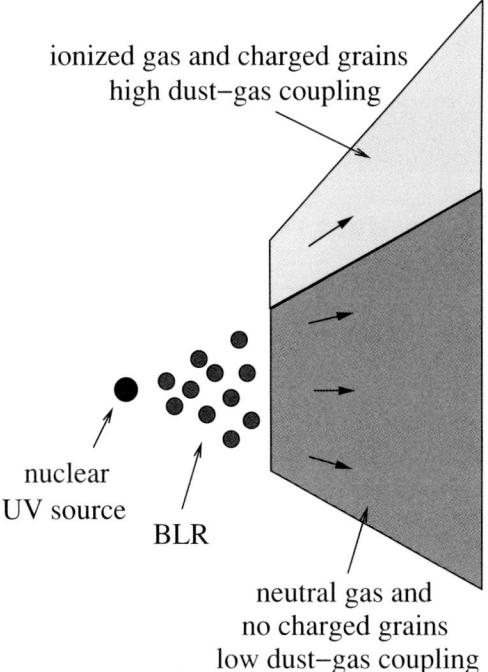

ionized gas and charged grains
high dust–gas coupling

nuclear
UV source
BLR

neutral gas and
no charged grains
low dust–gas coupling

Figure 3: Sketch showing the distribution of the two different phases of the gas and dust discussed in the text relative to the BLR.

cases for the dependence of gas density on distance are presented in Maiolino & Woltjer (in prep.). The gas temperature strongly depends on the phase of the gas: in the transition region, where the gas is mostly atomic, T $\sim$ 7000 K, while in the molecular phase the temperature is of the order of a few 100 K (Maloney et al. 1996). Here I assume an average temperature of $\sim$ 1000 K. Again, a more complete discussion with different temperatures is given in Maiolino & Woltjer. Fig. 4 shows the luminosity-distance plane where the loci of equilibrium velocities of dust grains (as given by Eq. 3) corresponding to the various regimes discussed above are plot.

In Fig. 4 we also report the expected properties of the dust on the same L-R plane corresponding to the v_{eq} in that region. As illustrated by the diagram, the most interesting consequence of dust destruction is that at high, QSO-like luminosities most of the circumnuclear region is free of dust and the obscuring medium which should produce a t ype-2 QSO survives only at large distances from the nucleus, hence giving a lower covering factor. At low, Sy-like luminosities a much wider region is available for the obscuring medium. This effect can explain the shortage of t ype 2 QSO with respect to the Seyfert population.

The scenario discussed above also predicts the existence of other dust phases. The region close to the dust-free medium, where the equilibrium velocity is large enough to produce grain fragmentation, we expect dust made

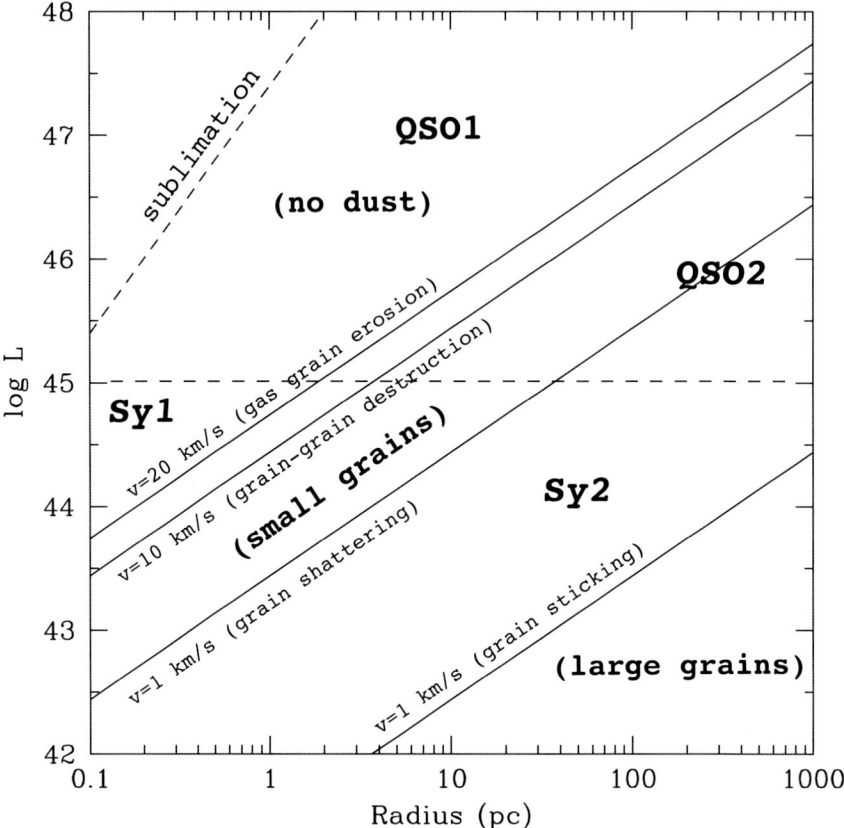

Figure 4: Diagram showing the behavior of dust in the luminosity-distance diagram. Note that at high luminosities dust can survive only at large distances from the nucleus, implying that the ratio QSO2/QSO1 should be lower than the ratio Sy2/Sy1.

mostly of small grains. Evidence for an excess of small grains in the circumnuclear medium of some AGNs has indeed been found, as traced by a steeply rise of their extinction curve in the extreme UV (Crenshaw et al. 2001).

   On the other hand the scenario also predicts the existence of regions with a dominant population of large grains due to sticking and coagulation. Evidence for large grains in the circumnuclear region of AGNs has also been found (Maiolino et al. 2001a, 2001b, Laor & Draine 1993, Hines et al. 2001). Here I shortly discuss only the evidence coming from the ratio between broad infrared lines. A dust composition biased in favor of large grains causes a flattening of the extinction curve in the optical, while at longer wavelengths the extinction curve bends to the same slope as the "standard" extinction. This is illustrated in Fig. 5, where the "large grains" curve is obtained by increasing the maximum grain size of the "standard" distribution given in Mathis et al. (1977) to $a_{max} = 1\ \mu$m and by setting the exponent of the distribution law to

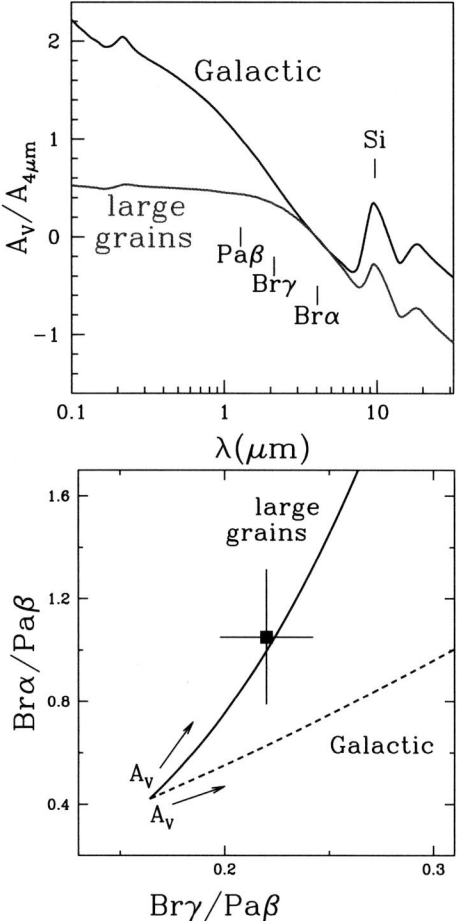

Figure 5: Top panel: extinction curve derived for a distribution of "large grains" $(n(a) \propto a^{-2.5},\ a_{min} = 0.005\ \mu m$ and $a_{max} = 1\ \mu m)$, compared to the Galactic standard extinction curve. Bottom panel: expected trend of the line ratios $Br\alpha/Pa\beta$ and $Br\gamma/Pa\beta$ in the case of absorption by assuming the t wo extinction curves given in the upper panel. The point with errorbars gives the value measured in an obscured Sy2 (NGC 2992).

-2.5 $(n(a) \propto a^{-2.5})$. The apparent knee at about $\lambda \sim 2\ \mu m$ can be sampled by means of three infrared (broad) lines: $Pa\beta$ (1.28 μm), $Br\gamma$ (2.16 μm) and $Br\alpha$ (4.05 μm). In the bottom panel of Fig. 5 I show the expected variation of the ratios $Br\alpha/Pa\beta$ and $Br\gamma/Pa\beta$ in case of reddening by using the standard Galactic extinction curve and the "large grains" curve shown in the upper panel. The point with errorbars gives the ratio for an obscured Sy (NGC 2992) for which both line ratios could be determined (Lutz et al. 2002, and Lutz et al. in prep). The value is in good agreement with the "large

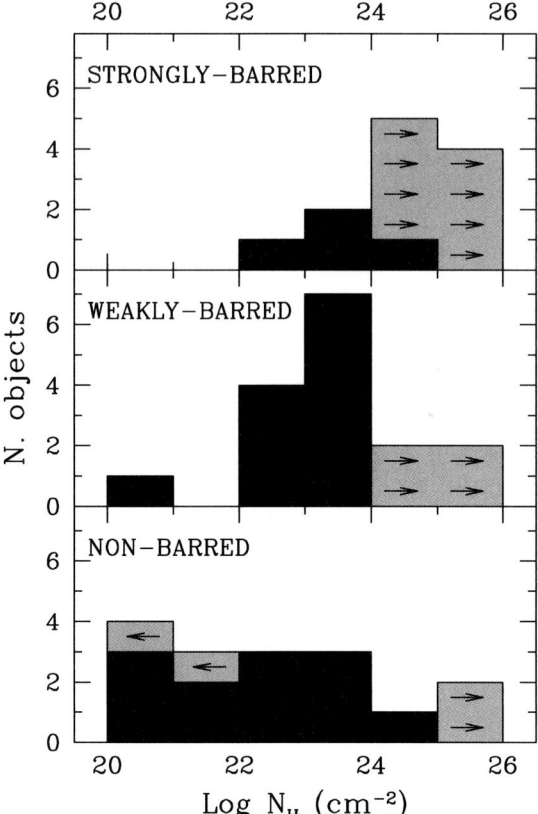

Figure 6: Distribution of absorbing column density N_H as a function of bar strength in the host galaxy.

grains" curve and inconsistent with the Galactic curve. Note that the mixed dust-gas case, which could also explain the observation, does not apply, since the Broad Line Region is dust-free (Laor & Draine 1993).

Finally, note that the diagram given in Fig. 4 is correct only if the dust is located in a cloud directly illuminated by the nuclear source. Yet, the regions which are in the shade of the dusty clouds do not suffer the consequences of the radiation pressure, and the dust properties will be the same as those in dense clouds in the galactic disk, i. e. dominated by large grains (see Maiolino & Natta 2002, for a review).

4   AGN obscuration at high redshift

The upper limit on the fraction of obscured QSO given in Sect. 2 only applies to the local Universe, since it is based on the space density of ULIRGs, which is measured locally. Also the argument based on the ratio between infrared and UV-optical luminosities to infer the covering factor is based on local objects. At higher redshift it is not possible to derive similar constraints on the obscuration and average covering factor of dust, at least with the currently available data.

However, indirect evidences suggest that the obscuration at higher redshift might be higher. Indeed at high redshift interactions and non-axisymmetric morphologies are more common and these are likely to drive gas into the nuclear region and, therefore, increase the obscuration. Evidence for this effect is actually observed in local low-luminosity AGNs. Indeed, the obscuration of Seyfert nuclei as traced by the N_H measured in the X-rays strongly correlates with the presence of a stellar bar in their host galaxies. This is illustrated in Fig. 6 where the distribution of N_H of a sample of Sy2s is shown as a function of the bar strength in the host galaxy (from Maiolino et al. 1999). Note that most of the Compton thick Sy2s are hosted in strongly barred galaxies.

In the same direction goes the finding by Levenson et al. (2001) that Sy2s associated with starburst activity tend to be more obscured in terms of N_H than Sy2s hosted in quiescent systems. This trend appears to continue to totally unobscured AGNs, i.e. Sy1s, indeed the latter appear to be generally hosted in quiescent galaxies (Maiolino et al. 1995). This correlation between AGN obscuration and star formation is probably indirect and to ascribe, again, to the effects of axisymmetric potentials discussed above: non-axisymmetric potentials both boost the star formation in the host galaxy and drive gas into the nuclear region to obscure the AGN (Maiolino et al. 1997). However, regardless of the origin of the link between star formation and obscuration, this also has important implications on the obscuration of AGNs at high redshift. Indeed, at high redshift the average star forming activity of galaxies is much higher than locally. As a consequence, the average obscuration of the associated AGNs is likely to be higher.

An average higher obscuration of AGNs at higher redshift appear to be required also by the X-ray background. Indeed, recent synthesis models of the X-ray background require that the fraction of obscured AGNs increases at high redshift in order to fit the observational constraints (Gilli et al. 2001). The recent identification campaigns of the deep Chandra fields seem to confirm this scenario (Giacconi et al. 2001).

5   Buried active galactic nuclei

The upper limit on the fraction of obscured AGNs inferred from the space density of ULIRGs in Sect. 2 only applies to high luminosity, QSO-like AGNs. Lower-luminosity, Seyfert-like nuclei ($L_{bol} \sim 10^{10} L_\odot$) can be accommodated in most of normal and starburst galaxies, which have similar IR luminosities. For Seyfert nuclei the problem is just opposite as for QSOs, in the sense that the fraction of known Sy2s might actually be a lower limit of the real fraction of obscured Seyferts, since the latter are more difficult to identify. An example of this problem is given by the nearby starburst galaxy NGC 4945. This edge-on galaxy hosts a nuclear powerful starburst whose superwind has created

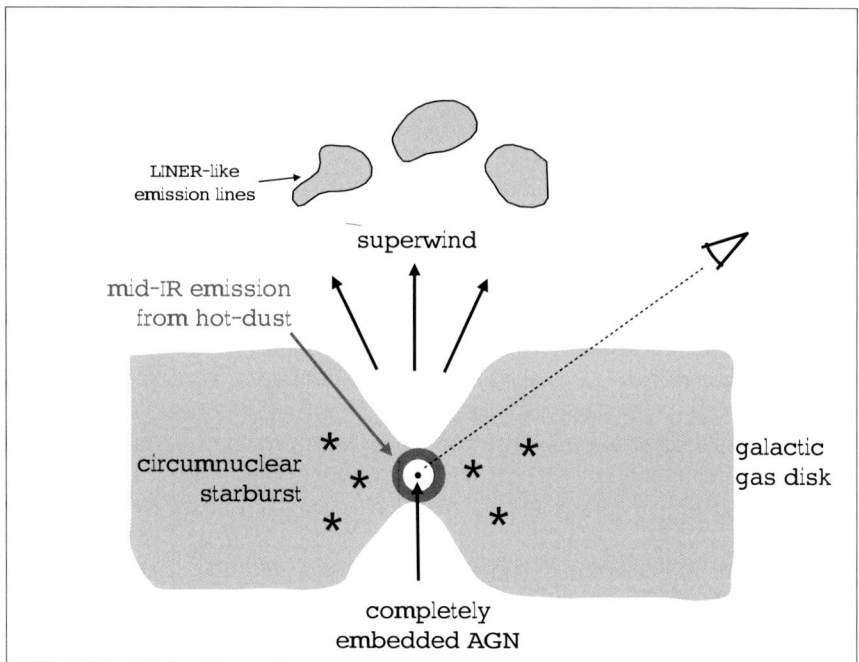

Figure 7: Sketch illustrating the nuclear geometry of NGC 4945 as inferred by the observational data.

a cavity. In the cavity a few gaseous knots have a LINER-like spectrum (Moorwood et al. 1996), meaning that no significant UV photons from the nucleus reach these clouds. The galaxy appears as a normal starburst at any wavelength from the mid-IR (at low angular and spectral resolution) to the UV (Maiolino et al. 2000, Marconi et al. 2000). Yet, hard X-rays observations have revealed a powerful, heavily obscured ($N_H \sim 5 \cdot 10^{24}$ cm^{-2}) AGN hosted in the nucleus of this galaxy (Done et al. 1996, Guainazzi et al. 2000). However, the AGN is obscured not only along our line of sight, but also in the direction of the cavity, since the clouds in the cavity do not show the AGN-like spectrum expected if ionized by the UV radiation associated to the powerful AGN. NGC 4945 is the case of an AGN which (at variance with the classical unified model) is obscured in all directions, probably by a shell of dust and gas. Fig. 7 shows a sketch of the geometry of the nuclear region as inferred by the observational data discussed above.

The fraction of these "buried" AGNs in unknown, since they are elusive in optical surveys, and even in the near-IR. They can be detected in the hard X-rays, as discussed above. However, if the absorbing gaseous column is larger than $\sim 10^{25}$ cm^{-2} (several Sy2s are characterized by N_H in excess of this value, Maiolino et al. 1998), then the AGN would be totally obscured even in the hard X-rays, and if the X-ray reflection efficiency is low, then the nucleus cannot be identified even through the scattered X-ray radiation. This

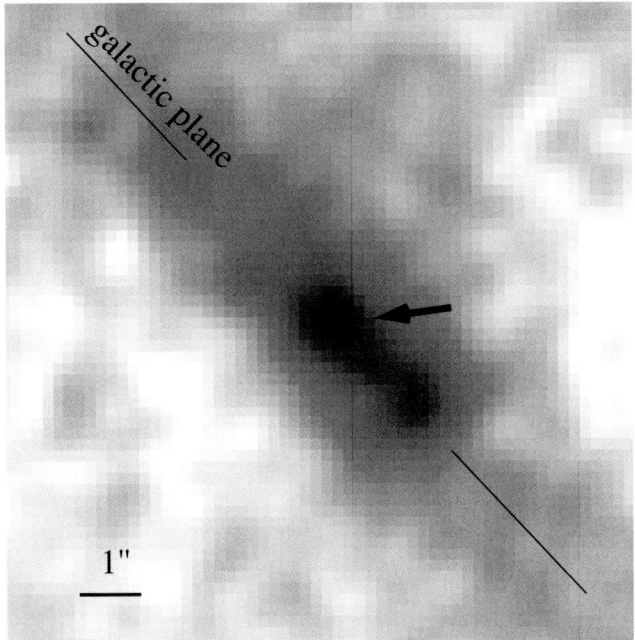

Figure 8: Mid-IR (N-band) image of NGC 4945 (from Krabbe et al. 2001). The arrow indicates the nuclear unresolved source which is most likely the counterpart of the buried AGN.

possibility is frustrating, since it would imply that a population of totally obscured AGNs might exist, but has been missed by any survey.

A possibility to identify these "totally" obscured AGNs is to look for their infrared radiation. Indeed, the shell of dust should reprocess the absorbed photons into near- and mid-IR thermal radiation. In the mid-IR, the size of the warm emitting dust is as large as a few pc, i.e. much more extended than the X-ray source and, presumably, more extended than the Compton thick medium. Therefore, high angular resolution mid-IR observations might be a tool to unveil starburst-embedded AGNs. In Fig. 8 I show the mid-IR (N-band) image of NGC 4945 obtained with an angular resolution of 1″ (Krabbe et al. 2001). Besides, the extended emission, on the disk of the galaxy, which traces the nuclear starburst, the image shows a nuclear point-like source. As discussed in Krabbe et al. (2001) this unresolved nuclear source is the mid-IR counterpart of the AGN. This result indicates that high angular resolution mid-IR observations are actually an excellent tool to identify buried AGNs, which are elusive at other wavelengths.

We have started a program of mid-IR imaging observations of a sample of starburst galaxies aimed at detecting starburst-embedded AGNs, which have been elusive at other wavelengths, as point-like mid-IR sources. So far we have identified a few candidate buried AGNs. Notably, on the starburst for which we have found indications for a buried AGN is the starburst prototype

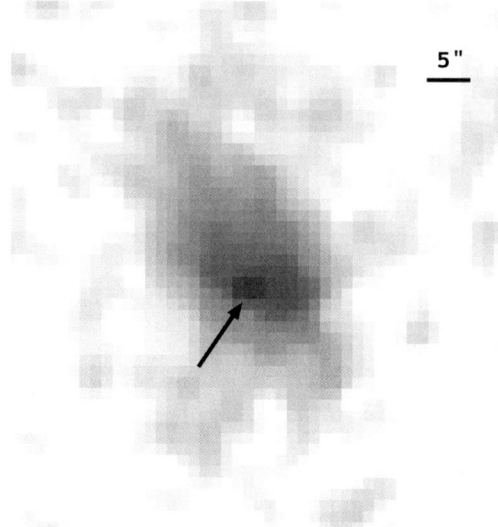

Figure 9: SCUBA-JCMT image at 450 μm of NGC 4945. The emission is clearly extended, similarly to what observed in the mid-IR image. The arrow indicates the location of the nuclear source.

NGC 253. Interesting enough, the putative AGN is undetected in the hard X-rays (Pietsch et al. 2001).

6   The relative contribution of starburst and AGN

The finding that some (possibly several) starburst galaxies host an AGN does not imply that the latter dominates the bolometric luminosity. More generally, evidence for an association between starburst and AGN have been presented by various authors (Heckman et al. 1997, Gonzalez-Delgado et al. 2001, Maiolino et al. 1995, Oliva et al. 1998), but what is the relative role played by the two phenomena in powering the bolometric luminosity has been a hotly debated issue. The uncertainty is even larger in obscured AGNs, where the primary radiation in not observed directly and its reprocessed component in the IR is mixed with the starburst contribution.

The relative contribution of AGN and starburst can be disentangled in the mid-IR, where ground-based observations have an angular resolution high enough to isolate the contribution of the nuclear component. However, generally the bulk of the luminosity is emitted in the far-IR (60–100 μm), where the luminosity is measured through IRAS data, which only give the total luminosity of the system due to the poor angular resolution (similarly ISO has not provided data with significantly improved resolution at these wavelengths). However, SCUBA-JCMT observations at 450 μm have a resolution

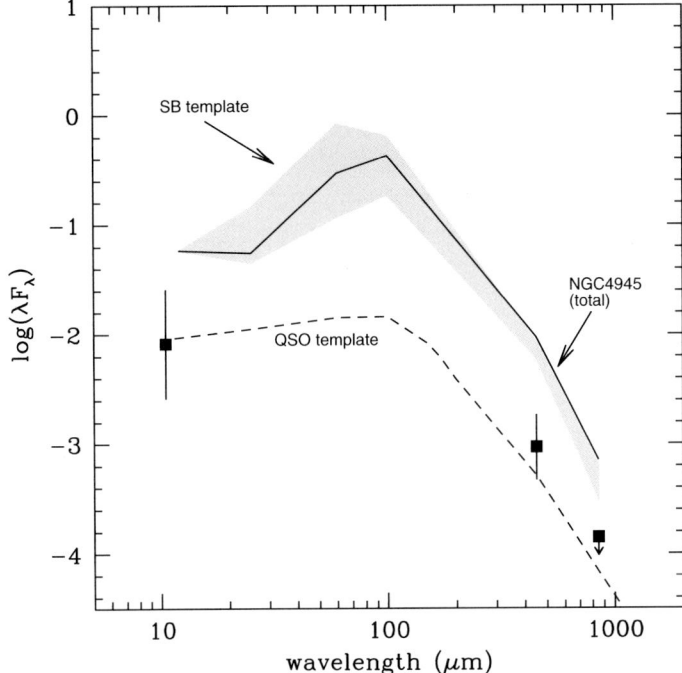

Figure 10: Spectral energy distribution of NGC 4945 (solid line) compared to the range of SEDs of starburst galaxies (shaded area). The dots give the photometric points of the nucleus alone. The dashed line is a fit to the nuclear photometric points with a QSO template.

high enough ($\sim 7''$) to identify the nuclear component. The 450 μm flux is linked to the luminosity peak at ~ 100 μm and is unaffected by dust extinction. In Fig. 9 I show the 450 μm SCUBA image of NGC 4945 (Maiolino et al. in prep). The 450 μm emission is extended, and has the same morphology as the 10 μm image in Fig. 8, therefore indicating that total luminosity is dominated by the circumnuclear starburst. The image also show a weak nuclear source, but which account for less than 10 % of the total flux at that wavelength. In Fig. 10 I report the total IR spectral energy distribution of NGC 4945 and the nuclear components measured in the mid-IR (corrected for extinction, Krabbe et al. 2001) and by SCUBA. The dashed line is the average IR SED of a sample of QSOs (Haas et al. 2000), which fits fairly well the nuclear photometric points. The shaded area gives the spread of SEDs of a sample of starburst and normal galaxies (Dunne & Eales 2001), which clearly encompasses the total infrared SED of NGC 4945. Summarizing, the infrared ($\sim$ bolometric) luminosity of NGC 4945 is dominated by the starburst, while the AGN contributes for less than 10 %. This analysis has been applied to other Seyfert 2 galaxies for which both SCUBA and mid-IR images are available. The analysis of the data is still in progress, but most of them appear dominated by starburst activity (Maiolino et al. in prep.).

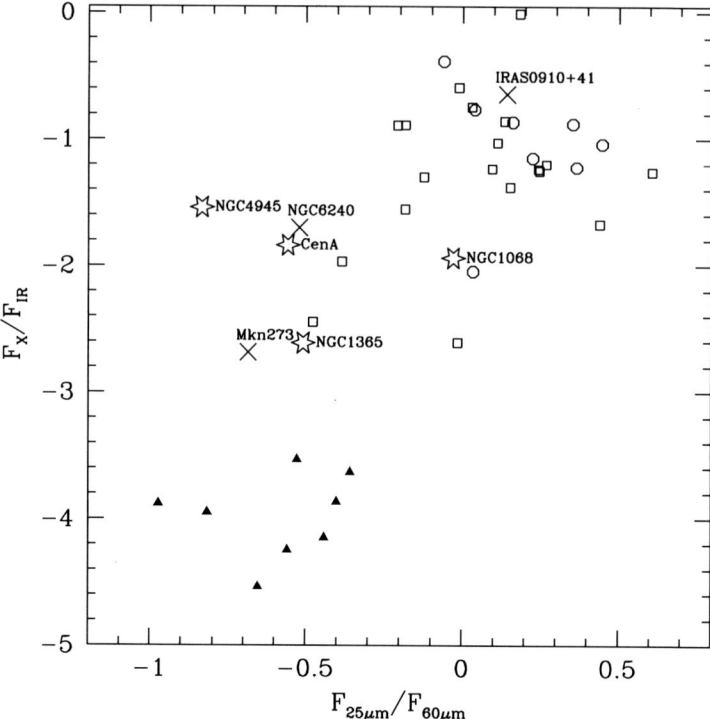

Figure 11: Diagram showing the ratio of (absorption corrected) hard X-ray flux (2–10 ke V)to far-IR flux versus the 60 μm to 25 μm IRAS flux ratio. Filled triangles are starburst galaxies; hollow squares are Sy1; hollow circles are QSOs; hollow stars are Sy2s whose luminosity appears dominated by the starburst component, according to the analysis discussed in the text; crosses are a few candidate type 2 QSOs hosted in ultraluminous infrared galaxies. The hard X-ray luminosity of NGC 1068 (totally reflection dominated) was derived by assuming a reflection efficiency of 1%.

It is important to note that several of these starburst-dominated Sy2s are characterized by a ratio of infrared luminosity to hard X-ray luminosity (2–10 keV, corrected for absorption) is similar to that found in some candidate QSO2s (e. g. Vignati et al. 1999), suggesting that some of the latter might also be mostly powered by star formation. This is shown in Fig. 11, where the ratio of (absorption corrected) hard X-ray flux (2–10 keV) to far-IR flux versus the 25 μm to 60 μm IRAS flux ratio is plot for various classes of objects. Filled triangles are starburst galaxies, hollow squares are Sy1s and hollow circles are QSOs. Hollow stars are Sy2s whose luminosity appears dominated by the starburst component, according to an analysis analogous to that discussed above. Their F_X/F_{IR} ratio partly overlaps with type 1 Sys and QSOs, while the $F_{60\mu m}/F_{25\mu m}$ ratio spans a wide range, from starburst-like to QSO-like values. Some ULIRGs suspected to host a type 2 QSO, such as NGC 6240 and Mkn 273, have the same properties as Sy2s dominated by star formation, suggesting that also the former are mostly powered by starburst activity.

7 Conclusions

Rather than giving a review on the unified model and on the obscured AGNs I have discussed some new issues concerning this class of objects.

I have shown that at high, QSO-like luminosities the fraction of obscured AGNs is significantly lower than at low, Seyfert-like luminosities or, equivalently, that the covering factor of the circumnuclear dust is lower at high luminosities.

I have shown that this effect might be a consequence of the strong radiation pressure impinging on the dust in the circumnuclear region. At high luminosities the dust is driven to high drift velocities with respect to the gas ($v_{eq} > 10$ km s^{-1}). This causes destruction of the dust due to erosion by the gas and by grain-grain collisions. As a consequence, QSOs tend to have less dust in the circumnuclear region with respect to low-luminosity AGNs. This model also predicts a phase of the dust dominated by small grains (due to dust shattering) and a phase dominated by large grains (due to dust sticking and coagulation); evidence for both phases has been found in AGNs.

The constraints on the maximum fraction of obscured AGNs only applies to the local universe. At high redshift probably the average obscuration of AGNs increases, due to the enhanced rate of interactions and distorted morphologies which are likely to drive gas into the nuclear region. Evidence for a tight connection between non-axisymmetric morphologies and obscuration is found in Seyfert 2 galaxies, supporting the scenario of higher obscuration at higher redshift. This is also expected by recent models of the hard X-ray background.

I have shown that, at low luminosities, Seyfert nuclei may be hosted in "classical" starburst galaxies and may miss the typical signatures of AGNs because obscured by dust and gas in all directions. Although the evidence for the existence for this class of "elusive" AGNs is clear, their fraction is unknown. High angular resolution mid-IR observations may be a powerful tool to identify this class of totally embedded AGNs.

I have also discussed the issue of the relative contribution of AGN and starburst in systems hosting both components. By combining SCUBA and mid-IR data of a sample of Seyfert 2 galaxies I could isolate the contribution of the nuclear AGN. In general, the latter appear to contribute only to a minor fraction of the bolometric luminosity, implying that the starburst component dominates. Based on a similar L_{IR}/L_X ratio, the bolometric luminosity of some candidate QSO2s appear also dominated by an associated starburst component.

Acknowledgments

The results presented in this paper were obtained in collaboration with other researchers and, more specifically, L. Woltjer, M. Salvati, A. Marconi, G. Risaliti, E. Oliva, A. Krabbe, T. Böker, and P. van der Werf. This work was partially supported by the University and Research (MURST) under grant Cofin00–02–36 and by the Italian Space Agency (ASI) under grant 1/R/27/00.

References

Chang, C. A., Schiano, A. V. R., Wolfe, A. M. 1987, ApJ 322, 180

Collin-Souffrin, S., Dumont, A. M. 1990, A&A 229, 292

Crawford, C. S., Fabian, A. C., Gandhi, P., Wilman, R. J., Johnstone, R. M. 2001, MNRAS 324, 427

Crenshaw, D. M., Kraemer, S. B., Bruhweiler, F. C., Ruiz, J. R. 2001, ApJ 555, 633

Dominik, C., Tielens, A. G. G. M. 1997, ApJ 480, 647

Done, C., Madejski, G. M., Smith, D. A. 1996, ApJ 463, L63

Dunne, L., Eales, S. A. 2001, MNRAS 327, 697

Genzel, R. et al. 1998, ApJ 498, 579

Giacconi, R. et al. 2001, ApJ 551, 624

Gilli, R., Salvati, M., Hasinger, G. 2001, A&A 366, 407

Gonzàlez-Delgado, R. M., Heckman, T., Leitherer, C. 2001, ApJ 546, 845

Guainazzi, M., Matt, G., Brandt, W. N., Antonelli, L. A., Barr, P., Bassani, L. 2000, A&A 356, 463

Haas, M. et al. 2000, A&A 354, 453

Heckman, T. M., Gonzalez-Delgado, R., Leitherer, C., et al. 1997, ApJ, 482, 114

Hines, D. C., Schmidt, G. D., Wills, B. J., Smith, P. S., Sowinski, L. G. 1999, ApJ 512, 145

Hines, D. C. et al. 2001, ApJ 563, 512

Jones, A. P., Tielens, A. G. G. M., Hollenbach, D. J. 1996, ApJ 469, 740

Kim, D.-C., Sanders, D. B. 1998, ApJS 119, 41

Krabbe, A., Böker, T., Maiolino, R. 2001, ApJ 557, 626

Laor, A., Draine, B. T. 1993, ApJ 402, 441

Laor, A., Brandt, N. 2002, ApJ in press (astro-ph/0201038)

Levenson, N. A., Weaver, K. A., Heckman, T. M. 2001, ApJ 550, 230

Lutz, D. et al. 1998, ApJ 505, L103

Lutz, D. et al. 2002, in *Issues in unification of AGNs*, eds. R. Maiolino, A. Marconi, N. Nagar, ASP Conf Ser. 258, 39

Maiolino, R., Thatte, N., Alonso-Herrero, A., Lutz, D., Marconi, A. 2000, in *Imaging the Universe in Three Dimensions*, eds. W. van Breugel, J. Bland-Hawthor, ASP Conf. Ser. 195, 307

Maiolino, R., Ruiz M., Rieke G. H., Keller L. D. 1995, ApJ 446, 561

Maiolino, R., Rieke, G. H. 1995, ApJ 454, 95

Maiolino, R., Ruiz, M., Rieke, G. H., Papadopoulous, P. 1997, ApJ 485, 552

Maiolino, R. et al. 1998, A&A 388, 781

Maiolino, R., Risaliti, G., Salvati, M. 1999, A&A 341, L35

Maiolino, R. et al. 2001a, A&A 365, 37

Maiolino, R., Marconi, A., Oliva, E. 2001b, A&A 365, 37

Maiolino, R., Salvati, M., Marconi, A., Antonucci, R. R. J. 2001c, A&A 375, 25

Maiolino, R., Natta, A. 2001, in *The evolution of galaxies. II-Basic building blocks*, ed. M. Sauvage et al. (Kluwer), in press

Maloney, P. R., Hollenbach, D. J., Tielens, A. G. G. M. 1996, ApJ 466, 561

Marconi, A. et al. 2000, A&A 357, 24

Mathis, J. S., Rumpl, W., Nordsieck, K. H. 1977, ApJ 217, 425

Moorwood, A. F. M., vand der Werf, P. P., Kotilainen, J. K., Marconi, A., Oliva, E. 1996, A&A 308, L1

Netzer, H. 1987, MNRAS, 225, 55

Norman, C. et al. 2002, ApJ in press (astro-ph/0103198)

Oliva, E., Origlia, L., Maiolino, R., Moorwood, A. F. M. 1999, A&A 350, 9P

Peres-Garcia, A. M., Rodriguez-Espinosa, J. M. 2001, ApJ 557, 39

Pietch, W. et al. 2001, A&A 365, L174

Scoville, N., Norman, C. 1995, ApJ 451, 510 (SN95)

Stern, D. et al. 2002, ApJ in press (astro-ph/0111513)

Tacconi, L. J., Genzel, R., Blietz, M., Cameron, M., Harris, A. I., Madden, S. 1994, ApJ 426, L77

Tielens, A. G. G. M., McKee, C. F., Seab, C. G., Hollenbach, D. J. 1994, ApJ 431, 321

Veilleux, S., Sanders, D. B., Kim, D.-C. 1999, ApJ 522, 139

Vignati, P. et al. 1999, A&A 349, L57

ASTRONOMISCHE GESELLSCHAFT: Reviews in Modern Astronomy **15**, 199–218 (2002)

Cosmological Evolution of AGN – A Radioastronomer's View

Silke Britzen

Landessternwarte Heidelberg
Königstuhl, 69117 Heidelberg, Germany
SBritzen@lsw.uni-heidelberg.de

Max-Planck-Institut für Radioastronomie
Auf dem Hügel 69, 53121 Bonn, Germany

Abstract

AGN come in different flavours, such as radio galaxies, BL Lacertae objects, quasars, etc. Still, according to unified theories, they belong to the same parent population. They differ in properties such as optical spectra and radio loudness, and these properties are used to classify the sources. In addition, source-specific pecularities are observed for the blazars (BL Lacs and FSRQ, Urry & Padovani 1995), e.g., flux-density variability from radio to gamma-rays. Superluminal motion of jet components is a frequently observed phenomenon in AGN. However, the key to understanding all these phenomena is not yet found. In addition to these unsolved problems, it is still unclear whether we see properties that evolve with cosmic epoch. Since AGN are our principal probes of the universe on large scales, understanding them is essential to studying the evolution of the universe. The best we can do is to study possible correlations between different phenomena in statistically valid samples of AGN in order to find the physical mechanisms that drive these sources. A possible solution to these questions might come from the observation that flux-density variability often coincides with morphological changes seen in the jets of AGN using interferometric techniques in the radio regime. An example is gamma-ray outbursts and component ejections seen in PKS 0420–014 (Britzen et al. 2000). The CJF – the just completed VLBI survey of 293 Caltech-Jodrell Bank flat-spectrum sources performed at 5 GHz – offers the unique opportunity to study such correlations between jet structures and flux-density variability. We here present first results of the analysis of the completed survey and discuss implications for unified theories and cosmology.

1   Introduction

The discovery by VLBI (very long baseline interferometry) of well-collimated, one-sided, apparently superluminal radio jets on parsec scales has revealed the dominant effects of relativistic beaming on the appearance of these objects (e. g., Eckart et al. 1986; Witzel 1987; Witzel et al. 1988; Readhead 1993; Zensus & Pearson 1987), and has motivated the development of the so-called "unified theories" of quasars and radio galaxies (Orr & Browne 1982; Urry & Padovani 1995). Realistic physical models can be tested and constrained through in-depth studies of prototypical objects, in particular when VLBI observations are combined with information from other spectral regimes, e. g., optical, X-ray, and γ-ray wavebands. While detailed studies of individual objects are undoubtedly important for understanding the origin and collimation of the jets and their emission mechanisms, a full understanding will only come from study of large, well-defined samples that can be subjected to statistical analysis. Taking note of the controversy in the literature over, for instance, the physical nature of jet components, pattern and bulk speeds, and unified schemes, the next step is to improve both the amount and the quality of the observational data available for consideration.

The newly completed VLBI survey "CJF" (see section 2) yields an unprecedented basis for a statistical analysis of jets in AGN. In the ideal case, one would aim to obtain as many observations (e. g., every week) for as many sources as possible (e. g., a few thousand). However, telescope time does not allow the performance of such an ample project. In section 3 we show as an example such detailed studies for one of the CJF-sources: S5 1803+784. There I show how detailed jet component motion can be studied on the basis of frequent and numerous (44) observations.

For the CJF, we believe that for the unambiguous determination of the jet component position and motion parameters, it is necessary to have at least three observing epochs, spread over roughly 4 years, for each source. These observations are now complete and a progress report will be given in this article.

With our analysis we specifically want to adress the following topics:

- the superluminal motion statistics for the complete sample
- the jet component paths (bending, multiple paths) as dependent on AGN class
- search for evolution with redshift
- test of the beaming hypothesis
- a possible correlation between X-ray prominence and radio properties

Future projects include the investigation of a possible correlation between radio- and optical variability and radio morphology.

2   The CJF

CJF, defined by Taylor et al. (1996) is a complete flux-limited sample of 293 flat-spectrum radio sources, drawn from the 6 cm and 20 cm Green Bank Surveys (Gregory & Condon 1991; White & Becker 1992) with selection criteria as follows: $S(6 \text{ cm}) \geq 350$ mJy, $\alpha_{20}^{6} \geq -0.5$, $\delta(1950) \geq 35°$, and $|b^{\text{II}}| \geq 10°$. This sample is mostly a superset of the flat-spectrum sources in the Pearson-Readhead Survey (Pearson & Readhead 1981) based on the 6 cm MPI-NRAO 5 GHz surveys (see Kühr et al. 1981), the First Caltech-Jodrell Bank Survey (CJ1: Polatidis et al. 1995; Thakkar et al. 1995; Xu et al. 1995) and the Second Caltech-Jodrell Bank Survey (CJ2: Taylor et al. 1994; Henstock et al. 1995). Optical identifications have been made for 97 % of the CJF sample, and redshifts obtained for 92 % of the objects (e. g., Stickel & Kühr 1994; Stickel et al. 1994; Xu et al. 1994; Vermeulen & Taylor 1995; Vermeulen et al. 1996). The redshifts of the presently identified quasars in the CJF range from $z = 0.263$ to $z = 3.886$, with 10 or more quasars per redshift interval of 0.2 in the range $z = 0.6$–2.6. This provides us with the opportunity to investigate important cosmological questions, such as the evolution of AGN with cosmic epoch. The CJF is now known to contain some 25 galaxies and 20 BL Lac objects at $z > 0.6$, enough to allow a meaningful comparison of their motion statistics with those from quasars at the same redshift and luminosity. This provides the opportunity for an important first-order test of AGN unification models. There are similar numbers of both galaxies and BL Lac objects at lower redshifts in the CJF; these can also be compared amongst themselves to form the basis for determining trends in cosmological evolution.

Continued VLBI observations of the CJF sources have been performed since 1990. Subsamples were observed in several global VLBI observations and in VLBA snapshot runs at 6 cm between March 1990 and November 2000. These observations are now complete; the last epoch, for a subsample of 34 sources, has been obtained in December 2000.

All sources and epochs have been analyzed in the same standardized way, details concerning the data analysis can be found in Britzen et al. (1999, 2001a).

3   S5 1803+784 – Detailed VLBI studies

With a time sampling of up to one observation per month, it is possible to determine the trajectories and velocities of the VLBI jet components with much higher accuracy, as required in particular with regard to the complex motion patterns seen in an increasing number of radio sources (Krichbaum et al. 1994, Zensus 1997).

Geodetic and astronomical VLBI data of S5 1803+784 obtained at various epochs between 1979 and 1985 indicated that the brightest jet component in observations at 8.4 GHz at 1.4 mas is stationary (Schalinski et al. 1988). This finding has been confirmed in geodetic X-band observations between

1983 and 1987 (Schalinski et al. 1988, Witzel et al. 1988), as well as in 5 GHz VLBI observations between 1979 and 1985. Several authors confirm this constant separation, leading to a subluminal velocity of $\beta_{app} < 0.74\ h^{-1}$ (Cawthorne et al. 1993). Britzen & Krichbaum (1995) present 43 X-band maps obtained from geodetic VLBI campaigns (from 1986.21 to 1992.34). They discuss significant position shifts observed for the brightest jet component with displacements between $r \sim 0.7$ mas and $r \sim 1.5$ mas. Obviously the improved time sampling of the geodetic VLBI data led to the detection of systematic position variations for components regarded as stationary on the basis of the less frequent earlier observations. In addition, evidence is presented for t wo jet components separating from the core with an apparent speed of $\beta_{app} \sim 6$ ($H_0 = 100\ h\ \mathrm{km\ s^{-1}\ Mpc^{-1}}$, $q_0 = 0.5$). Further analysis of these datasets yields evidence for a third componet moving at similar velocities, but which has been ejected earlier from the core. Fig. 1 shows 1803+78 as a complex, possibly bent, jet with eight jet components between 0.2 and 6 mas. The former "stationary" component is marked with filled black squares (component A in Fig. 1).

In Fig. 7 (upper right) we see a similar motion scenario for this source at 5 GHz and the blending effect with a new jet component "merging" with the bright component (marked with circle).

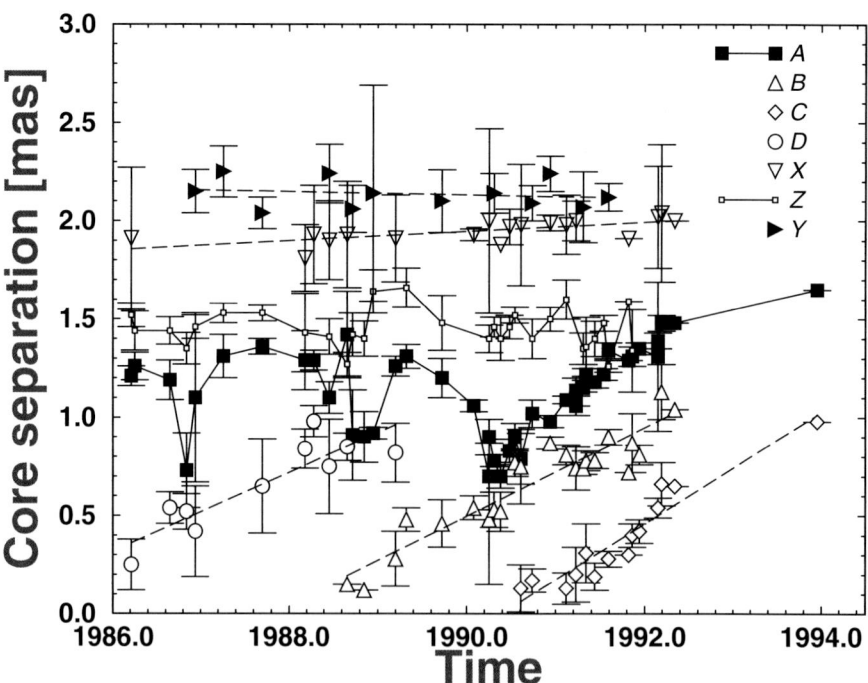

Figure 1: The jet components in S5 1803+784 and the kinematics of their core separation as a function of time between 1986 and 1994.

4 Superluminal motion statistics of a complete sample

A major goal of the CJF survey was to make a statistical study of the apparent component speeds in order to understand the general properties of the phenomena rather than the details of an individual object. The CJF sample is suited to study the velocity distribution, as well as to develop population models that test for the dependence on jet bending and accelerations, and whether a single Lorentz factor applies to all jet components in a given source. The great majority of CJF sources show a parsec-scale jet on one side of a flat-spectrum, compact core. Many of the observed jets are curved, and in some cases quasi-oscillatory trajectories or ridge lines have been observed. Components show a wide range of apparent velocities – different components in the same source can show different velocities; components can accelerate, decelerate, merge, or split; and in some cases a stationary component can coexist with moving components. In the following, I report on first results for the now complete survey.

4.1 Proper motion

Our results presented here are based on the analysis of 402 jet component motions determined for 177 quasars, 107 for 40 galaxies, 58 for 23 BL Lac objects, and 30 for 12 still unclassified sources. 42 sources can not be included into the analysis presented here, since they either have no redshift yet, were too faint in the observations, or clearly need another epoch to clarify jet component motion.

Only in rare cases is *one single* motion value representative for jet component motion in a given source; in general, a broad range of velocities can be measured. The extracted values (based on three or four snapshot observations) can not cover the complete history of jet component motion, but can reproduce the motion of the brightest and dominant jet components. In many cases brightening or dimming of the jet components and/or reference component (the "core") from one epoch to another complicates the assignment of component identities.

From inspection of Fig. 2, we find that the internal proper motions cover a wide range of positive and negative values. The most important scenarios that lead to negative values include true "backwards" motion, blending effects between components (see Fig. 7 (upper right) for the component closest to the core (C0) apparently merges with component C1), and mis-identifications of the core (sources like 4C39.25 where the brightest component is not the core).

Despite the large spread in the data, quasars and BL Lac objects show a tendency to have the highest proper motions (mean value: 0.053 ± 0.089 mas/year[1] and 0.053 ± 0.065 mas/year respectively) and the galaxies the

---

[1] We give mathematical statistical errors throughout this article. Based on the internal results we receive, we find that this statistical error overestimates the real physical errors significantly. A careful error analysis taking this into account is in preparation.

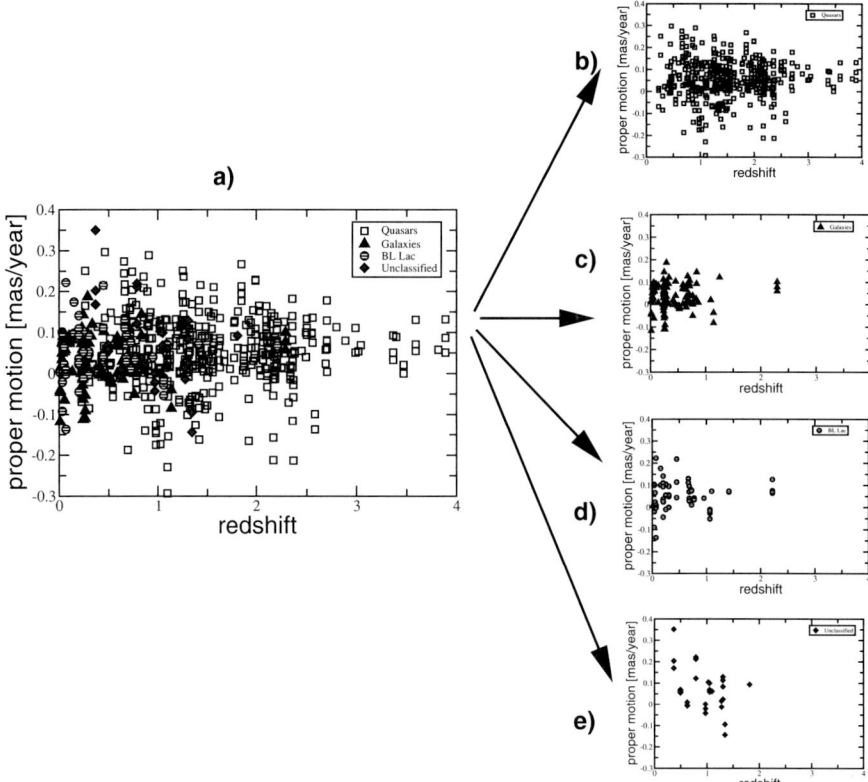

Figure 2: All jet component proper motion values for all sources are shown as function of redshift. In addition we show this relation separately for the different classes: a) quasars, b) galaxies, c) BL Lac objects, and d) unclassified objects.

slowest proper motions (mean value: 0.035 ± 0.058 mas/year). The unclassified subgroup that most likely consists of quasars reveals fast proper motions (mean value: 0.067 ± 0.098 mas/year). Given the large spread, it is difficult to draw conclusions concerning a possible evolution with redshift – we find the maximum proper motion for the quasar [components] decreases with increasing redshift, while the median increases. The proper motions for the galaxies and BL Lac objects appear to rise with redshift, but this is largely controlled by only a small number of high-redshift objects.

4.2   The β_{app}-redshift relation

Fig. 3 shows the apparent velocities calculated from the observed proper motions using $H_0 = 65$ km s^{-1} Mpc^{-1} and a deceleration parameter of $q_0 = 0.5$. In Fig. 3, quasars have highest β_{app} (mean value: 2.89 ± 4.19), the galaxies show the lowest values (mean value: 0.81 ± 1.58), and the BL Lac objects have values in between (mean value: 1.36 ± 1.80).

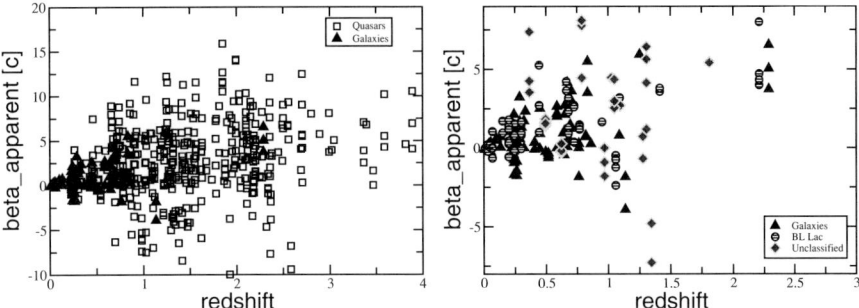

Figure 3: The β_{app}-z relation is shown. The figure on the left includes only quasars and galaxies, the figure on the right shows all classes except the quasars.

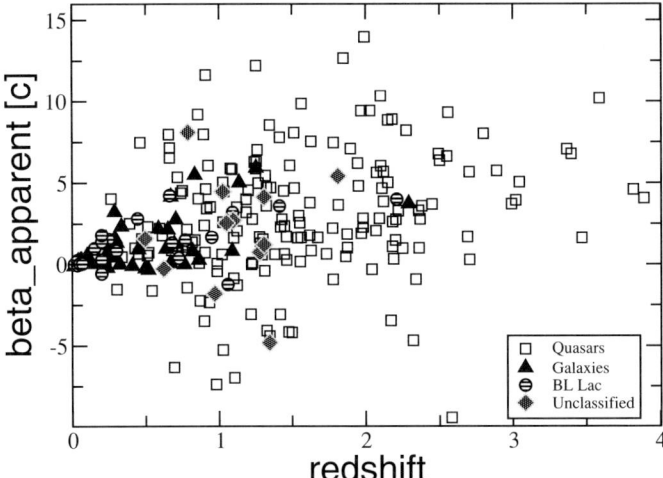

Figure 4: The β_{app}-z relation for the brightest jet components in the sources of all classes.

In Fig. 4 we show the β_{app}-redshift relation for the brightest jet component only. The brightest component should be the maximally beamed component in a jet and can yield some representative value for motion in a given source. In addition, the brightest component can be determined with higher accuracy and traced more reliably across the epochs. Again the quasar population exhibits the highest values (mean value: 2.95 ± 3.90), the galaxies reveal slower motion (mean value: 1.28 ± 1.62), but even slower are the BL Lacs (mean value: 1.22 ± 1.45). The unclassified objects again show values indicating that this group consists of a mix of the other populations (mean value: 2.00 ± 3.27). Interestingly, the brightest galaxy components move faster than the less bright galaxy components. It appears that the observed apparent velocities show a slight increase with redshift.

4.3   Class-specific pecularities in jet structures

As expected from unified theories, the different AGN classes have different morphologies in VLBI observations. In quasars, as we have seen, the jet components tend to move faster. Within the time covered by the observations, the jet components clearly separate from the core (or approach the core in case of negative values). Two t ypical cases of jet component motion in quasars are shown in Fig. 5.

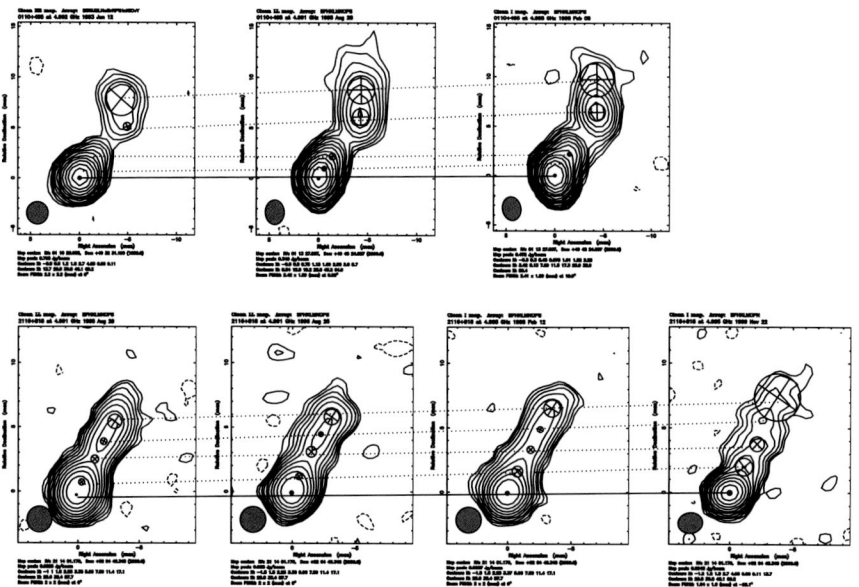

Figure 5. Three epochs of the quasar 0110+495 (top) and four epochs of the quasar 2116+818 (bottom) are shown. The reference point (presumably the core) is marked by a solid line. The jet component identification and separation from the core is marked by the dotted lines. The individual jet component positions and sizes are indicated in this and following figures by encircled crosses.

In BL Lac objects, the component identification is more problematic. In quite a number of BL Lac objects, jet components appear at similar core separations but different position angles from epoch to epoch. Therefore it is difficult to decide whether these are indeed the same components. One possible solution could be some sort of superposition between outward motion and rotation. We show t wo examples of BL Lac morphologies in Fig. 6. In addition the core separation as function of time and the position angle as function of core separation is shown in Fig. 7 and Fig. 8.

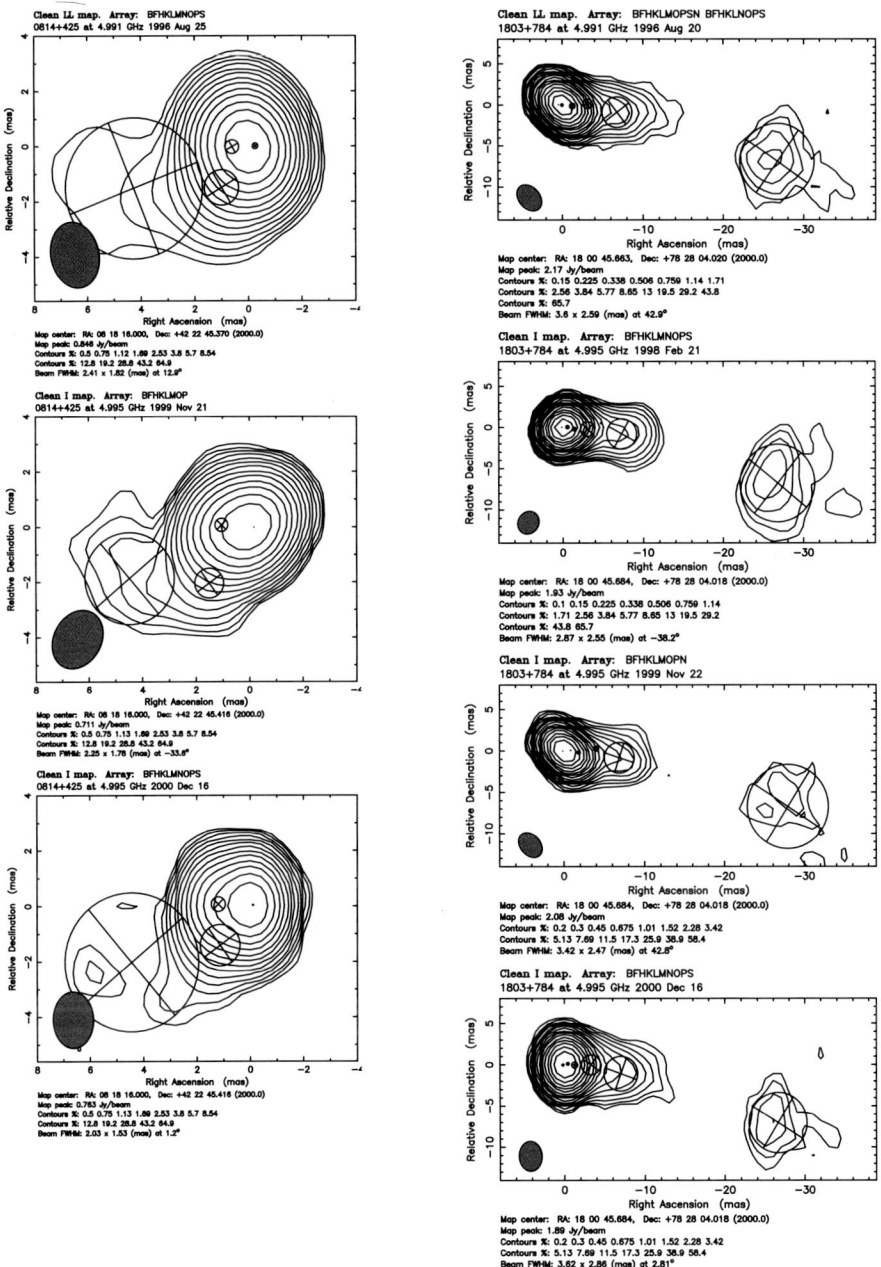

Figure 6: Two examples for jet component motion in BL Lac objects are shown: three epochs of 0814+425 (left), and four epochs for S5 1803+784 (right).

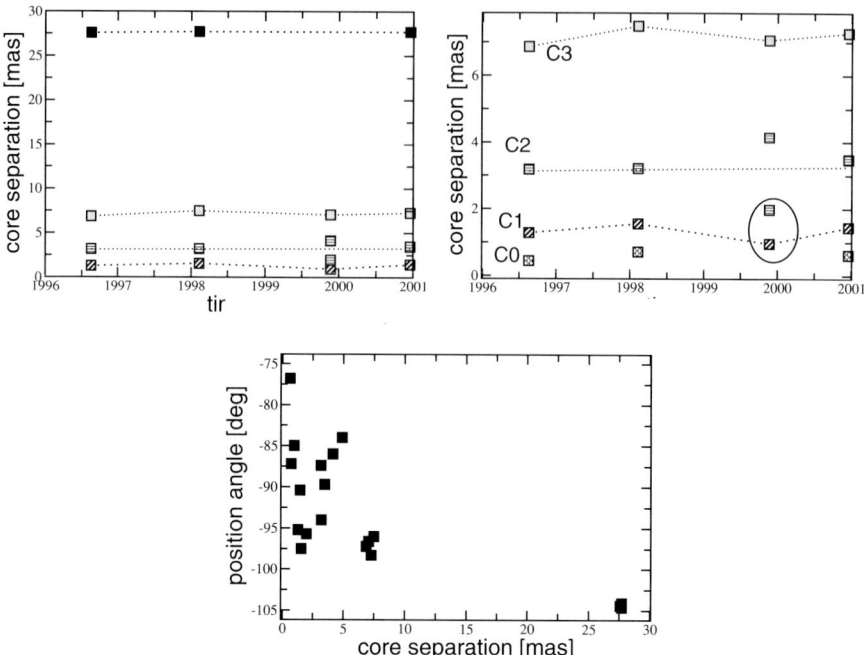

Figure 7: The upper figures show the core separation as function of time for all jet components in S5 1803+784 (left) and for the inner 8 mas (right) based on VLBI observations at 5 GHz. In comparison, Fig. 1 showed the components in 8.4 GHz observations. Component *C1* in this figure is most likely identical with the component labelled *A* in Fig. 1. The bottom figure shows the position angle as function of core separation for all jet components in S5 1803+784.

Radio galaxies tend to show more complex jet structures (e. g., Fig. 9), longer jets, and the identification of jet components is difficult compared to most quasars. Here, misidentifications can easily lead to much higher positive and lower negative velocities. In addition, for several galaxies the assignment as a galaxy is uncertain.

4.4   Parsec-scale jet curvature

Straight jets like in the quasar 2116+818 (Fig. 5, bottom) appear to be rare cases within the CJF. Most sources reveal curved jet structures, whereby different forms of curvature occur. 0110+495 (Fig. 5, top) is an example of curvature typically observed in quasars. The dominant trend of component motion effect is outward motion along the main jet path. The BL Lac objects especially show quite strongly bent jet paths. Jet components in these sources move on curved trajectories with significantly changing position angle from epoch to epoch. The outward motion, usually displayed in core separation against time plot, is affected by this "rotation", as we can see in Fig. 8.

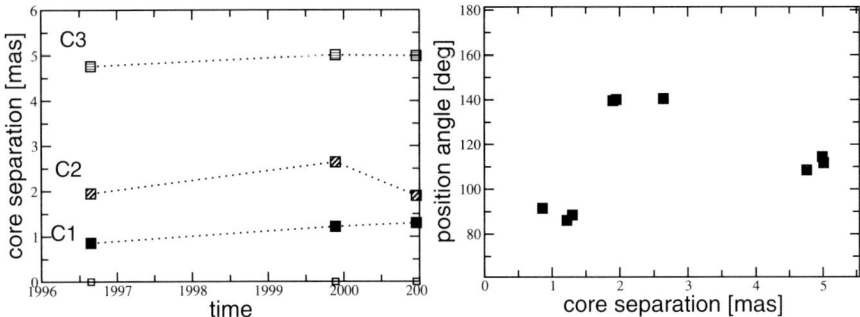

Figure 8: The two figures show the core separation as function of time for three components in 0814+425 (left), and the position angle versus core separation for all three components in 0814+425.

Two typical examples are shown here (0814+425 and S5 1803+784). In the latter we see a clearly marked cosinus in the plot of position angle versus redshift (Fig. 7, bottom), indicating that jet components might move on helical trajectories (this is supported by mm-VLBI results for the inner region, Britzen & Krichbaum 1995). A similar result is shown for 0814+425. Several BL Lac sources show this sort of behavior, and this is reflected in a tendency towards higher values for the observed pc-scale curvature (determined from smallest and largest value for position angle observed). The mean value for this curvature is $26° \pm 23°$ for the quasars and $37° \pm 26°$ for the BL Lac objects.

4.5 Flux-density variability of core and jet

The vast majority of the CJF sources are either quasars or blazars, detectable in all wavebands of the electromagnetic spectrum accessible for astrophysical investigations. Their flux density is variable on a wide range of time-scales in all energy bands, and the events at the shortest timescales have not yet been resolved (Wagner & Witzel 1995 and references therein). A number of highly compact sources (all members of the CJF) show flux density changes within two hours (Witzel et al. 1986; Wagner & Witzel 1995 and references therein). This intraday variability (IDV) is the most amazing aspect of radio variability. In contrast to measurements at shorter wavelengths, direct imaging of milliarcsecond jet structures is possible in the radio regime by VLBI. Evidence has been presented that an observed millimeter flux-density excess can in fact be associated with a newly emerging VLBI component (e.g., Britzen et al. 2000 for PKS 0420–014). In this case, the millimeter flux-density increase is followed by a rise in flux at lower radio frequencies. VLBI maps obtained at later epochs show a new emitted jet component close to the core.

Based on the CJF, we can perform a check of the radio flux variability, and find which part of the source – jet or core – contributes predominantly to the variability. We find evidence that suggests that BL Lac objects show the highest core+jet variability (mean value: $25 \pm 14\%$), the quasars show

Figure 9: A 5 GHz VLBA image of 1943+546 is shown as an example for the complex structures in radio galaxies.

weaker variability (mean value: $19 \pm 10\,\%$), and the galaxies, as expected, are less variable sources. The core is the most variable component for all classes. In galaxies with total flux-density variability less than BL Lacs and quasars, the jet components contribute $49 \pm 22\,\%$ of the variability. In quasars and BL Lacs – with higher total flux-density and flux-density variability – $36 \pm 17\,\%$ and $36 \pm 19\,\%$ of the variability, respectively, comes from the jet components.

5   Do we see cosmological evolution?

The statistical distribution of apparent velocities as determined from pc-scale structures in sources over a wide range of redshifts is a property of quasars that, one hopes, might show little cosmological evolution. It is determined by astrophysical processes in the nuclear region only, and consequently is unlikely to change as a result of varying environmental conditions on galactic and intergalactic scales. The hope therefore is that superluminal motion studies may yield the derivative of the standard rod for use as a cosmological probe: a standard β_{app} distribution. According to Vermeulen & Cohen (1994), the study of the β_{app} distribution as a function of redshift might be used to constrain the deceleration parameter q_0. They find a fairly well-defined upper envelope in the β_{app}-z diagram as traced out by the core-selected quasars. The envelope seems to rise as a function of redshift if $q_0 = 0.05$, but it could well be flat if $q_0 = 0.5$. On the assumption that evolution of the Lorentz factors in parsec-scale jets is unlikely, the result could be turned into the conclusion that the universe is not far from closed. We perform a similar analysis based on a larger, and more uniform sample: the brightest jet components in the quasar sample of the CJF (176 sources). Fig. 10 shows β_{app} both for $q_0 = 0.5$ and $q_0 = 0.05$. We find that in the case of $q_0 = 0.05$, β_{app} rises more steeply with increasing redshift for the positive values. However, the negative values become even more negative. The spread in both distributions is very large and this effect is enhanced by using $q_0 = 0.05$.

Outward motion of components is not the only effect observed in the sources; curvature together with projection effects might lead to a spreading of the apparent velocities. We might have to disentangle these effects first, before a determination of the deceleration parameter is possible.

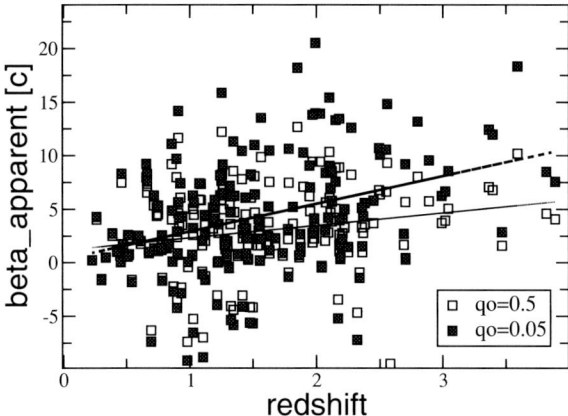

Figure 10: β_{app} as a function of the redshift for the brightest components in the quasar sample for $q_0 = 0.5$ (open squares, linear regression: thin line) and for $q_0 = 0.05$ (filled squares, linear regression: thick line). In both cases is $H_0 = 65$ km s^{-1} Mpc^{-1}. Note the open squares reproduce Fig. 4.

6 ROSAT observations: the soft X-ray properties of the CJF

Many models accounting for the observed broadband spectra of blazars have been developed. Most of them attribute the emission at radio through optical wavelengths to synchtrotron radiation, and X-ray through γ-ray emission to Compton scattering (e. g., Marscher 1980; Königl 1981; Sikora et al. 1994). The models differ in the location and structure of the acceleration and emission region(s). Flux and spectral variations observed in blazars can be caused by shock waves propagating along the jet (Marscher & Gear 1985; Hughes et al. 1989a; Hughes et al. 1989b; Valtaoja et al. 1992). Some authors concentrate on a single shock front, while others model an inhomogeneous jet where particle acceleration and radiation are assumed to occur continuously along the jet. The problem is so complex that we are still far from a realistic description of the phenomenon, but the temporal evolution of such flares is qualitatively similar from one model to the next: outbursts occur first at IR-optical frequencies and then propagate to successively longer wavelenghts (Marscher & Gear 1985; Valtaoja et al. 1992). Extensive VLBI monitoring studies of AGN at centimeter and millimeter wavelengths, when combined with broadband total flux density observations, can be used to determine the overall physical conditions in parsec-scale radio jets. Here is an excellent opportunity to directly compare the theoretical predictions with observational data, despite the complexity and the large number of free parameters of the theoretical models.

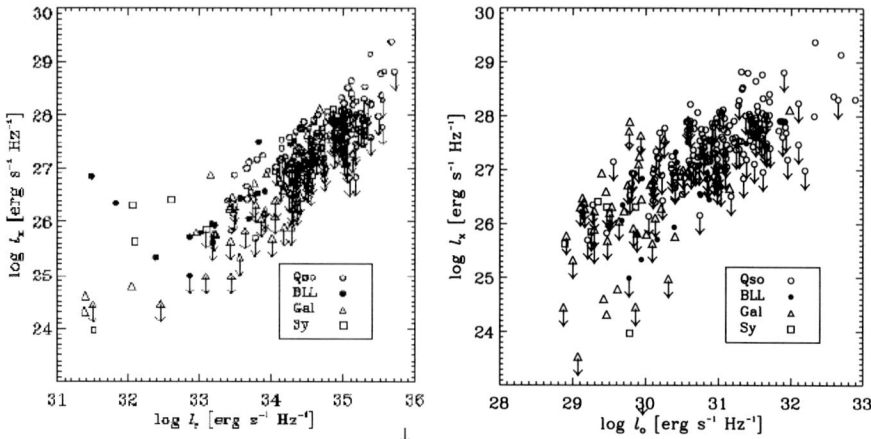

Figure 11: The monochromatic X-ray luminosity (at 2 keV) as a function of the radio luminosity (left) and as a function of the optical luminosity (right). The arrows denote the upper limits of non-detected objects.

The CJF sources have been observed within the ROSAT All-Sky survey and/or in pointed PSPC observations (Britzen et al. 2001b). In total we have 188 quasars, of which 87 were non-detected in the X-ray survey; 59 galaxies (including Seyferts), of which 37 were undetected; 43 BL Lacs (8 non-detections); and 9 objects which have not yet been classified. The highest rate of non-detection is amongst the galaxies, whereas most of the BL Lacs have been found as strong X-ray emitters.

Interestingly, most of the objects (the galaxies included) exhibit a nearly linear relation between X-ray and radio luminosities. Far above this general trend in Fig. 11 (left) we find the three extreme BL Lac objects Mrk 421, Mrk 501 (both at low radio luminosities), and 3C 66A. The three Seyfert galaxies 2116+818, 0402+379, and 0309+411 also show excess X-ray emission at low radio luminosities. A similar behavior is seen in the plot of the X-ray vs. the optical luminosities (Fig. 11, right).

Comparing the observed and the predicted X-ray fluxes by assuming the observed X-rays to be of inverse Compton origin, we can compute the beaming or Doppler factor δ_{IC} for the CJF sources and can compare this Doppler factor with other beaming indicators derived from the VLBI observations, such as the value of the apparent expansion velocity.

6.1   Doppler factors: phase or pattern speeds?

The observed super-luminal motions strictly require only that some "phase" or "pattern" speed of a wave traveling along the jet is relativistic, but there are strong arguments also for the bulk velocity of the radiating plasma to be relativistic, with associated forward beaming of the emitted radiation (e.g., Witzel et al. 1988; Eckart et al. 1989).

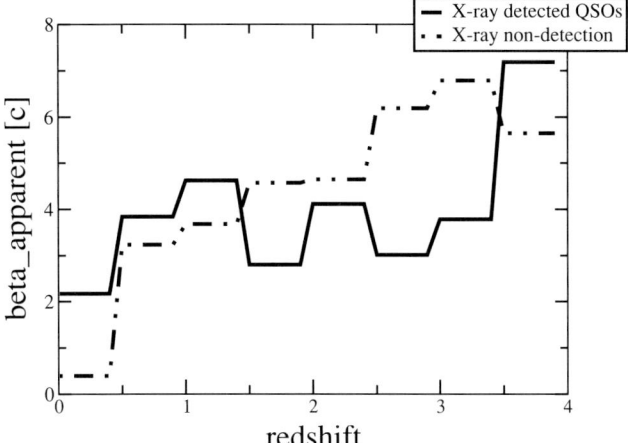

Figure 12: The behaviour of the ROSAT detected and non-detected quasars in β_{app} versus redshift are shown. The data are binned in intervals of 0.5 in redshift.

In Fig. 12 we show the β_{app}-redshift plot for the ROSAT detected and non-detected quasars, binned in intervals of 0.5 in redshift. While the detected population shows more or less a constant β with redshift (from redshift 3 onwards the numbers are not significant due to lack of sources), the non-detected populations shows an increase with redshift. This will have to be investigated in more detail but might be evidence for the existence of two different populations of quasars.

The Doppler factors can be derived via the standard synchrotron self-Compton (SSC) argument, from equipartition arguments, and from the apparent velocities determined from VLBI observations. The comparison of Doppler factors calculated on the basis of velocity and X-ray information may answer the question whether the pattern and the bulk velocities are different.

Fig. 13 shows that apparent expansion speeds and Doppler factors have similar average numerical values. This can be taken as evidence that the bulk motion causing the beaming also causes the superluminal expansion, and that it does not require different pattern and bulk velocities.

6.2 Correlation between X-ray prominence and large-scale radio structure

In Britzen et al. (2001b) we describe the VLA structure of the sources. The major part of the information on the large scale structure comes from VLA information by T. Pearson (http://www.astro.caltech.edu/tjp/cj/). VLA maps were not available for all sources and since the information on the extended structure has been collected from the literature, this data can not be homogeneous in quality. In Fig. 14 (left) we show the observed X-ray flux-density as a function of the complexity of the extended radio structure

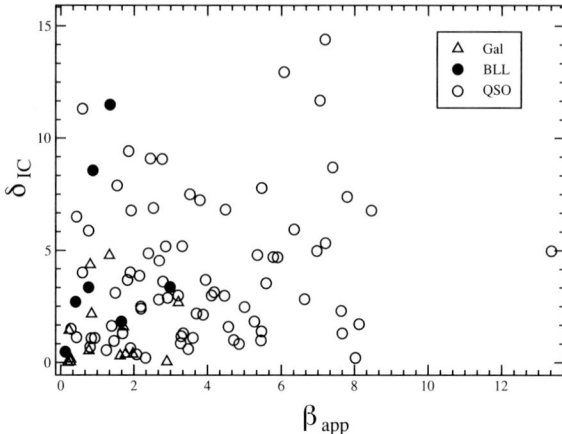

Figure 13: The figure shows the relation between β_{app} and δ_{IC}.

(the numbers from 0 to 5 serve as a measure for the complexity of the large-scale structure, 0 meaning point-like and 5 very complex extended structure). The X-ray flux-density of those sources detected by ROSAT seems to increase with the complexity of the extended structure. Except for t wo objects (0014+813, 1246+586), the point-like sources reveal relatively low X-ray fluxes (below 1.7). Fig. 14 (right) shows the distribution of sources according to their large scale radio structure. The black bars denote the detected population. 101 of the 293 CJF sources reveal a point-like VLA structure. Among those, 71 have not been detected by ROSAT suggesting that X-ray selected samples preferentially have extended VLA structure.

7   Conclusions

Flux-limited samples of AGNs are thought to contain a particularly complicated form of selection bias, since they are expected to include not only sources of high intrinsic luminosity but also lower luminosity sources whose emission is Doppler boosted by virtue of their orientation with respect to the observer. The exact composition of such a sample is therefore highly dependent on certain aspects of the parent population. These include the intrinsic dispersion of jet Lorentz factors, the distribution of jet orientations of the sources we detect, and the intrinsic (non-boosted) luminosity function (Britzen et al. 2001a). Our investigations show that there is a clear and very intriguing correlation between jet velocity and observed luminosity: there is a lack of sources with high velocity at low luminosity. Recent simulations by Lister & Marscher (1997) show that this effect can be obtained as a sort of Malmquist bias from the interplay among Lorentz-factor distributions weighted to low values and the shape of the unbeamed luminosity function and its cosmological evolution; they also raise the possibility that the Lorentz factor might be positively correlated with the unbeamed luminosity.

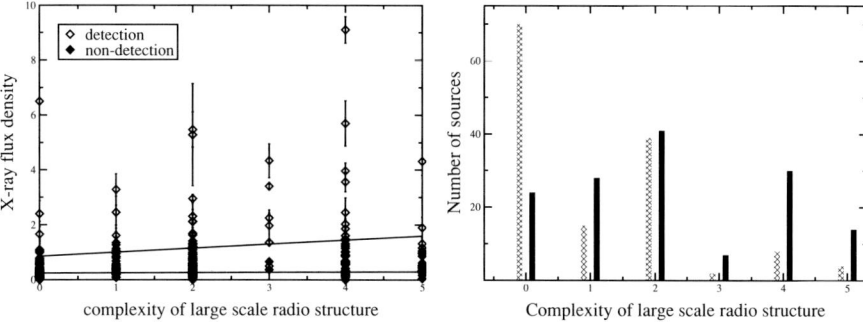

Figure 14: In the figure (left) we show the X-ray flux-density of the CJF sources (in units of 10^{-12} erg cm^{-2}s^{-1}, sources detected by ROSAT as open diamonds, thin line represents a linear regression; non-detections are shown as filled diamonds, thick line represents the linear regression) as function of the complexity of extended radio structure. The figure on the right shows the distribution of the sources according to the complexity of extended structure (detected sources as filled bar, non-detections as crosses).

At present, the following results and conclusions have been drawn from the ongoing multi-wavelength analysis of the CJF sample sources:

• We find evidence that jet component motion usually can not be described by *one single* velocity for all jet components. In most cases, jet components move with different velocities, they can take different paths in the jet.

• Quasars and BL Lac objects show similar proper motion values, although the spread is very large. There seems to be slight evidence for a decrease in the upper envelope of quasars in proper motion with redshift, but the median seems to rise with redshift.

• We find smaller apparent velocities for BL Lacertae objects than for quasars. It was noted before by other authors that the expansion velocities of BL Lacertae objects are smaller than those of quasars, which was explained either by assuming that they are extremely aligned sources, but having the same Lorentz factor as the other superluminal sources (e.g., Roberts & Wardle 1987; Cohen 1989), or by assuming that they have smaller Lorentz factors (e.g., Mutel et al. 1990). Statistical results from the study of motions in large samples of superluminal sources have been discussed among others by Ghisellini et al. (1993), Vermeulen & Cohen (1994), and Vermeulen & Taylor (1995). These previous studies have been hampered by the fact that they relied on smaller and/or inhomogeneously sampled data collections, i.e., taken from literature searches. For a sample of 81 flat-spectrum objects – a subsample of the CJF sample – Vermeulen & Taylor (1995) found no evidence for intrinsically different populations of galaxies, BL Lac objects, and quasars, which is in contrast to other reports (see Gabuzda 1995) and the results presented here. Ghisellini et al. (1993) instead find that the mean apparent speed of BL Lacertae objects is similar to that of quasars, but smaller than that of HPQs (highly polarized quasars).

Analyses of this kind so far have quite clearly demonstrated the need for larger samples, as is available now for the first time with the CJF.

• Curvature is a common phenomenon in AGN; most CJF sources show curvature of some degree. We find that BL Lac objects tend to show more bending than quasars or galaxies. This tendency is supported by unified models and different viewing angles for BL Lac objects, quasars and galaxies.

• It is debatable whether the components are discrete objects following ballistic paths; the apparent motions that we measure are probably phase velocities of a moving disturbance and may not be simply related to the velocity of the underlying jet flow. The occurrence of different component speeds, as in 3C 120 or Cygnus A – if not interpreted as geometric effect – might be caused by variable pattern speed (Vermeulen & Cohen 1994). For example, some theoretical studies predict interacting instability modes that can produce such variable pattern speeds (Hardee 1995). However, based on a comparison of Doppler factors calculated from the data presented here and ROSAT observations, we here show that no additional pattern speeds are required to explain the observed X-ray observations.

Ghisellini et al. (1993) derive Doppler factors for about 100 sources with known VLBI structures by comparing predicted and observed X-ray fluxes in the synchrotron self-Compton model. The main results agree with those from other beaming indicators (superluminal motion and core-to-extended flux-density ratio) and support a simple kinematic model of ballistic motion of knots in relativistic jets. The derived Doppler factors are largest for core-dominated quasars, intermediate for BL Lacertae objects, and smallest for lobe-dominated radio galaxies and quasars. For a subsample of 39 superluminal sources, they find that apparent expansion speeds and Doppler factors correlate and have similar average numerical values. This is consistent with our results and can be taken as evidence that the bulk motion causing the beaming also causes the superluminal expansion, and that it does not require different pattern and bulk velocities. However, their corresponding Lorentz factors are on average about 10, with no significant differences between core- and lobe-dominated quasars and BL Lacs, quite in contrast to our results presented here.

• We find a nearly linear correlation between X-ray luminosities and radio luminosities, as well as between optical and X-ray luminosities. We find a tendency for a different distribution in the β_{app}-z relation for the quasars detected by ROSAT and those not detected. Whether these are actually different quasar populations will have to be studied in more detail.

• We find a positive correlation between the complexity of the large scale radio structure and the X-ray flux of the ROSAT-detected objects. Most non-detections are point-like sources on VLA scales. But this analysis has to be taken with some caution, since the VLA information was collected from the literature and is thus not homogeneous in map quality.

Within the so called "unified" models of AGNs, the diversity in properties of extragalactic radio sources is attributed to orientation effects (e. g., Scheuer & Readhead 1979; Orr & Browne 1982). The known properties of radio-loud AGN are key to identifying the correct parent and beamed populations. Selection effects introduced by relativistic beaming have to be incorporated. A fundamental test of the proposed unification of blazars and radio galaxies will be whether the number ratios of the populations agree with the relativistic beaming hypothesis. The ratio of these two samples depends only on the critical angle dividing blazars and radio galaxies, which in turn depends on the amount of beaming. We plan for the future to test unified schemes via these population ratios of the complete sample.

Acknowledgements

I am particularly grateful to R. M. Campbell for many helpful suggestions and critical comments on the manuscript that helped to improve the paper substantially. I am indebted to A. Witzel, S. Wagner, T. P. Krichbaum, and J. A. Zensus for stimulating discussions on the subjects of variability and superluminal motion. Thanks are owed to my colleagues from the CJF collaboration and the CERES-network for their support. I am grateful to R. Vermeulen for sharing data prior to publication. I wish to acknowledge support by the Claussen-Simon-Stiftung. Part of this work has been supported by the European Comission, TMR Programme, Research Network Contract ERBFMRXCT96-0034 "CERES". This research has made use of the NASA/IPAC Extragalactic Database (NED) which is operated by the Jet Propulsion Laboratory, California Institute of Technology, under contract with the National Aeronautics and Space Administration.

References

Britzen, S., Brinkmann, W., Gliozzi, M., et al. 2001b, in preparation

Britzen, S., Krichbaum, T. P. 1995, in: R. Lanotte, G. Bianco (eds.) Proc. 10th Working meeting on European VLBI for Geodesy and Astrometry. Centro Di Geodesia Spaziale, Matera, Italy, 172

Britzen, S., Vermeulen, R. C., Taylor, G. B., et al. 1999, in: L. O. Takalo & A. Sillanpää (eds.) Proceedings of the BL Lac phenomena conference. San Francisco: Astron. Soc. Pac., PASP Conf. Ser. 159, 431

Britzen, S., Vermeulen, R. C., Taylor, G. B., et al. 2001a, in preparation

Britzen, S., Witzel, A., Krichbaum, T. P., et al. 2000, A&A 360, 65

Cawthorne, T. V., Wardle, J. F. C., Roberts, D. H., et al. 1993, ApJ 416, 496

Cohen, M. H. 1989, in L. Maraschi, T. Maccacaro, M.-H. Ulrich (eds.): BL Lac Objects. Springer-Verlag, Berlin, 13

Eckart, A., Hummel, C. A., Witzel, A. 1989, MNRAS 239, 381

Eckart, A., Witzel, A., Biermann, P., et al. 1986, A&A 168, 17

Gabuzda, D. C. 1995, Proc. Nat. Acad. Sci. USA 92, 11393

Ghisellini, G., Padovani, P., Celotti, A., Maraschi, L. 1993, ApJ 407, 75

Gregory, P. C., Condon, J. J. 1991, ApJS 75, 1011

Hardee, P. E., Clarke, D. A., Howell, D. A. 1995, ApJ 441, 644

Henstock, D. R., Browne, I. W. A., Wilkinson, P. N., et al. 1995, ApJS 100, 1

Hughes, P. A., Aller, H. D., Aller, M. F. 1989a, ApJ 341, 68

Hughes, P. A., Aller, H. D., Aller, M. F. 1989b, ApJ 341, 54

Königl, A. 1981, ApJ 243, 700

Krichbaum, T. P., Standke, K. J., Graham, D. A., et al. 1994, IAU Symp. 159: Multi-Wavelength Continuum Emission of AGN, 159, 187

Kühr, H., Witzel, A., Pauliny-Toth, I. I. K., Nauber, U. 1981, A&AS 45, 367

Lister, M. L., Marscher, A. P. 1997, ApJ 476, 572

Marscher, A. P. 1980, ApJ 235, 386

Marscher, A. P., Gear, W. K. 1985, ApJ 298, 114

Mutel, R. L., Philips, R. B., Su, B., Bucciferro, R. R. 1990, ApJ 352, 81

Orr, M. J. L., Browne, I. W. A. 1982, MNRAS 200, 1067

Pearson, T. J., Readhead, A. C. S. 1981, ApJ 248, 61

Polatidis, A. G., Wilkinson, P. N., Xu, W., et al. 1995, ApJS 98, 1

Readhead, A. C. S. 1993, in R. J. Davis, R. S. Booth (eds.) Sub-Arcsecond Radio Astronomy. Cambridge: Cambridge Univ. Press, 173

Roberts, D. H., Wardle, J. F. C. 1987, in: Superluminal radio sources. Cambridge: Cambridge Univ. Press, 193

Schalinski, C. J., Alef, W., Witzel, A., Campbell, J., Schuh, H. 1988, in J. M. Reid, J. M. Moran (eds.): The impact of VLBI on astrophysics and geophysics, Proc. 129th IAU Symp., Dordrecht, Kluwer Acad. Publ., 359

Scheuer, P. A. G., Readhead, A. C. S. 1979, Nature 277, 182

Sikora, M., Begelman, M. C., Rees, M. J. 1994, ApJ 421, 153

Stickel, M., Kühr, H. 1994, A&AS 103, 349

Stickel, M., Meisenheimer, K., Kuehr, H. 1994, A&AS 105, 211

Taylor, G. B., Vermeulen, R. C., Pearson, T. J., et al. 1994, ApJS 95, 345

Taylor, G. B., Vermeulen, R. C., Readhead, A. C. S., et al. 1996, ApJS 107, 37

Thakkar, D. D., Xu, W., Readhead, A. C. S., et al. 1995, ApJS 98, 33

Urry, C. M., Padovani, P. U. 1995, PASP 107, 803

Valtaoja, E., Teräsranta, H., Urpo, S., et al. 1992, A&A 254, 71

Vermeulen, R. C., Cohen, M. H. 1994, ApJ 430, 467

Vermeulen, R. C., Taylor, G. B. 1995, AJ 109, 1983

Vermeulen, R. C., Taylor, G. B., Readhead, A. C. S., Browne, I. W. A. 1996, AJ 111, 1013

Wagner, S. J.., Witzel, A. 1995, ARA&A 33, 163

White, R. L., Becker, R. H. 1992, ApJS 79, 331

Witzel, A. 1987, in: J. A. Zensus, T. J. Pearson (eds.): Superluminal radio sources, Cambridge: Cambridge Univ. Press, 83

Witzel, A., Heeschen, D. S., Schalinski, C. J., Krichbaum, T. P. 1986, Mitt. Astron. Ges. 65, 239

Witzel, A., Schalinski, C. J., Biermann, P. L., et al. 1988, A&A 206, 245

Xu, W., Lawrence, C. R., Readhead, A. C. S., Pearson, T. J. 1994, AJ 108, 395

Xu, W., Readhead, A. C. S., Pearson, T. J., et al. 1995, ApJS 99, 297

Zensus, J. A. 1997, ARA&A 35, 607

Zensus, J. A., Pearson, T. J. (eds.) 1987, Cambridge: Cambridge Univ. Press

ASTRONOMISCHE GESELLSCHAFT: Reviews in Modern Astronomy **15**, 219–238 (2002)

The Epochs of Early-Type Galaxy Formation in Clusters and in the Field

Daniel Thomas, Claudia Maraston, and Ralf Bender

Universitäts-Sternwarte München
Scheinerstraße 1, D-81679 München, Germany
daniel@usm.uni-muenchen.de

Abstract

We compute new population synthesis models of Lick absorption line indices with variable α/Fe ratios and use them to derive average ages, metallicities, and [α/Fe] element enhancements for a sample of 126 field and cluster early-type galaxies. Calibrating the models on galactic globular clusters, we show that any population synthesis model being based on stellar libraries of the Milky Way is intrinsically biased towards super-solar α/Fe ratios at metallicities below solar. We correct for this bias, so that the models presented here reflect constant α/Fe ratios at all metallicities. The use of such unbiased models is essential for studies of stellar systems with sub-solar metallicities like (extragalactic) globular clusters or dwarf galaxies.

For the galaxy sample investigated here, we find a clear correlation between [α/Fe] and velocity dispersion. Zero-point, slope, and scatter of this correlation turn out to be independent of the environment. Additionally, the [α/Fe] ratios and mean ages of elliptical galaxies are well correlated, i.e. galaxies with high α/Fe ratios have also high average ages. This strongly reinforces the view that the [α/Fe] element enhancement in ellipticals is produced by short star formation timescales rather than by a flattening of the initial mass function. With a simple chemical evolution model, we translate the derived average ages and α/Fe ratios into star formation histories. The more massive the galaxy, the shorter is its star formation timescale, and the higher is the redshift of the bulk of star formation, independent of the environmental density. We show that this finding is incompatible with the predictions from hierarchical galaxy formation models, in which star formation is tightly linked to the assembly history of dark matter halos.

1 Introduction

The stellar population properties of early-type galaxies represent a key challenge for theories of galaxy formation. In hierarchical models of galaxy formation, more massive galaxies have longer assembly timescales and therefore younger mean ages (Kauffmann 1996). Without taking metallicity effects into account, this results in more massive galaxies having bluer colors in contradiction to the observational evidence (Bower, Lucey, & Ellis 1992). It is well known, that metallicity can be traded for age. Therefore, by assuming that metallicities steeply increase with the mass of the galaxy, the blueing due to age effects can be masked by metallicity. In this way hierarchical galaxy formation models are able to produce the correct slope of the color magnitude relation (Kauffmann & Charlot 1998). It remains open, however, if the correct ages of early-type galaxies are predicted. A meaningful test of the model must include the direct comparison of predicted and observed ages.

A powerful tool to derive ages and metallicities are absorption line indices (Worthey 1994). Under the assumption that the stellar populations of ellipticals do not exhibit a significant spread in metallicity (Maraston & Thomas 2000), a combination of the Lick indices $H\beta$, $Mg\,b$, and $\langle Fe \rangle =$ (Fe5270 + Fe5335)/2 (Faber et al. 1985) serves to disentangle age and metallicity. Still, this approach is hampered by the fact that the data of elliptical galaxies lie below the models in the $Mg\,b$-$\langle Fe \rangle$-index diagram (e.g., Worthey, Faber, & González 1992; Davies, Sadler, & Peletier 1993). More precisely, the Mg lines observed in elliptical galaxies are stronger – at a given Fe line strength – than predicted by population synthesis models. As a consequence, Mg line indices yield higher metallicities and younger ages than the Fe line indices, which indicates the presence of $[\alpha/Fe]$ enhanced stellar populations. This interpretation gets empirical support from the work of Maraston et al. (2002), who find that metal-rich globular clusters of the Bulge with independently known super-solar $[\alpha/Fe]$ ratios exhibit the same pattern in the $Mg\,b$-$\langle Fe \rangle$ and Mg_2-$\langle Fe \rangle$ diagrams. A non-ambiguous derivation of ages therefore requires population synthesis models with variable α/Fe ratios, and the consideration of the 3-dimensional parameter space of Balmer, Mg, and Fe lines.

The paper is organized as follows. In Section 2 we present the main ingredients of our population synthesis models, whose calibration on globular clusters is shown in Section 3. The ages, metallicities, and α/Fe ratios of early-type galaxies are derived and presented in Section 4. A comparison of these results with the predictions from models of hierarchical galaxy formation is shown in Section 5. The main conclusions of this paper are discussed and summarized in Sections 6 and 7.

2   New Population Synthesis Models

The classical input parameters for population synthesis models are age and metallicity. In this paper, we introduce the element abundance ratio α/Fe as a third input parameter. For the construction of the α/Fe-enhanced simple stellar population (SSP) models we use the SSP models of Maraston (1998, 2002) as the base models, which we then modify according to the desired α/Fe ratio. In the following we summarize the procedure and introduce the main input parameters. For a more detailed presentation of the model we refer to Thomas, Maraston, & Bender (2002).

2.1   The base SSP model

The underlying solar-scaled SSP models are presented in Maraston (1998, 2002). In these models, the fuel consumption theorem (Renzini & Buzzoni 1986) is adopted to evaluate the energetics of the post main sequence phases. The input stellar tracks (solar abundance ratios) with metallicities from $1/200$ to 2 solar, are taken from Cassisi, Castellani, & Castellani (1997), Bono et al. (1997), and S. Cassisi (1999, private communication). The tracks with 3.5 solar metallicity are taken from the solar-scaled set of Salasnich et al. (2000). Lick indices are computed by adopting the fitting functions of Worthey et al. (1994).

2.2   Element abundance variations

We construct models with super-solar α/Fe ratios by increasing the abundances of the α-elements (i. e. N, O, Mg, Ca, Na, Ne, S, Si, Ti) and by decreasing those of the Fe-peak elements (i. e. Cr, Mn, Fe, Co, Ni, Cu, Zn), such that total metallicity is conserved (Trager et al. 2000a). The abundances of Carbon and all elements heavier than Zinc are assumed not to vary. It is important to notice that super-solar α/Fe ratios at constant total metallicity are accomplished mainly through a depletion of the Fe-peak element abundances, because total metallicity is made up predominantly by oxygen and the other α-elements.

2.3   Effects on absorption line indices

The abundance variations of individual elements in the stellar atmospheres certainly impact on the observed absorption-line strengths of a stellar population. This effect represents the principal ingredient in the present α/Fe-enhanced models. The variations of the Lick absorption line indices owing to the element abundance changes described in the previous section are taken from Tripicco & Bell (1995; hereafter TB95).

TB95 computed model atmospheres and synthetic spectra along a 5 Gyr-old isochrone with solar metallicity, alternately doubling the abundances of the elements C, N, O, Mg, Fe, Ca, Na, Si, Cr, Ti, Mn, Ni, and V. Note that total metallicity is not conserved but slightly increased. The impact of the abundance variations on the temperature of the isochrone is on purpose not considered by TB95. All models are based on the same isochrone with fixed T_{eff} and $\log g$ distributions, so that the abundance effects are isolated at a given temperature and surface gravity.

On the model atmospheres defined by these parameters, TB95 measure the absolute Lick index value I_0^{TB95} and the index change $\Delta I(i)^{\text{TB95}}$, for the variation of the abundance of element i. From these numbers one obtains the fractional index change (*response function*) $R_{0.3}(X_i) = \Delta I(i)^{\mathrm{TB95}}/I_0^{\mathrm{TB95}}$ of the index I due to the enhancement of the abundance of element i by 0.3 dex.

Following Trager et al. (2000a), the total fractional change δI^{TB95} of the index I when enhancing all α-elements and depressing all Fe-peak elements can then be written as the product of the fractional changes due to individual element abundance variations:

$$\delta I^{\mathrm{TB95}} = \left\{ \prod_i [1 + R_{0.3}(i)]^{([X_i/\mathrm{H}]/0.3)} \right\} - 1 \qquad (1)$$

In this equation, $[X_i/\mathrm{H}]$ is the change of the abundance ratios of element i over Hydrogen relative to the solar value. This equation assumes that the percentage index change is constant for each step of 0.3 dex in abundance, which assures that index values approach zero gracefully (Trager et al. 2000a).

The index variation ΔI of the index I is then given by the product of the total fractional index changes δI^{TB95} from TB95 and the value of the index I. Note, however, that δI^{TB95} is calculated in TB95 for $Z = Z_\odot$, so that this procedure is in principle only valid for metallicities reasonably close to solar. Indeed, with this prescription the α/Fe ratios of galactic globular clusters cannot be reproduced, the fractional index changes given by TB95 turn out to be by far too small at metallicities $Z \ll Z_\odot$. As a first-order approximation, we therefore assume that the *absolute* index change calculated by TB95 at solar metallicity is conserved when going to lower metallicities. At super-solar metallicities we adopt the *fractional* index changes of TB95. The index variation ΔI for the index I for given element abundance variations $[X_i/\mathrm{H}]$ is then

$$\Delta I = \left\{ \begin{array}{ll} \delta I^{\mathrm{TB95}} \times I & \text{if } Z \geq Z_\odot \quad \text{(fractional)} \\ \delta I^{\mathrm{TB95}} \times I_0^{\mathrm{TB95}} & \text{if } Z < Z_\odot \quad \text{(absolute)} \end{array} \right. \qquad (2)$$

As response functions for metallicities different from solar are not available, this is the most straightforward approximation at present. It is further supported by the fact that the α/Fe ratios we derive for galactic globular clusters from their Lick indices Mg_2, $\langle \mathrm{Fe} \rangle$, and $\mathrm{H}\beta$ are in very good agreement with independent spectroscopic high-resolution measurements as shown in Section 3.

It should also be emphasized that at metallicities relevant for early-type galaxies and metal-rich globular clusters, i.e. $-0.5 \leq [Z/\mathrm{H}] \leq 0.5$, the difference between fractional response and absolute response in Eqn. 2 has only a marginal effect on the resulting SSP models (see Thomas et al. 2002).

TB95 compute the variations of the individual Lick indices on model atmospheres with well defined values of temperature and gravity. These values are chosen to be representative of the three evolutionary phases dwarfs ($T_{\mathrm{eff}} = 4575\ K$, $\log g = 4.6$), turnoff ($T_{\mathrm{eff}} = 6200\ K$, $\log g = 4.1$) and giants ($T_{\mathrm{eff}} = 4255\ K$, $\log g = 1.9$), on the 5 Gyr, solar metallicity isochrone used by the authors.

We separate the turnoff region from the dwarfs on the Main Sequence at 5000 K, independent of age and metallicity. Note that the impact on the final model from varying this temperature cut-off as a function of age and metallicity is negligible (Thomas et al. 2002). We assign the Sub Giant Branch phase to the turnoff because of the very similar T_{eff} and g. The evolutionary phase 'giants' consist of the Red Giant Branch, the Horizontal Branch, and the Asymptotic Giant Branch phases. The Lick indices are computed for each evolutionary phase separately, and modified according to the response functions presented in Eqn. 2. The total integrated index of the SSP is then

$$I_{\text{SSP}} = \frac{I^{\text{D}} \times F_{\text{C}}^{\text{D}} + I^{\text{T}} \times F_{\text{C}}^{\text{T}} + I^{\text{G}} \times F_{\text{C}}^{\text{G}}}{F_{\text{C}}^{\text{D}} + F_{\text{C}}^{\text{T}} + F_{\text{C}}^{\text{G}}} , \qquad (3)$$

where I^{D}, I^{T}, I^{G} are the integrated indices of the three phases, and F_{C}^{D}, F_{C}^{T}, F_{C}^{G} are their continua. It can be easily verified that Eqn. 3 is mathematically equivalent to

$$I_{\text{SSP}} = \Delta \left(1 - \frac{\sum_i F_{\text{L}}^i}{\sum_i F_{\text{C}}^i} \right) , \qquad (4)$$

which defines integrated indices (in EW) of SSPs. In Eqn. 4 F_{L}^i and F_{C}^i are the fluxes in the line and the continuum (of the considered index), for the i-th star of the population, Δ is the line width.

2.4   The α/Fe bias of stellar libraries

In population synthesis models, the link between Lick absorption line indices and the stellar parameters temperature, gravity, and metallicity is provided by stellar libraries that inevitably reflect the chemical history of the Milky Way[1]. This implies that every population synthesis model suffers from a bias in the α/Fe ratio, i.e. the model does not reflect solar α/Fe at all metallicities (Borges et al. 1995). From the abundance patterns of the Milky Way stars (see the review by McWilliam 1997 and references therein), we know that the underlying α/Fe ratio is solar at $Z = Z_\odot$ and increases with decreasing metallicity to $[\alpha/\text{Fe}] \approx 0.3$ at $Z \leq Z_\odot/10$.

Table 1: The α/Fe Bias in the Milky Way

$[Z/H]$	-2.25	-1.35	-0.33	0.00	0.35	0.67
$[\alpha/\text{Fe}]$	0.40	0.30	0.10	0.00	0.00	0.00

In this paper we construct SSP models for different α/Fe with the aid of the index response functions as described in the previous sections. With this

---

[1] In our (and most) population synthesis models this link is given by the so-called fitting functions of Worthey et al. (1994).

procedure it is straightforward to correct for the bias given by the chemical history of our Galaxy (see Table 1). The values in Table 1 are in excellent agreement with the bias derived by Maraston et al. (2002) with the aid of the fitting functions of Borges et al. (1995). By correcting for this bias we present for the first time SSP models that have a constant α/Fe ratio at all metallicities. The use of such models is particularly important for the interpretation of metal-poor stellar systems like (extragalactic) globular clusters or dwarf spheroidal galaxies.

2.5   The effect of α/Fe-enhanced stellar tracks

In principle the element abundance variations in a star affect also the star's evolution and the opacities in the stellar atmosphere, hence the effective temperature. A fully consistent α/Fe-enhanced SSP model should therefore consider α-enhanced stellar evolutionary tracks. The impact of non-solar α/Fe ratios on stellar evolution, however, was very controversially discussed in the literature (see discussion in Trager et al. 2000a). In particular, we are still missing a homogeneous set of stellar tracks with non-solar abundance ratios using consistent opacities, that are computed for a large range in metallicities. Salasnich et al. (2000) computed α/Fe-enhanced stellar tracks for metallicities $Z > 1/2\ Z_\odot$, while Bergbusch & VandenBerg (2001) recently published α/Fe-enhanced stellar tracks for metallicities $Z < 1/2\ Z_\odot$. It would be valuable to have one set of models that includes sub-solar metallicities for calibration purposes on globular clusters and super-solar metallicities for the application on early-type galaxies.

Nevertheless, in Thomas et al. (2002) we additionally compute models in which the α/Fe-enhanced stellar tracks of Salasnich et al. (2000) are included at metallicities above 1/2 solar. The lower opacities of the α/Fe enhanced tracks lead to hotter isochrones, so that with these models we derive unreasonably high ($\sim$ 30 Gyr) ages for early-type galaxies. Unfortunately, a calibration of the model with the α/Fe-enhanced tracks of Salasnich et al. (2000) on globular clusters is not possible, because metallicities below 1/2 solar are missing. We therefore do not consider α/Fe-enhanced stellar tracks in the models presented here.

3   Calibration on Globular Clusters

In this section we show the calibration of our α/Fe-enhanced SSP models on galactic globular clusters. We derive [α/Fe] ratios and metallicities from the Mg_2 and Fe5270/Fe5335 indices of the globular clusters observed by Covino, Galletti, & Pasinetti (1995). These are then compared with independent spectroscopic measurements from individual stars in these clusters, taken from the compilations by Carney (1996) and Salaris & Cassisi (1996). We restrict our study to the data of Covino et al. (1995), because they give very consistent measurements of the indices Fe5270 and Fe5335 for all globulars (except NGC 6356), which is not the case for the data of Trager et al. (1998).

We do not use the Hβ index, but assume a fixed age of 12 Gyr for the following reason. In Maraston, Greggio, & Thomas (2001) it is shown that, because of the appearance of hot horizontal branch stars at very low metallicities, Hβ does not monotonically decrease with increasing age for [Fe/H] ≤ -1, but has a minimum at $t \approx 12$ Gyr. As a consequence, the age determination through Hβ is ambiguous at very low metallicities. Moreover, assuming the galactic globular clusters to be uniformly old, their Hβ indices and in particular the strong increase of Hβ with decreasing metallicity can be perfectly reproduced with our models (Maraston & Thomas 2000). Note also that the galactic globular clusters are found to be coeval independent of their metallicities (Rosenberg et al. 1999; Piotto et al. 2000). The ages derived from color-magnitude diagrams lie between 9 and 14 Gyr (VandenBerg 2000). The exact assumed age does not impact on the α/Fe ratios derived here, because of the mild derivatives with age of the Mg and Fe indices at old ages.

3.1 Mg$_2$ and Fe indices

Fig. 1 shows data of galactic globular clusters from Covino et al. (1995) as triangles and squares in the Mg$_2$-Fe5335 plane. Squares are those clusters for which independent measurements of [α/Fe] are given in the literature. SSP models of fixed age (12 Gyr) and metallicities $-2.25 \leq$ [Z/H] ≤ 0.67 are overplotted. The dotted line are the underlying SSP models of Maraston (1998, 2002). These models – like other SSP models in the literature (Buzzoni, Gariboldi, & Mantegazza 1992; Buzzoni, Mantegazza, & Gariboldi 1994; Worthey 1994; Tantalo et al. 1996; Vazdekis et al. 1996; Kurth, Fritze-von Alvensleben, & Fricke 1999; and others) – are biased in α/Fe, i.e. they reflect super-solar α/Fe ratios at sub-solar metallicities (see Section 2.4). Therefore, at the lowest metallicities, the models predict weaker Fe-indices than observed in globular clusters. The dashed and solid lines are the unbiased SSP models of this paper for [α/Fe] = 0.0 and [α/Fe] = 0.3 at all metallicities, respectively. The globular cluster data lie between these two models, indicating super-solar α/Fe ratios in agreement with independent spectroscopic measurements of single stars in these clusters (see next section).

We note that it is very unlikely that the excess of Fe measured in the globular clusters can be explained by anomalies in the horizontal branch morphologies, i.e. clusters with larger Fe have relatively red horizontal branches for their metallicities. The clusters considered here do not show such an effect. The HBR parameter defined by Lee (1990), which essentially is a measure for the fraction of horizontal-branch stars on the blue side of the RR Lyrae region, indicates blue horizontal branches for all metal-poor clusters in the Covino et al. (1995) sample (Harris 1996).

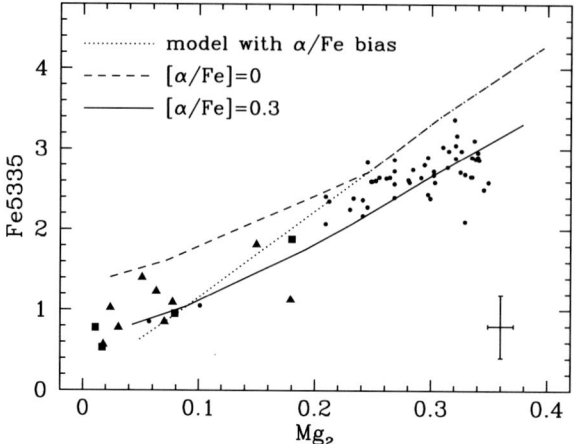

Figure 1: Index-index diagram Fe5335 vs. Mg_2. Squares and triangles are galactic globular cluster data from Covino et al. (1995). Typical error bars for these data are shown in the lower-right corner of the diagram. Squares are those clusters for which independent measurements of $[\alpha/Fe]$ are given in Carney (1996) and Salaris & Cassisi (1996). The circles are early-type galaxies from Beuing et al. (2002). The lines are 12 Gyr SSP models in the metallicity range $-2.25 \leq [Z/H] \leq 0.67$. The dotted lines are the underlying SSP models of Maraston (1998, 2002), which are biased in α/Fe, i.e. they reflect super-solar α/Fe ratios at sub-solar metallicities (see Section 2.4). The dashed and solid lines are the unbiased SSP models of this paper for $[\alpha/Fe] = 0.0$ and $[\alpha/Fe] = 0.3$ at all metallicities, respectively.

3.2   The α/Fe ratios

In Fig. 2 we plot the $[\alpha/Fe]$ ratios derived for the Covino et al. (1995) clusters versus their total metallicities $[Z/H]$ (filled symbols). Both these quantities are determined with our models described in Section 2. The filled squares are those clusters for which the $[\alpha/Fe]$ ratios are known from independent spectroscopic measurements of individual stars. These values are plotted as open squares. The bottom panel shows the deviation of our determinations from the literature: $\Delta[\alpha/Fe] = [\alpha/Fe]_{ThisWork} - [\alpha/Fe]_{lit}$. The left and right panels show the results for the Fe5270 and Fe5335 indices, respectively.

Within the errors, the values for $[\alpha/Fe]$ derived with our models are in good agreement with independent spectroscopic measurements in single stars. Both Fe indices yield consistent α/Fe ratios, although the errors (particular in Fe5270) are uncomfortably large. A more accurate calibration would certainly require data of better quality. For the entire Covino et al. (1995) sample we find $[\alpha/Fe] \approx 0.2 - 0.4$, independent of metallicity. This result strongly supports the view that the galactic globular cluster population formed early in a short (< 1 Gyr) star formation episode, so that no significant trend of age and α/Fe with metallicity is detectable.

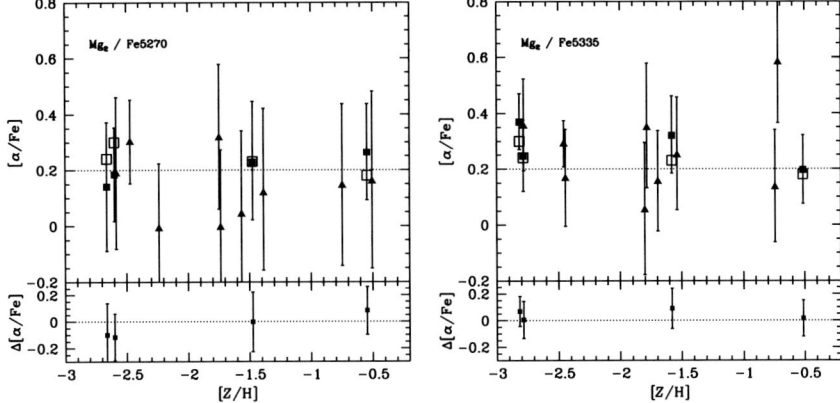

Figure 2: Abundance ratios $[\alpha/\text{Fe}]$ as a function of metallicity $[Z/\text{H}]$ for galactic globular clusters. Filled triangles and squares are the Covino et al. (1995) sample for which we determined $[\alpha/\text{Fe}]$ and $[Z/\text{H}]$ with our SSP models using Mg_2 and Fe5270 (left panel) or Fe5335 (right panel). Squares are those clusters for which independent measurements of $[\alpha/\text{Fe}]$ are given in Carney (1996) and Salaris & Cassisi (1996). The literature values are plotted as open squares. The bottom panel shows the deviation of our determinations from the literature: $\Delta[\alpha/\text{Fe}] = [\alpha/\text{Fe}]_{\text{ThisWork}} - [\alpha/\text{Fe}]_{\text{lit}}$.

For the application of the models to early-type galaxies a calibration at solar and super-solar metallicities is desirable. For a sample of bulge clusters with metallicities up to solar (Maraston et al. 2002; Puzia et al. 2002), we derive abundance ratios $0.2 \le [\alpha/\text{Fe}] \le 0.4$, which are in good agreement with independent spectroscopic element abundance measurements in these clusters (see Maraston et al. 2002 and references therein).

4 Key Parameters of Early-Type Galaxies

4.1 Data sample

We analyze a sample of 126 early-type galaxies, 71 of which are field and 55 cluster objects, containing roughly equal fractions of elliptical and lenticular (S0) galaxies. The sample is constructed from the following sources: 41 Virgo cluster and field galaxies (González 1993), 32 Coma cluster galaxies (Mehlert et al. 2000), and 53 mostly field galaxies (highest quality data from Beuing et al. 2002) selected from the ESO–LV catalog (Lauberts & Valentijn 1989). In the latter sample, objects with a local galaxy surface density $\text{NG}_T > 9$ are assumed to be cluster galaxies. NG_T is given in Lauberts & Valentijn (1989) and is the number of galaxies per square degree inside a radius of one degree around the considered galaxy.

In Fig. 3 we show as filled circles the Lick absorption line indices $H\beta$, $\text{Mg}\,b$, and $\langle\text{Fe}\rangle = (\text{Fe}5270 + \text{Fe}5335)/2$ of the sample measured within $1/10\,r_e$. Overplotted are our models with $[\alpha/\text{Fe}] = 0.0$ as dotted lines and $[\alpha/\text{Fe}] = 0.3$ as solid lines for the metallicities $[Z/\text{H}] = 0.0, 0.35, 0.67$ and for the ages $t = 2, 3, 5, 10, 15$ Gyr (see the labels in the top-right, and bottom panels).

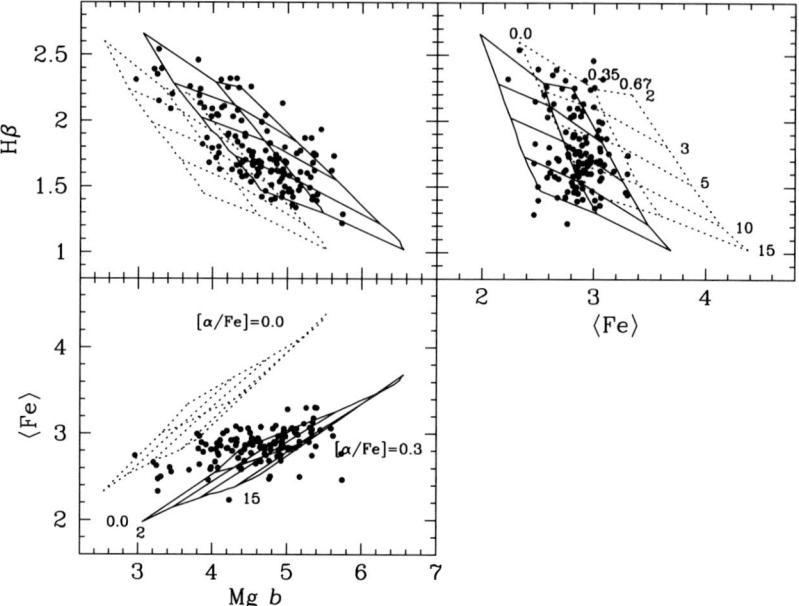

Figure 3: Three-dim. parameter space of the Lick indices Hβ, Mg b, and $\langle$Fe$\rangle$ = (Fe5270 + Fe5335)/2. The filled circles are the early-type galaxies analyzed in this paper, taken from González (1993), Mehlert et al. (2000), and Beuing et al. (2002). Index values are measured within 1/10 r_e. Our models with [α/Fe] = 0.0 and [α/Fe] = 0.3 are plotted as dotted and solid lines, respectively. Models of constant ages $t = 2, 3, 5, 10, 15$ Gyr and metallicities [Z/H] = 0.0, 0.35, 0.67 are shown (see labels in the top right, and bottom panels).

Note that the models with [α/Fe] = 0.0 are identical with the original SSP models of Maraston (1998, 2002), as we do not assume any α/Fe bias at these metallicities (see Table 1). The Mg b indices of the α/Fe-enhanced model are higher, and the $\langle$Fe$\rangle$ indices are lower, while Hβ increases only marginally. At metallicities close to solar, the fractional index changes are only very little dependent on age and metallicity. Therefore, the shape of the model grid is almost invariant under changes of α/Fe.

By means of our α/Fe-enhanced population synthesis models, we can now uniquely determine the average ages, metallicities, and α/Fe ratios from the observed absorption indices Hβ, Mg b, and $\langle$Fe$\rangle$.

4.2 The α/Fe ratios

In Fig. 4 we plot the element abundance ratio [α/Fe] as functions of velocity dispersion σ (left panel) and mean age (right panel). Grey and black symbols are field and cluster early-type galaxies, respectively. Triangles are lenticular, circles elliptical galaxies. Squares are the Coma cD galaxies NGC 4874 and NGC 4889.

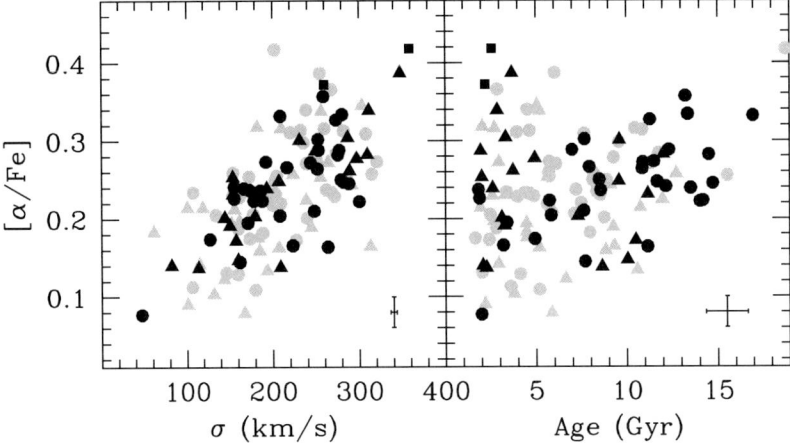

Figure 4: Element abundance ratio [α/Fe] as a function of velocity dispersion σ (measured within 1/10 r_e) and mean age. Grey and black symbols are field and cluster early-type galaxies, respectively. Triangles are lenticular, circles elliptical galaxies. The two squares are the Coma cD galaxies NGC 4874 and NGC 4889. Mean ages and abundance ratios are derived with [α/Fe] enhanced SSP models described in Section 2 from the indices Hβ, Mg b, and ⟨Fe⟩ = (Fe5270 + Fe5335)/2 measured within 1/10 r_e. Typical error-bars are given in the bottom-right corners. Galaxy data are taken from González (1993), Mehlert et al. (2000), and Beuing et al. (2002).

As anticipated qualitatively by Fisher, Franx, & Illingworth (1995), the [α/Fe] ratio and σ are well correlated, in agreement with the study of Trager et al. (2000b). Additionally, we find that zero-point, slope, and scatter of this relation are the same for field and cluster galaxies. This result extends the discovery that the Mg-σ relation is independent of environmental density (Bernardi et al. 1998; Colless et al. 1999), and may provide a deeper understanding of the origin for the Mg-σ relation.

The increase of [α/Fe] as a function of galaxy mass can be explained by either a flattening of the initial mass function (IMF) or by a shortening of the star formation timescale with increasing galaxy mass (e. g., Matteucci 1994; Thomas, Greggio, & Bender 1999). The additional consideration of average ages helps to disentangle this degeneracy. The right panel of Fig. 4 shows that the [α/Fe] element enhancement correlates with the average age of the galaxy, such that objects with higher α/Fe tend to be older. If IMF variations were the main cause for the observed α/Fe ratios, we would not expect to find such a trend. In particular, the lack of old objects with low α/Fe in all environments could not be easily understood. This non-detection strongly supports the conclusion that formation timescales rather than IMF variations are the driving mechanism for the [α/Fe]-σ relation.

We conclude that the depth of the potential well, measured through the velocity dispersion, defines the star formation timescales in early-type galaxies and hence their α/Fe ratios, independent of the environment. If α/Fe ratios are the main driver for the Mg-σ relation, its independence of the environment is then easily understood through the link between potential well and star formation timescale.

From Fig. 4 it can be seen that the correlation between mean $[\alpha/\text{Fe}]$ and mean age is well defined for all cluster ellipticals. For the two Coma cD galaxies and roughly 15 per cent of the field ellipticals and lenticular galaxies we derive ages below 5 Gyr and $[\alpha/\text{Fe}] \geq 0.3$. Naively interpreted, the coexistence of low ages and high α/Fe could imply that the majority of stars in these objects were formed recently on a short timescale. However, this is very implausible. More likely explanations are:

- Metal-poor subcomponent. Old, metal-poor stellar populations develop hot horizontal branches, which lead to rather strong Balmer absorption despite the old age. A composite stellar population that includes a small fraction of metal-poor stars can explain the Balmer absorption of these objects (H$\beta \approx 2$) without invoking young ages (Maraston & Thomas 2000). In this case the derivation of young average ages is an artifact, the galaxy is about 15 Gyr old, formed its stars on a short timescale and is therefore α/Fe-enhanced.

- Recent star formation. The object formed most of its stars at high redshift and suffered very recently (~ 300 Myr ago) from a minor (a few per cent in mass) star formation episode. As a consequence, the object has a very young V-light averaged age, but also a very low average α/Fe because of the Fe enrichment from Type Ia supernovae of the underlying old population (Thomas et al. 1999). In this case, the seemingly large α/Fe derived here is an artifact, as composite stellar populations with a major old and a very small very young subcomponent can mimic the existence of α/Fe-enhancement (Kuntschner 2000). This effect comes from the different partial time derivatives of the Mg and the Fe indices at ages below 1 Gyr. We note the caveat, however, that the fitting functions, and therefore also the population synthesis models, are not valid at such low ages (Buzzoni et al. 1994; Worthey et al. 1994).

4.3   The ages

The correlation of α/Fe with both σ and average age – essentially valid for elliptical galaxies – implies a relation between average age and velocity dispersion. Both in clusters and in the field, more massive objects are older.

This result is better illustrated in Fig. 5, in which the age distributions of elliptical galaxies with velocity dispersions above and below 200 km/s are shown as the shaded and grey histograms, respectively. More than 50 per cent of the low-mass ellipticals, but only 10 per cent of the massive ellipticals, have average ages younger than 5 Gyr. The trend of massive ellipticals being older is in agreement with a recent study by Poggianti et al. (2001a) of a large

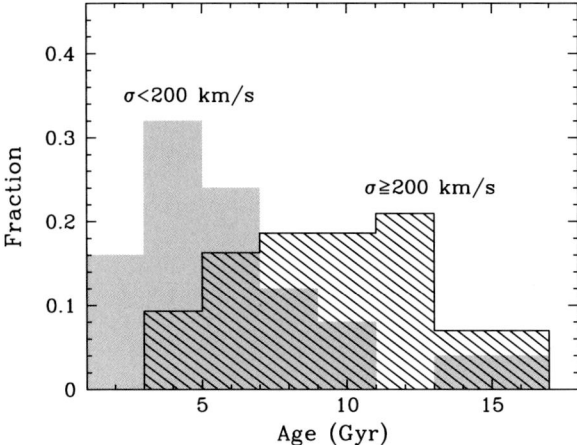

Figure 5: Age distributions of elliptical galaxies with velocity dispersions above (shaded histogram, 43 objects) and below (grey histogram, 25 objects) 200 km/s. The histograms include both cluster and field galaxies.

number of Coma cluster galaxies. We add the important conclusion that this correlation is an intrinsic property of elliptical galaxies and does not depend on the environment.

It is interesting to note that lenticular galaxies do not follow this trend. There is a considerable fraction of S0 galaxies with $\sigma > 200$ km/s, for which we derive relatively young average ages (< 5 Gyr). This might indicate recent star formation episodes in such objects as concluded by Poggianti et al. (2001b) or the existence of metal-poor subpopulations (Maraston & Thomas 2000; see the previous section).

4.4 Star formation histories

The relations shown in Fig. 4 can be used to constrain the epochs of the main star formation episode and the star formation timescales for objects as a function of their velocity dispersions. Assuming a Gaussian distribution for the star formation rate, we calculate the chemical enrichment of α and Fe-peak elements for an initially primordial gas cloud. The delayed enrichment from Type Ia supernovae is taken into account using the prescription of Greggio & Renzini (1983; see Thomas, Greggio, & Bender 1998, 1999 for more details). The simulations are done for a set of different star formation histories (Gaussians) varying the star formation timescale (width) and the lookback time of the maximum star formation (peak). For this set of star formation histories we compute the V-light averaged ages and $[\alpha/\text{Fe}]$ ratios of the resulting composite stellar population today. These can be directly compared with the observationally derived values plotted in Fig. 4. Finally, the Gaussians are linked to velocity dispersion using the relation between σ and $[\alpha/\text{Fe}]$ (left panel of Fig. 4).

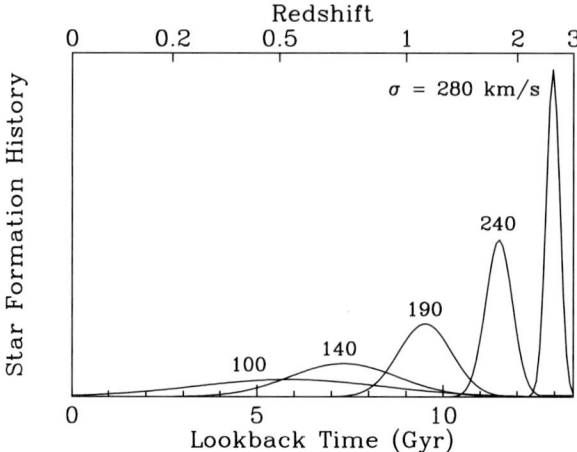

Figure 6: Star formation rates as functions of lookback time for early-type galaxies with different velocity dispersions σ between 100 km/s and 300 km/s. The star formation histories are derived from the mean ages and $[\alpha/\mathrm{Fe}]$ ratios shown in Fig. 4 (see text). Redshifts assume $\Omega_m = 0.2$, $\Omega_\Lambda = 0.8$, $H_0 = 65$ km/s/Mpc.

The resulting star formation rates as a function of lookback time are shown in Fig. 6. The upper y-axis gives the corresponding redshifts assuming $\Omega_m = 0.2$, $\Omega_\Lambda = 0.8$, and $H_0 = 65$ km/s/Mpc. The more massive the galaxy, the shorter is its formation timescale and the higher is its formation redshift. Galaxies with $\sigma < 140$ km/s exhibit significant star formation at redshifts $z < 1$. The star formation histories shown in Fig. 6 are in principle derived for cluster ellipticals, but are also valid for the bulk of field ellipticals and S0 galaxies, as only roughly 15 per cent deviate from the $[\alpha/\mathrm{Fe}]$-age relation in Fig. 4.

Most interestingly, objects are observed at high redshifts whose properties are consistent with the abundance ratios and star formation histories derived in this paper. At least part of the SCUBA sources at $2 \leq z \leq 3$ turn out to be star forming galaxies with extremely high star formation rates up to 1000 $M_\odot$/yr (de Mello et al. 2002; Smail et al. 2002). They are difficult to detect in the optical rest-frame because they are enshrouded in dust. These objects are likely to be the precursors of the most massive ellipticals ($\sigma \approx 300$ km/s) forming in a violent star formation episode at high redshift. Ly-break galaxies, instead, exhibit more moderate star formation rates of the order a few 10 $M_\odot$/yr (Pettini et al. 2001), and may therefore be the precursors of less massive ellipticals ($\sigma \approx 200$ km/s). In a recent work, Pettini et al. (2002) analyze deep Keck spectra of the lensed Ly-break galaxy cB58 at redshift $z = 2.73$ (Seitz et al. 1998). They find the *interstellar medium* to be significantly α/Fe-enhanced, which is in good agreement with the α/Fe-enhancement derived in this paper for the *stellar population* of local elliptical galaxies.

5 Hierarchical Galaxy Formation

The element abundance ratios $0.2 \leq [\alpha/\text{Fe}] \leq 0.4$ derived in this paper for massive ($\sigma > 200$ km/s) early-type galaxies require star formation timescales below ~ 1 Gyr (see Fig. 6). In models of hierarchical galaxy formation, however, star formation in ellipticals typically does not truncate after 1 Gyr, but continues to lower redshift (Kauffmann 1996). It is therefore questionable if hierarchical clustering would lead to significantly α/Fe-enhanced giant ellipticals (Bender 1996). In a more quantitative investigation, Thomas et al. (1999) compute the chemical enrichment in mergers of evolved spiral galaxies, and show that such a merger does not produce significantly super-solar α/Fe ratios. This prediction is confirmed observationally by the study of Maraston et al. (2001), who find that the newly formed globular clusters in the merger remnant NGC 7252 and the integrated light of the galaxy are indeed not α/Fe-enhanced.

So far, semi-analytic models have not considered this constraint. Thomas (1999) demonstrates that the average star formation history of cluster ellipticals within the hierarchical formation scheme leads to $[\alpha/\text{Fe}] \sim 0.04$, a value that clearly lies below the observational estimates in Fig. 4. In Thomas & Kauffmann (1999), we follow this aspect in more detail by taking the star formation histories of individual Monte Carlo realizations of elliptical galaxies into account. This allows us to explore the distribution and the scatter of the α/Fe ratios among the galaxies as they are predicted by the semi-analytic models. We calculate the chemical enrichment of closed box systems for the star formation rate predicted by semi-analytic models. Both modes of star formation, the quiescent and the merger induced burst components, are taken into account. A universal Salpeter IMF slope $x = 1.35$ is assumed. The simulations in this analysis are for a cold dark matter power spectrum with $\Omega = 1$, $H_0 = 50$ km/s/Mpc, and $\sigma_8 = 0.67$. For more details please refer to Thomas & Kauffmann (1999) and Thomas (1999).

In Fig. 7, we plot the resulting $[\alpha/\text{Fe}]$ ratios of elliptical galaxies as a function of their V-magnitudes. The observationally derived $[\alpha/\text{Fe}]$ ratios from Fig. 4 are indicated as the grey shaded area. The scatter in α/Fe predicted by the hierarchical models is large because of the relatively large variety of formation histories comprising star formation timescales from 10^9 to 10^{10} yr.

The figure reveals a striking failure of the hierarchical model: More luminous ellipticals are predicted to have lower α/Fe ratios. The opposite is observed. As a consequence, the α/Fe ratios predicted by the hierarchical models for massive elliptical galaxies are significantly below the observed values. These problems are directly related to the hierarchical clustering scheme, in which the largest objects form last and have therefore more extended star formation histories. This hierarchical paradigm leads also to the prediction of an age-mass anti-correlation (Kauffmann & Charlot 1998), which again stands in clear contradiction to the observational evidence that early-type galaxies with larger velocity dispersions tend to have higher average ages (see Section 4.2).

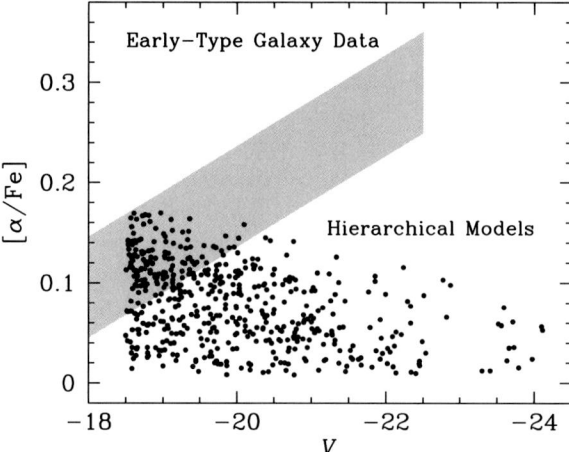

Figure 7: Absolute V-magnitude vs. $[\alpha/\mathrm{Fe}]$ ratio. Points are early-type galaxies modeled in the framework of hierarchical galaxy formation (Thomas & Kauffmann 1999). Halos with circular velocities $V_c = 1000$ km/s are considered, corresponding to cluster environments. The grey shaded area indicates the range of $[\alpha/\mathrm{Fe}]$ values derived in this paper from observed absorption line indices (see Fig. 4).

6 Discussion

As recently summarized by Peebles (2002), the arguments of this paper are accompanied by a number of evidences from the tightness and redshift evolution of the scaling relations of early-type galaxies. Both push the formation ages of the stellar populations in large elliptical galaxies to redshifts $z > 2$ (e. g., Bower et al. 1992; Bender, Burstein, & Faber 1993; Renzini & Ciotti 1993; Bender, Ziegler, & Bruzual 1996; Kodama, Bower, & Bell 1999; Ziegler et al. 1999). These findings are not accomplished by standard hierarchical models.

At least part of the reason for this failure may be connected to the way how non-baryonic dark matter and baryonic matter are linked. In current models, star formation is tightly linked to the assembly history of dark matter halos, so that galaxies with longer assembly times also form stars on longer timescales. In a recent paper, Granato et al. (2001) present a promising – even though more heuristic – approach, in which star formation is enhanced in massive systems. The resulting 'anti-hierarchical baryonic collapse' leads to higher formation redshifts and shorter formation timescales of the stellar populations in massive objects. A modification of this kind is certainly the right step towards reconciling hierarchical models with the constraints set by the stellar populations properties of early-type galaxies. Still, this modification may not be sufficient to harmonize hierarchical galaxy formation with the conclusion of Gerhard et al. (2001), who argue that the very high core densities of the halos of elliptical galaxies push also the collapse of their dark matter halos (not only of their baryonic matter) to high redshifts.

7   Summary

In this paper we derive the central average ages, metallicities, and α/Fe ratios for 126 early-type galaxies from both cluster and field environments.

For this purpose we develop new population synthesis models with variable α/Fe ratios, that allow for the derivation of these stellar population parameters from the Lick absorption line indices Hβ, Mg b, and $\langle$Fe$\rangle$. The models are based on the SSP models of Maraston (1998, 2002). The effect from varying the α/Fe ratio is calculated with the Tripicco & Bell (1995) index response functions, following the method introduced by Trager et al. (2000a). At sub-solar metallicities we calibrate our models on galactic globular clusters, putting particular attention to the derivation of α/Fe ratios. The base SSP models of Maraston (1998, 2002) – like all population synthesis models up to now – rely on stellar libraries that reflect the chemical history of the Milky Way. They are therefore biased towards super-solar α/Fe ratios at sub-solar metallicities. We correct for this bias, so that our new SSP models reflect constant α/Fe ratios at all metallicities.

For the sample of early-type galaxies investigated here, we find that α/Fe ratios correlate well with velocity dispersion σ, independent of the environmental density. Elliptical galaxies and the majority of lenticular galaxies additionally exhibit a good correlation of α/Fe ratio with average age. These results strongly support the view that the increase of α/Fe with increasing velocity dispersion is due to a decrease of star formation timescales rather than due to a flattening of the IMF. We show that more massive ellipticals have higher average ages and higher α/Fe ratios, because of earlier formation epochs and shorter formation timescales of their stellar populations. This observational result is not matched by current models of hierarchical galaxy formation, mainly because they predict too extended star formation histories for massive ellipticals.

Acknowledgments

DT thanks the organizers of the JENAM 2001 conference for the invitation to a highlight talk. We would like to acknowledge Santi Cassisi for providing a large number of stellar evolutionary tracks, and Beatriz Barbuy, Laura Greggio, Claudia Mendes de Oliveira, Dörte Mehlert, and Manuela Zoccali for very interesting and stimulating discussions. DT and CM give sincere thanks to Claudia Mendes de Oliveira, Beatriz Barbuy, and the members of the Instituto Astronomico e Geofisico of São Paulo for their kind hospitality. The the BMBF, the DAAD, and the "Sonderforschungsbereich 375-95 für Astro-Teilchenphysik" of the DFG are acknowledged for financial support.

References

Bender, R. 1996, in New Light on Galaxy Evolution, ed. R. Bender & R. L. Davies (Dordrecht: Kluwer Academic Publishers), 181

Bender, R., Burstein, D., & Faber, S. M. 1993, ApJ 411, 153

Bender, R., Ziegler, B. L., & Bruzual, G. 1996, ApJ 463, L51

Bergbusch, P. A., & VandenBerg, D. A. 2001, ApJ 556, 322

Bernardi, M., et al. 1998, ApJ 508, 143

Beuing, J., Bender, R., Mendes de Oliveira, C., Thomas, D., & Maraston, C. 2002, A&A, submitted

Bono, G., Caputo, F., Cassisi, S., Castellani, V., & Marconi, M. 1997, ApJ 489, 822

Borges, A. C., Idiart, T. P., de Freitas Pacheco, J. A., & Thévenin, F. 1995, AJ 110, 2408

Bower, R. G., Lucey, J. R., & Ellis, R. S. 1992, MNRAS 254, 589

Buzzoni, A., Gariboldi, G., & Mantegazza, L. 1992, AJ 103, 1814

Buzzoni, A., Mantegazza, L., & Gariboldi, G. 1994, AJ 107, 513

Carney, B. W. 1996, PASP 108, 900

Cassisi, S., Castellani, M., & Castellani, V. 1997, A&A 317, 10

Colless, M., Burstein, D., Davies, R. L., McMahan, R. K., Saglia, R. P., & Wegner, G. 1999, MNRAS 303, 813

Covino, S., Galletti, S., & Pasinetti, L. E. 1995, A&A 303, 79

Davies, R. L., Sadler, E. M., & Peletier, R. F. 1993, MNRAS 262, 650

de Mello, D., Wiklind, T., Leitherer, C., & Pontoppidan, K. 2002, in The Evolution of Galaxies II. Basic Building Blocks, ed. M. Sauvage, G. Stasinska, L. Vigroux, D. Schaerer, & S. Madden (Dordrecht: Kluwer), in press

Faber, S. M., Friel, E. D., Burstein, D., & Gaskell, D. M. 1985, ApJS 57, 711

Fisher, D., Franx, M., & Illingworth, G. 1995, ApJ 448, 119

Gerhard, O., Kronawitter, A., Saglia, R. P., & Bender, R. 2001, AJ 121, 1936

González, J. 1993, Phd thesis, University of California, Santa Cruz

Granato, G. L., Silva, L., Monaco, P., Panuzzo, P., Salucci, P., De Zotti, G., & Danese, L. 2001, MNRAS 324, 757

Greggio, L., & Renzini, A. 1983, A&A 118, 217

Harris, W. E. 1996, AJ 112, 1487

Kauffmann, G. 1996, MNRAS 281, 487

Kauffmann, G., & Charlot, S. 1998, MNRAS 294, 705

Kodama, T., Bower, R. G., & Bell, E. F. 1999, MNRAS 306, 561

Kuntschner, H. 2000, MNRAS 315, 184

Kurth, O. M., Fritze-v. Alvensleben, U., & Fricke, K. J. 1999, A&AS 138, 19

Lauberts, A., & Valentijn, E. A. 1989, The Surface Photometry Catalogue of the ESO-Upsalla Galaxies (Garching: ESO)

Lee, Y.-W. 1990, ApJ 363, 159

Maraston, C. 1998, MNRAS 300, 872

Maraston, C. 2002, MNRAS submitted

Maraston, C., et al. 2002, in preparation

Maraston, C., Greggio, L., & Thomas, D. 2001, Ap&SS 276, 893

Maraston, C., Kissler-Patig, M., Brodie, J. P., Barmby, P., & Huchra, J. 2001, A&A 370, 176

Maraston, C., & Thomas, D. 2000, ApJ 541, 126

Matteucci, F. 1994, A&A 288, 57

McWilliam, A. 1997, ARA&A 35, 503

Mehlert, D., Saglia, R. P., Bender, R., & Wegner, G. 2000, A&AS 141, 449

Peebles, P. J. E. 2002, in A New Era in Cosmology, ed. N. Metcalfe & T. Shanks, ASP Conference Series (Dordrecht: Kluwer Academic Publishers), in press

Pettini, M., Rix, S. A., Steidel, C. C., Adelberger, K. L., Hunt, M. P., & Shapley, A. E. 2002, ApJ, in press, astro-ph/0110637

Pettini, M., Shapley, A. E., Steidel, C. C., Cuby, J., Dickinson, M., Moorwood, A. F. M., Adelberger, K. L., & Giavalisco, M. 2001, ApJ 554, 981

Piotto, G., Rosenberg, A., Saviane, I., Zoccali, M., & Aparicio, A. 2000, in Ap&SS Library, Vol. 255, The evolution of the Milky Way: stars versus clusters, ed. F. Matteucci & F. Giovannelli (Dordrecht: Kluwer Academic Publishers), 249

Poggianti, B., et al. 2001a, ApJ 562, 689

Poggianti, B., et al. 2001b, ApJ 563, 118

Puzia, T., et al. 2002, in preparation

Renzini, A., & Buzzoni, A. 1986, in Spectral evolution of galaxies, ed. C. Chiosi & A. Renzini (Dordrecht: Reidel), 135

Renzini, A., & Ciotti, L. 1993, ApJ 416, L49

Rosenberg, A., Saviane, I., Piotto, G., & Aparicio, A. 1999, AJ 118, 2306

Salaris, M., & Cassisi, S. 1996, A&A 305, 858

Salasnich, B., Girardi, L., Weiss, A., & Chiosi, C. 2000, A&A 361, 1023

Seitz, S., Saglia, R. P., Bender, R., Hopp, U., Belloni, P., & Ziegler, B. 1998, MNRAS 298, 945

Smail, I., Ivison, R. J., Blain, A. W., & Kneib, J.-P. 2002, MNRAS, in press, astro-ph/0112100

Tantalo, R., Chiosi, C., Bressan, A., & Fagotto, F. 1996, A&A 311, 361

Thomas, D. 1999, MNRAS 306, 655

Thomas, D., Greggio, L., & Bender, R. 1998, MNRAS 296, 119

Thomas, D., Greggio, L., & Bender, R. 1999, MNRAS 302, 537

Thomas, D., & Kauffmann, G. 1999, in Spectrophotometric dating of stars and galaxies, ed. I. Hubeny, S. Heap, & R. Cornett, Vol. 192 (ASP Conf. Ser.), 261

Thomas, D., Maraston, C., & Bender, R. 2002, MNRAS, submitted

Trager, S. C., Faber, S. M., Worthey, G., & González, J. J. 2000a, AJ 119, 164

Trager, S. C., Faber, S. M., Worthey, G., & González, J. J. 2000b, AJ 120, 165

Trager, S. C., Worthey, G., Faber, S. M., Burstein, D., & González, J. J. 1998, ApJS 116, 1

Tripicco, M. J., & Bell, R. A. 1995, AJ 110, 3035

VandenBerg, D. A. 2000, ApJS 129, 315

Vazdekis, A., Casuso, E., Peletier, R. F., & Beckmann, J. E. 1996, ApJS 106, 307

Worthey, G. 1994, ApJS 95, 107

Worthey, G., Faber, S. M., & González, J. J. 1992, ApJ 398, 69

Worthey, G., Faber, S. M., González, J. J., & Burstein, D. 1994, ApJS 94, 687

Ziegler, B. L., Saglia, R. P., Bender, R., Belloni, P., Greggio, L., & Seitz, S. 1999, A&A 346, 13

ASTRONOMISCHE GESELLSCHAFT: Reviews in Modern Astronomy **15**, 239–258 (2002)

Modelling the Spectral Energy Distribution
of Galaxies
from the Ultraviolet to Submillimeter

Cristina C. Popescu [1,2,3] and Richard J. Tuffs [4]

[1] The Observatories of the Carnegie Institution of Washington
813 Santa Barbara Street, Pasadena, CA 91101, USA
popescu@ociw.edu

[4] Max-Planck-Institut für Kernphysik
Saupfercheckweg 1, 69117 Heidelberg, Germany

Abstract

We present results from a new modelling technique which can account for the observed optical/NIR–FIR/submm spectral energy distributions (SEDs) of normal star-forming galaxies in terms of a minimum number of essential parameters specifying the star-formation history and geometrical distribution of stars and dust. The model utilises resolved optical/NIR images to constrain the old stellar population and associated dust, and geometry-sensitive colour information in the FIR/submm to constrain the spatial distributions of young stars and associated dust. It is successfully applied to the edge-on spirals NGC 891 and NGC 5907. In both cases the young stellar population powers the bulk of the FIR/submm emission. The model also accounts for the observed surface brightness distribution and large-scale radial brightness profiles in NGC 891 as determined using the Infrared Space Observatory (ISO) at 170 & 200 μm and at 850 μm using SCUBA.

1 Introduction

Historically, almost all our information about the current and past star-formation properties of galaxies has been based upon spatially integrated measurements in the ultraviolet (UV), visible and near-infrared (NIR) spectral regimes. However, star-forming galaxies contain dust which absorbs some

---

[2] Otto Hahn Fellow of the Max-Planck-Institut für Astronomie, Königstuhl 17, 69117 Heidelberg, Germany

[3] Research Associate, The Astronomical Institute of the Romanian Academy, Str. Cuţitul de Argint 5, Bucharest, Romania

fraction of the emitted starlight, re-radiating it predominantly in the far-infrared (FIR)/sub-millimeter (submm) range. The true significance of this process even for "normal" (i. e. non-starburst) galaxies has been revealed by observations of a representative sample of late-type Virgo cluster galaxies with the ISOPHOT instrument on board the Infrared Space Observatory (ISO). These showed the dust emission to typically account for 50 percent of the bolometric output of these systems, with a spectral peak generally lying between 100 and 250 μm (Tuffs et al. 2002; Popescu et al. 2002).

In view of this, the measurement of current and past star-formation in galaxies – and indeed of the universe as a whole – requires a quantitative understanding of the role different stellar populations play in powering the FIR/submm emission. For this both optical and FIR/submm data needs to be used, as they contain complementary information about the distribution of stars and dust.

On the one hand, optical data probes the colour and spatial distribution (after correction for extinction) of the photospheric emission along sufficiently transparent lines of sight. This is particularly useful to investigate older, redder stellar populations in galaxian disks with scale heights larger than that of the dust. On the other hand, grains act as test particles probing the strength and colour of ultraviolet (UV)/optical interstellar radiation fields. This constitutes an entirely complementary constraint to studies of photospheric emission. In the FIR, grains are moreover detectable over almost the full range of optical depths present in a galaxy. At least part of this regime is inaccessible to direct probes of starlight, especially at shorter wavelengths, even for face-on systems. This particularly applies to light from young stars located in, or close by, the dust clouds from which they formed, since a certain fraction of the light is locally absorbed. Furthermore, there is at least a possibility that most of the remaining UV and even blue light from young stars that can escape into the disk might be absorbed by diffuse dust there. A combined analysis of the whole UV/optical/FIR/submm spectral energy distribution (SED) of galaxies seems to be a promising way to constrain the problem.

Here we present the results from a new tool for modelling the optical/FIR/submm SED of galaxies (Popescu et al. 2000b, Misiriotis et al. 2001). Our model predicts the SED as a function of intrinsic SFR, star formation history (from population synthesis calculations) and dust content. This tool includes solving the radiative-transfer problem for a realistic distribution of absorbers and emitters, considering realistic models for dust (taking into account the grain-size distribution and stochastic heating of small grains) and the contribution of H II regions. The model addresses the fundamental and controversial question of optical thickness in the disk of galaxies and can account for the detailed UV/optical/FIR/submm SED with an absolute minimum of independent physical variables, each strongly constrained by data.

2   An overview of SED modelling in the literature

The SED models of galaxies have been built using the tools and results from previous studies of the key ingredients required by such models: radiative transfer codes and dust emission models. Radiative transfer codes were initially developed to account for the UV-optical/NIR appearance of galaxies, their extinction properties, and the role the spatial distribution of stars and dust play in shaping the SED. These codes used either analytical methods (Kylafis & Bahcall 1987) or Monte Carlo simulations (Witt & Gordon 1996, 2000, Ferrara et al. 1999, Gordon et al. 2001). In parallel, dust models have been built to model extinction curves and the heating and emission of grains in different environments. These include the studies of Mathis et al. (1977), Draine & Lee (1984), Draine & Anderson (1985), Dwek (1986), Guhathakurta & Draine (1989), Désert et al. (1990), Laor & Draine (1993), Weingartner & Draine (2001).

Applications of such methods to study the SED of galaxies have been initiated, as expected, in the optical/NIR regime. Thus, in their fundamental work, Kylafis & Bahcall (1987) applied a radiation transfer modelling technique to edge-on systems, where the scale height of the stars and dust extinction can be directly constrained. The technique was subsequently applied to several edge-on galaxies by Xilouris et al. (1997, 1998, 1999). Radiative transfer codes in combination with observations of nearby edge-on galaxies were also used by Ohta & Kodaira (1995), Kuchinski et al. (1998), and Matthews & Wood (2001).

A further step in the SED modelling was to include the FIR-submm spectral regime in the overall analysis. Different tools have been proposed starting with the pioneering works of Xu & Buat (1995) and Xu & Helou (1996). They used radiative transfer codes for an assumed "sandwich" configuration of dust and stars and considered in detail the relative contribution of the non-ionising UV photons and the optical photons in heating the grains. However these calculations did not incorporate a model for the dust grain emission, nor for radial variations in the absorbed radiation in the disk, and therefore could not account for the shape of the FIR SED.

Recently, there have been several works modelling the SED of galaxies from the UV to submm, which considered more realistic geometries and dust models, and which have started to make use of the recent observational data becoming available at longer FIR/submm wavelengths. In parallel with these self-consistent models a series of empirical or semi-empirical approaches for modelling the SED from UV to submm wavelengths have been adopted, mainly for statistical applications. Starting with the semi-empirical approaches, we should first mention the phenomenological model of Devriendt et al. (1999), which combines spectrophotometric and chemical evolution models with phenomenological extinction curves to account for the UV/optical/NIR SEDs. The FIR/submm SEDs are again obtained using a phenomenological model involving various dust component templates, where their mix is determined from the correlations of IRAS and submm colours with IRAS FIR luminosities. Obviously this is not a self-consistent model in which dust grains are heated by a radiation field calculated from radiative transfer. Such a model cannot describe the exact intrinsic distribution of stellar populations and dust, and seems to underestimate the luminosity of the cold dust in normal galaxies. For example, their predicted FIR SEDs of Virgo cluster galaxies are systematically warmer than those obtained by Popescu et al. (2002) using the new ISOPHOT data. Nevertheless this method can be used for an empirical characterisation of the SEDs of statistical samples, where proper calculations would entail prohibitively large computational times. Another semi-empirical treatment of the SEDs was presented by Dale et al. (2001). Essentially, a library of galaxy spectra was computed using three dust components and the model prediction of Desert et al. (1990) for different intensities of the radiation field, for a power law distribution of dust masses over radiation fields. The model reproduces the trend in the mid-infrared (MIR) colours and has been used to make predictions for the FIR wavelength range. The model is not constructed to fit the whole wavelength spectral range from UV to submm, and is based entirely on radiation fields with the colour of the local ISRF from our Galaxy. It is however known that the colour of the radiation fields changes dramatically within and between galaxies. Therefore the application of such a model is obviously limited to revealing trends in the diagnostic diagrams for the FIR regime, rather than giving a detailed SED model of individual galaxies.

On the other hand self consistent calculations of the SED over the entire spectral range have been developed slowly, due to the large amount of computational time required both for the radiative transfer and for the calculation of the probability distribution of temperatures of dust grains, especially of the small grains, which undergo large temperature fluctuations. To this should be added that it is only now becoming possible to obtain resolved images covering the MIR to FIR and submm spectral range, even for nearby galaxies. Such observations are needed for constraining SED models. Recent self-consistent SED modelling was presented by Silva et al. (1998). They applied photometric and chemical evolution models of galaxies to explain the SED of both normal and starburst galaxies. Their model considers radiative transfer in both the molecular clouds and in the diffuse ISM, and includes a consistent treatment of dust emission and stochastic heating of small grains. It can thus reproduce reasonably well the SED of the studied galaxies, in both the optical and FIR spectral range and is therefore suited to describe the general shape of the volume-integrated SED. The model has however more free parameters than the generally available observational constraints and the solution obtained may not be unique. The model is also less adequate in describing the intrinsic distributions of stars and dust because of its simplified geometry. For example the model does not consider different scale heights and scale lengths for the stellar and dust distribution. Furthermore, all stellar populations are constrained to have the same scale heights, including the young stellar population which is known to reside in a very thin stellar disk. The radiative transfer is only an approximation, and is rigorously applicable only to an infinite homogeneous medium, and there is no treatment of anisotropic scattering.

Bianchi et al. (2000) attempted to model NGC 6946 from the UV to FIR using a 3D Monte Carlo radiative-transfer code for a simplified geometry of emitters (a single stellar disk). They concluded that the total FIR output is consistent with an optically thick solution. However, their model did not consider a size distribution of grains and there is no consistent treatment of grain heating and emission. Accordingly, there is no calculation of stochastic heating for small grains, but rather a MIR correction calculated according to the model of Désert et al. (1990). Furthermore the model did not consider the contribution of localised sources within star-forming complexes. This resulted in a poor fit of the FIR SED and a failure to reproduce the IRAS flux densities.

Several studies have been dedicated to modelling the SED of starburst galaxies. Recent work on modelling the UV to submm emission was presented by Efstathiou et al. (2000) for starburst galaxies, which were treated as an ensemble of optically thick giant molecular clouds (GMCs) centrally illuminated by recently formed stars. The model has a proper treatment of dust emission, including stochastic heating of small grains, and it allows for the star-forming complexes to evolve with time, based on a physical model of the dusty H II region phase and of the supernova phase. The star-forming complexes are modelled assuming spherical symmetry, which, as remarked by the authors, may not be true in the later stages of the GMC evolution. Also there is no further transfer of radiation between GMCs, meaning that the stellar light is not allowed to escape the GMC. The FIR spectrum is fitted for an exponential decaying star formation rate, where the age and time constant of the star-burst (or in some cases of two star-bursts) are fitted parameters. This modelling technique successfully reproduced the observed SED of M82 and NGC 6090, though for the latter case there is an ambiguity between the existence of an older burst and a diffuse component not considered in the model. Obviously this method is suitable for starburst dominated galaxies and can bring interesting insights into the ages of the stellar populations, but cannot be applied to normal "quiescent" disk galaxies dominated by emission from the diffuse interstellar radiation field.

Bekki et al. (2000) studied the time evolution of the SED in starburst galaxies based on numerical simulations that can analyse simultaneously dynamical and chemical evolution, structural and kinematical properties. This method is rather unique and can be used to study mergers, for example, or other physical processes related to galaxy formation. However both the radiative transfer and dust emission are treated quite rudimentarily: scattering is not considered in the radiative transfer, there is no grain size distribution and no treatment of stochastic heating of small grains. Therefore, their model can be used to indicate trends in the SED evolution rather than to give quantitative predictions and is suitable for numerical simulations.

As a general remark, most of the work described above has been concentrated in fitting the volume-integrated emission of the galaxies at different spectral wavelengths and has not been constrained by morphological information from images. A more robust knowledge of the intrinsic distributions of stellar populations and dust, and ultimately of the SF history in galaxies can be gained if also the surface brightness distribution of optical and even FIR emission can be analysed. The method we present here constrains the problem by using the whole information in the brightness distribution. This approach has been only used in the optical/NIR regime, by Xilouris et al. (1997, 1998, 1999). With the ISO and SCUBA data we are now able to make predictions also for the appearance of the FIR/submm images and compare the predicted maps with the observed ones. In the future SIRTF observations will extend knowledge of the FIR morphology in galaxies and will bring new constraints for the SED modelling methods.

3   Our model

Star-forming galaxies are fundamentally inhomogeneous, containing highly obscured massive star-formation regions, as well as more extended structures harbouring older stellar populations which may be transparent or have intermediate optical depths to starlight. Accordingly, our model divides the stellar population into an "old" component (considered to dominate the output in B-band and longer wavelengths) and a "young" component (considered to dominate the output in the non-ionising UV).

3.1   The "old" stellar population

The "old" stellar population is generally observed to have scale heights of several hundred pc in rotationally supported spiral galaxies, a result which can be physically attributed to the increase of the kinetic temperature of stellar populations on timescales of order Gyr due to encounters with molecular clouds and/or spiral density waves. This means that the "old" stellar population can be constrained from resolved optical and near-IR images via the modelling procedure of Xilouris et al. (1999). The procedure uses the technique for solving the radiation transfer equation for direct and multiply scattered light for arbitrary geometries by Kylafis & Bachall (1987).

For edge-on systems these calculations completely determine the scale heights and lengths of exponential disk representations of the old stars (the "old stellar disk") and associated diffuse dust (the "old dust disk"), as well as a dustless stellar bulge. This process is feasible for edge-on systems since the scale height of the dust is less than that of the stars. For face-on systems the scale height cannot be fixed by the optical data alone, and must be determined through consideration of its effect on the FIR SED.

The extinction law of the dust can be directly determined from the observed optical/near-IR wavelength, since the calculation is done independently at each wavelength. In all cases so far, an extinction law consistent with that predicted from the graphite/silicate mix of Laor & Draine (1993) and the $a^{-3.5}$ grain size distribution of Mathis Rumpl & Nordsieck (1977) has been found (Xilouris et al. 1999).

3.2 The "young" stellar population

The "young" stellar population is also specified by an exponential disk, which we shall refer to as the "young stellar disk". Invisible in edge-on systems, its scale height is constrained to be 90 pc (the value for the Milky Way) and its scale length is equated to that of the "old stellar disk" in B-band. The emissivity of the "young stellar disk" is parameterised in terms of the current star formation rate (SFR) by relating the non-ionising UV emission to SFR using the population synthesis models of Bruzual & Charlot (2001) for $Z = Z_\odot$, a Salpeter initial mass function, a mass cut-off of $100\,\mathrm{M}_\odot$, and an exponential decrease of the SFR with time, with a time constant $\tau = 5$ Gyr.

A second exponential dust disk of grain mass M_dust – the "second dust disk" is associated with the young stellar population. This is needed to account for the observed submm emission from edge-on disk galaxies, which cannot be reproduced by models containing only the old dust disk determined from the optical images (Popescu et al. 2000b, Misioritis et al. 2001). It is constrained to have the same scale length and height as that of the young stellar disk. This second dust disk is assumed to be composed of grains with the same graphite/silicate and $a^{-3.5}$ grain size distribution as for the old dust disk. All the exponential disks for young and old stars and dust in the model are truncated at three times the exponential scale length. Because two disks of dust are required for the model, we refer to it as to the "two-dust-disk" model.

The current star-formation rate (SFR) and mass of the second dust disk (M_dust) are the first two primary free parameters of the model to determine the FIR/submm radiation. They both relate to the smooth distribution of stars and dust in the second disk. A third primary parameter, F, is included to account for inhomogeneities in the distributions of dust and stars in the young stellar disk. F is defined as the fraction of non-ionising UV which is locally absorbed in H II regions around the massive stars. It determines the additional likelyhood of absorbtion of non-ionising UV photons due to autocorrelations between an inhomogeneous distribution of young stars and parent molecular clouds. Astrophysically, this arises because at any particular epoch some fraction of the massive stars have not had time to escape the vicinity of their parent molecular clouds. Thus, F is related to the ratio between the distance a star travels in its lifetime due to it's random velocity and the typical dimensions of star-formation complexes.

3.3 Calculation of dust emission

The first step in the calculation of the dust emission is to determine the radiation energy density of the unabsorbed light versus wavelength from the non-ionising UV to the near-IR for the diffuse radiation field. This is done for trial values of the mass of dust in the second dust disk M_D by solving the radiation transfer equation for the observed optical wavelengths and three wavelengths in the non-ionising UV. (For edge on systems, M_D is independent of the dust mass in the old dust disk, which is fixed by the optical observations). The actual energy density distribution can then be determined for any trial combination of SFR and F without a further radiation transfer calculation. Both the colour and intensity of the radiation field vary with position in different ways according to the combination of SFR, M_D and F.

The FIR/submm emission for each combination of SFR, M_D and F is then calculated for graphite and silicate grains of size a immersed in the radiation field at each point of a grid of positions in the galaxy, including an explicit treatment of stochastic emission. Subsequently we integrate over the entire galaxy to obtain the FIR-submm SED of the diffuse disk emission. Prior to comparison with observed FIR-submm SEDs, an empirically determined spectral template for the H II regions, scaled according to the value of F, must be added to this calculated spectral distribution of diffuse FIR emission.

Due to the precise constraints on the distribution of stellar emissivity in the optical-NIR and the distribution and opacity of dust in the "old dust disk" yielded by the radiation transfer analysis of the highly resolved optical-NIR images, coupled with the simple assumptions for the distribution of the young stellar population and associated dust, our model has just three free parameters – SFR, F and $M_{\rm dust}$. These fully determine the FIR-submm SED, and allow a meaningful comparison with broad-band observational data in the FIR/submm, where, in particular for distant objects, typically only a few spectral sample points for the spatially integrated emission are available. The parameters are strongly coupled, but in general terms, $M_{\rm dust}$ is principally constrained by the submm emission, $SFR \times (1 - F)$ by the bolometric FIR-submm output and the factor F (in the absence of high resolution images and/or for edge-on systems) by the FIR colour.

4   Application to edge-on spiral galaxies

Modelling edge-on spiral galaxies has several advantages, mainly when investigating them in the optical band. One advantage is that, in this view of a galaxy, one can easily separate the three main components of the galaxy (i. e., the stellar disk, the dust and the bulge). Another advantage is that the dust is very prominently seen in the dust lane, and thus its scalelength and scaleheight can be better constrained. A third advantage is that many details of a galaxy that are evident when the galaxy is seen face-on (e. g., spiral arms), are smeared out to a large degree when the galaxy is seen edge-on (Misiriotis et al. 2000). Thus, a simple model with relatively few parameters can be used for the distribution of stars and dust in the galaxy. However, in edge-on galaxies it is very difficult to see localised sources (i. e., H II regions), in which the radiation can be locally absorbed and thus not contribute to

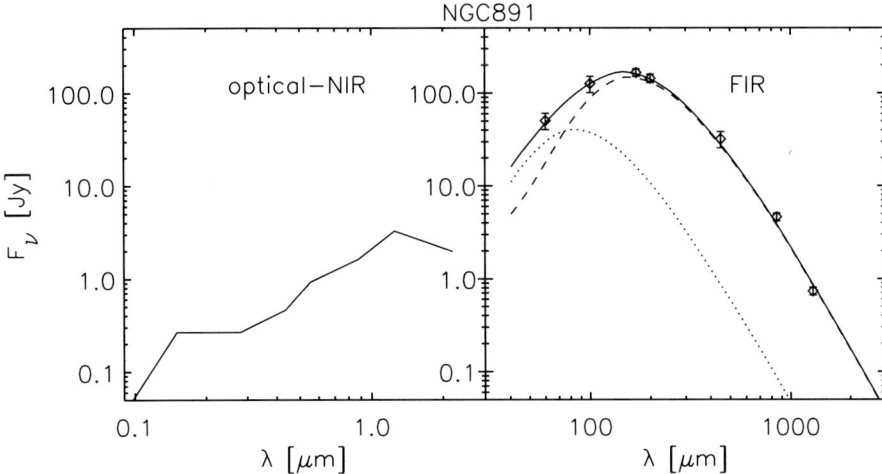

Figure 1: The predicted SED of NGC 891 from the "two-dust-disk" model with $SFR = 3.8\,\mathrm{M_\odot/yr}$, $F = 0.22$ and $M_{\mathrm{dust}} = 7 \times 10^7\,\mathrm{M_\odot}$ in the second disk of dust. LH panel: the intrinsic emitted stellar radiation (as would have been observed in the absence of dust). RH panel: the re-radiated dust emission, with diffuse and H II components plotted as dashed and dotted lines, respectively. The data (integrated over $\pm 225''$ in longitude), are from Alton et. al. 1998 (at 60, 100, 450 & 850 μm), Guélin et al. 1993 (at 1300 μm) and from Popescu et al. (2001) (at 170 & 200 μm).

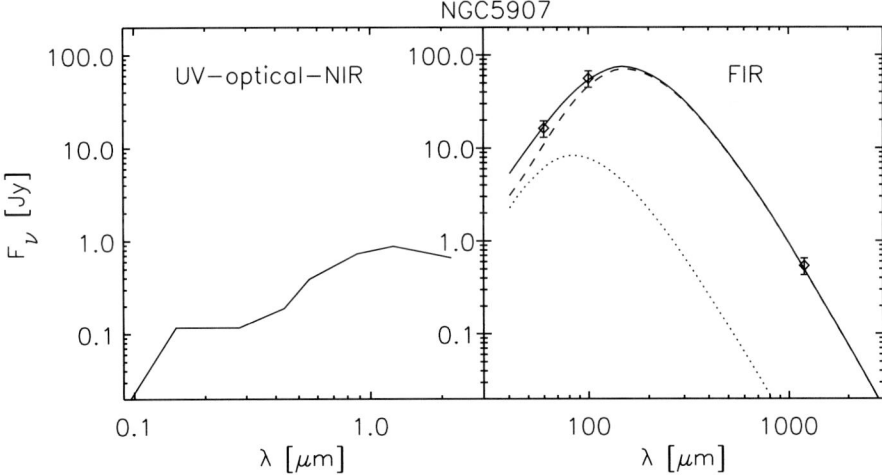

Figure 2: The predicted SED of NGC 5907 from the "two-dust-disk" model with $SFR = 2.2\,\mathrm{M_\odot/yr}$, $F = 0.10$ and $M_{\mathrm{dust}} = 4.5 \times 10^7\,\mathrm{M_\odot}$ in the second disk of dust. The legend is as in Fig. 1. The data are from Young et al. (1989) (at 60 & 100 μm), and from Dumke et al. (1997) (at 1200 μm).

the diffuse radiation field. Also, if the galaxy has a thin (young) stellar/dust disk, highly obscured by the dust lane in the plane of the galaxy, then this disk cannot be inferred from observations in the optical/NIR spectral range. As described in the previous section, our model makes use of the information in the FIR-submm regime to constrain this problem.

We first applied the above method to the well-known edge-on spiral galaxy NGC 891. This is one of the most extensively observed edge-on galaxies in the nearby universe, which makes it ideal for a verification of our modelling technique. We have also extended our SED modelling technique to four additional edge-on systems – NGC 5907, NGC 4013, UGC 1082 and UGC 2048 – with the aim of examining whether the features of the solution we obtained for NGC 891 might be more generally applicable. In this section we mainly show and discuss the results for NGC 891, and only briefly illustrate the solution for NGC 5907.

The "two-dust-disk" model can successfully fit the shape of the SED for both NGC 891 and NGC 5907. The best solution for NGC 891 (Fig. 1) has a central face-on V-band optical thickness $\tau_v^f = 3.1$ and a corresponding non-ionising UV luminosity $\sim 8.2 \times 10^{36}$ W. The luminosity of the diffuse component is 4.07×10^{36} W, which accounts for 69 % of the observed FIR luminosity, and the luminosity of the H II component is 1.82×10^{36} W, making up the remaining 31 % of the FIR luminosity. The best solution for NGC 5907 (Fig. 2) has a central face-on optical depth in the optical band $\tau_v = 1.4$. The total FIR-submm re-radiated luminosity of NGC 5907, obtained by integrating the "two-dust-disk" model SED, is 50.5×10^{35} W out of which 27.0×10^{35} W is attributed to heating from the young stellar population. Thus, about 40 % of the dust emission is powered by the old stellar population. The major difference between NGC 891 and NGC 5907, on the basis of the "two-dust-disk" model, is that the spectrum of the former apparently allows for the existence of a larger contribution from H II regions (see Misiriotis et al. 2001 for a detailed discussion) – F takes values of 0.22 and 0.10, respectively. Such small values of F are expected for "normal" galaxies, in contrast to starburst systems, where the FIR/submm SEDs peak shortwards of 100 μm, and one would anticipate that F would be closer to unity.

In both cases most of the luminosity comes from the diffuse component, and the main heating source is provided by the young stellar population. The relative contribution of optical and UV photons in heating the dust has been a longstanding question in the literature. Since we have a detailed calculation of the absorbed energy over the whole spectral range and at each position in the galaxy, we can directly calculate which part of the emitted FIR luminosity from each volume element of the galaxy is due to the optical and NIR photons, and which part is due to the UV photons. In this way we can also predict the contribution of different stellar populations in heating the dust as a function of FIR wavelength. Volume-integrated IR spectral components arising from re-radiated optical and UV light are presented in Figs. 3, 4 for the case of NGC 891. We note that the diffuse optical radiation field makes only a relatively small contribution to the total emitted dust luminosity. This is in

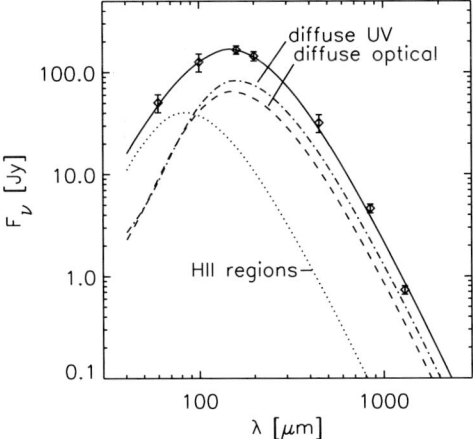

Figure 3: The absolute contribution of the three stellar components to the FIR emission of NGC 891 versus wavelength for the "two-dust-disk" model. Dashed-line: diffuse optical radiation (4000–22000 Å); dashed-dotted line: diffuse UV radiation (912–4000 Å); dotted-line: H II regions. The total predicted FIR SED is given by the solid line. The data points are as for Fig. 1.

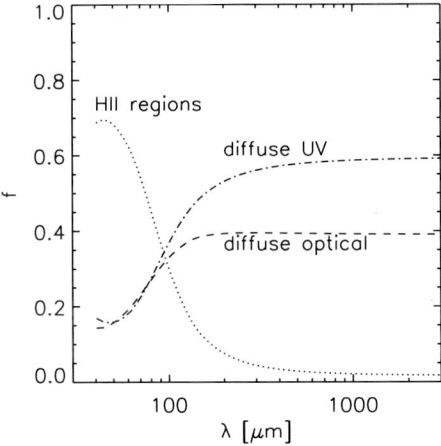

Figure 4: The fractional contribution of the three stellar components to the FIR emission of NGC 891 versus wavelength for the "two-dust-disk" model. The legend is as for Fig. 3 and the data points are as for Fig. 1.

qualitative agreement with various statistical inferences linking FIR emission with young stellar populations, in particular the FIR-radio correlation. Our analysis predicts the predominance of UV-powered grain emission even in the submm range, which, in turn, would predict a tighter FIR-radio correlation when the FIR luminosity integrated over the FIR-submm range is considered.

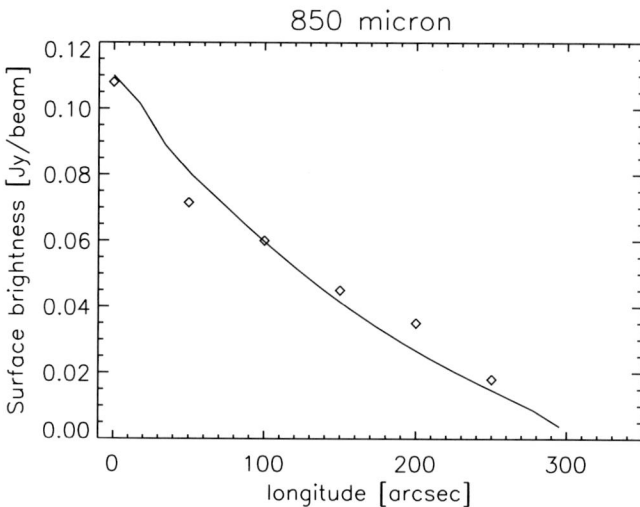

Figure 5: The averaged radial profile of NGC 891 at 850 μm from the diffuse component of the "two-dust-disk" model, plotted with the solid line. The profile is averaged over a bin width of 36″, for a sampling of 3″ and for a beam width of 16″, in the same way as the observed averaged radial profile from Alton et al. (2000) (plotted with diamonds).

This prediction has been recently confirmed by Popescu et al. (2002), using the new ISOPHOT observations of a complete sample of late-type Virgo cluster galaxies (Tuffs et al. 2002).

A more stringent test of the model is to compare its predictions for the morphology of the dust emission with spatially resolved maps. Because observed radial profiles were derived by Alton et al. (2000) using the SCUBA observations at 850 μm, we first attempt to calculate the radial profiles at this wavelength and compare it with the observations (Fig. 5), which are mainly sensitive to dust column density. We have found that in the case of the "two-dust-disk" model there is a very good agreement between the model predictions and the observations, where the observed profiles were mirrored for compatibility with the symmetry in our model. The predicted radial profile can be traced out to 300 arcsec radius (15 kpc), as also detected by the SCUBA.

Recent deep observations of NGC 891 with the ISOPHOT instrument at 170 and 200 μm (Popescu et al. 2001) offer a still stronger test of the model, as the FIR emission here depends on the distribution of both stellar luminosity and dust.

Simulated FIR maps were obtained using the actual pointing data to scan the diffuse disk model. The model map was then convolved with empirical PSFs derived from point source measurements. The comparison between the observed map (Fig. 6a) and the simulated one (Fig. 6b) show a remarkable

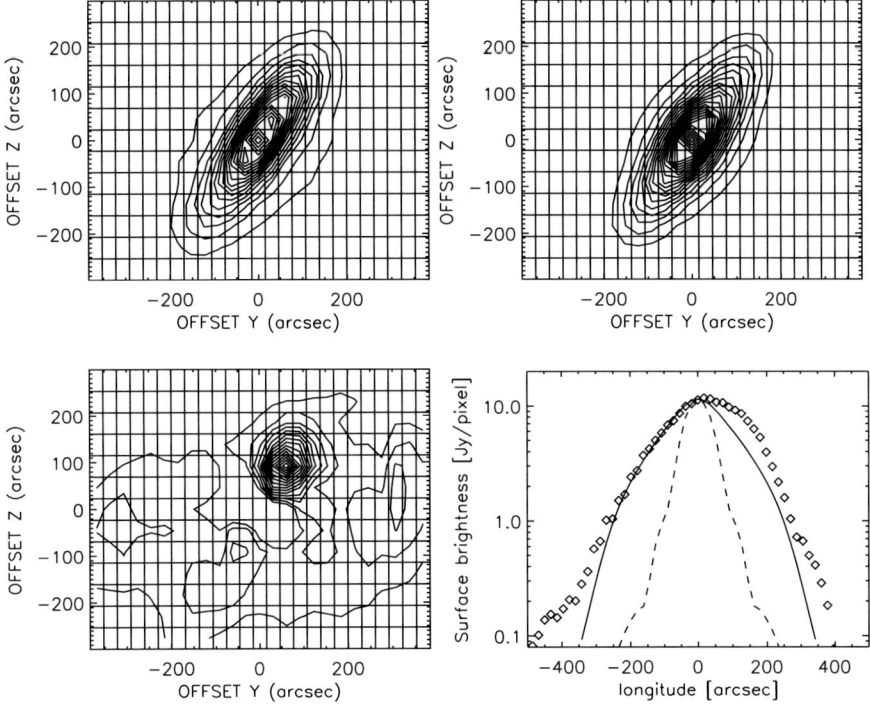

Figure 6:
a) Contour plot of the observed brightness distribution at 170 μm. The contours are plotted from 10.7 to 230.0 MJy/sr in steps of 12.2 MJy/sr. The grid indicates the actual measured sky positions sampled at 31 × 46 arcsec in the spacecraft coordinates Y and Z.
b) Contour plot of the simulated diffuse brightness distribution at 170 μm. The contours are plotted from 10.4 to 239.7 MJy/sr in steps of 10.4 MJy/sr.
c) Contour plot of the observed minus simulated diffuse brightness distribution at 170 μm. The contours are plotted from 2.2 to 46.1 MJy/sr in steps of 3.1 MJy/sr. The unresolved source from the Northern side of the galaxy and the faint extended source from the Southern side account for 8 % of the total integrated flux density, in agreement with our model prediction for the FIR localised sources at this wavelength.
d) The longitudinal profile (diamonds) observed by ISO at 170 μm (Popescu et al. 2001), with a common bin width and sampling interval of 18.4″. North is towards positive longitude. The solid line is the prediction from the diffuse component of the "two-dust-disk" model and the dashed line the projected beam profile (FWHM 1.8 arcmin).

agreement. To search for small differences between the model and the observations, not detectable in the maps due to the high dynamical range of the displayed data, we present in Fig. 6c the residual map of the difference between the observed and the simulated map. The main feature in the residuals is a localised, unresolved source in the Northern side of the disk, with a peak of 46.1 MJy/sr. This localised source is probably a giant molecular cloud complex – associated with one of the spiral arms, and not considered in the simulated map, which only includes the diffuse component of the model. At this FIR wavelength our model predicts a 11 % contribution from the star-forming complexes. The integration of the unresolved source gives a flux density of 9.4 Jy, which is 6 % of the total integrated flux density. Furthermore a faint extended source is seen in the southern side of the galaxy, of 4.0 Jy integrated flux density. This makes another 2 % of the total integrated emission. Thus the faint localised sources seen in the residual maps are in reasonable agreement with the prediction of our model, which reassures us that the template used in our model and scaled according to our model parameters, SFR and the F factor, are indeed a good representation for the galaxy. Apart from the two sources, a faint extended halo (at $\sim$1 % brightness level) is seen extending at large heights perpendicular to the disk. The interpretation of this halo component is beyond the scope of this paper.

A final comparison can be made between the projected radial profiles of the simulated and observed maps. Due to the larger longitudinal coverage of the ISO data, which embraces the outer, asymmetrical H I disk (Swaters et al. 1997), this time we did not mirror the observed profile. Again, a very good agreement between the model prediction and the observed profiles can be seen in Fig. 6d for the Southern side of the galaxy. The localised FIR source from the northern side of the residual map becomes prominent in the radial profiles as well. There also seems to be an excess of FIR emission at radii larger than 300 $''$, not reproduced by our model. We interpret this result as indicative of a dust disk larger than considered by our model, in which all dust disks are truncated at three scale lengths of the B-band stellar disk. A finer grid of models with varying truncation of the scale length may be needed to reproduce the faint FIR emission at large galactocentric radii.

5   The SFRs derived from SED modelling

To statistically evaluate our results for SFR, we compare the SFR characteristics of the 5 edge-on galaxies modelled by us with the larger sample of 61 galaxies with inclinations less than 75 degrees which were studied by Kennicutt (1998) on the basis of H_α measurements.

Fig. 7 depicts the relation between the disk-averaged surface density in SFR (Σ_{SFR}) as a function of the average gas surface density (Σ_g) for the 5 galaxies. The upper and lower limits for the SFRs are calculated such that the predicted SED is still consistent with the IRAS colours (within the 20 % IRAS error bars). Lower error bars are not given when the plotted SFRs

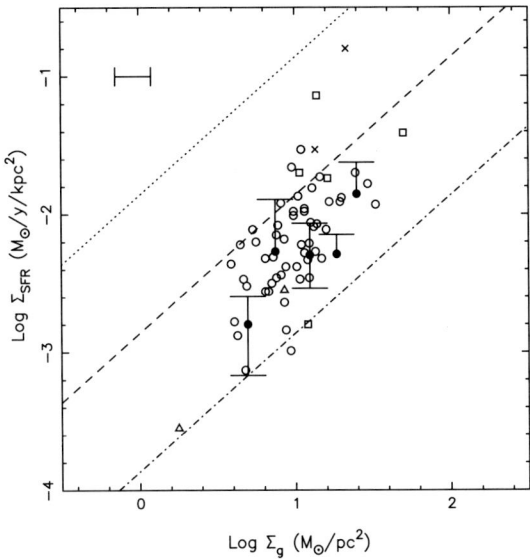

Figure 7: Disk-averaged SFR surface density (Σ_{SFR}) as a function of average gas surface density (Σ_g) for our galaxy sample and for the sample of Kennicutt (1998) of 61 normal disk galaxies with SFR determined from H_α measurements. The 5 galaxies from our sample are plotted as filled circles and the corresponding Σ_{SFR}, Σ_{SFR}^{min}, Σ_{SFR}^{max}, Σ_g and disk areas are listed in Table 3 of Misiriotis et al. (2001). The Σ_{SFR} are derived from the errors in the IRAS data, e. g. the upper and lower limits for the SFRs are calculated such that the predicted SED is still consistent with the IRAS colours (within the 20 % IRAS error bars). The SFR surface densities were calculated by averaging the SFRs over disks with an optically defined boundary (R_o) taken to be 3 times the intrinsic radial scalelength h_s determined from the radiation transfer modelling in the I band. The galaxies from the sample of Kennicutt are plotted as open circles (Sb, Sc, SBb, SBc), triangles (Sa), open squares (Unknown/Not Available), crosses (Irr). The dotted, dashed and dot-dashed lines represent star-formation efficiencies corresponding to consumptions of 100, 10 and 1 percent of the gas mass in 10^8 yr.

represent lower limits for the SFR (maximum values for the factors F, see discussion in Sect. 3 of Misiriotis et al. 2001). In these cases lower limits would be possible only if we allowed for different sources of uncertainties, like variations in the spectral shape of the template used for the H II regions. However this is hard to quantify, and in the following we assume that the errors of the SFR are given only by the uncertainties in the IRAS data. The plotted data are summarised in Table 3 of Misiriotis et al. (2001) and details on the calculations of the error bars and of the gas masses can be found in the same paper. The horizontal error bar corresponds to the uncertainty in the gas masses for which we adopted an average 0.2 dex error. The surface area of the disk was calculated for $R_o = (3 \pm 0.5)h_d$, where h_d is the intrinsic radial scalelength determined from the radiation-transfer modelling in the I band. In their analysis of surface photometry of the outer regions of spiral disks, Pohlen et al. (2000) show that the disk boundaries are typically in this range.

The points for the 5 galaxies in Fig. 7 lie within the area of the diagram occupied by the galaxies in the Kennicutt sample. The match is even better for those members of the Kennicutt sample with Hubble types Sb to Sc. This agreement is quite reassuring, bearing in mind the several factors which could introduce a systematic difference between the SFRs inferred for a sample of nearly face-on systems from H_α measurements compared with the present technique for edge-on systems based on an analysis of broad-band non-ionising UV re-radiated in the FIR-submm range.

Firstly, the H_α analysis is sensitive to the most massive stars and in particular to the assumed mass cut of $100\,M_\odot$. Whereas the FIR-submm modelling also assumes the same mass cut in the conversion of SFR to non-ionising luminosity, our model is less sensitive to this effect.

Secondly, whereas the H_α is sensitive to the star-formation history of the last 10^7 yr, our broad-band FIR-submm SED analysis samples approximately the last 10^8 yr. Thus, our analysis is consistent with the basic hypothesis (see Kennicutt 1998) for "normal" spiral galaxies of a steady star-formation activity. In principle, we could extend our analysis based on our determinations of the intrinsic populations and use the determined intrinsic colours to determine more accurately the SFR history of the galaxies.

The assumption of a steady-state star-formation rate is also broadly consistent with the timescales for the exhaustion of the current gas supply under the derived SFRs. The dotted, dashed and dot-dashed lines in Fig. 7 represent star-formation efficiencies corresponding to consumptions of 100, 10 and 1 percent of the gas mass in 10^8 yr.

Thirdly, the SFRs derived from H_α were corrected by a single factor for extinction, despite the varying orientations. As well as possibly affecting the vertical position of the galaxies on the plot, this may induce some scatter, especially if all the dust were diffusely distributed. The systematic effect may be expressed in terms of the factor F: an overestimation of the factor F is equivalent to an overestimation of the local extinction in the star-formation regions (statistically averaged over the population of $H\,II$ regions in a disk). Thus, while moving to higher factors F would move the points for the 5 galaxies towards lower SFRs, it would have the opposite effect for the SFRs determined from the H_α.

Lastly, we remark that NGC 891 does not appear as an exceptional system compared with the other 4 galaxies in our sample (and with Kennicutt's [1998] normal galaxy sample) on the basis of SFR normalised to disk area. Our work thus provides no evidence that this galaxy's exceptional layer of extraplanar H_α-emitting diffuse ionising gas (e. g. Hoopes et al. 1999) and surrounding X-ray-emitting hot gas (e. g. Bregman & Houck 1997) is attributable to unusual star-formation activity.

6 Discussion and Outlook

We have described a "two-dust-disk" model which can successfully account for the observed optical-FIR/submm characteristics of "normal" edge-on spiral galaxies in terms of three fundamental parameters – the SFR, F – the fraction of non-ionising UV absorbed locally in H II regions, and M_{dust} – the mass of a second dust disk associated with the young stellar population. We found these parameters to play the key role in determining the amplitude and shape of the FIR-submm SED in galaxies. Some other parameters were given larger attention in the literature, like for example clumpiness, different grain populations, metallicity.

Clumpiness has been extensively discussed by Witt & Gordon (1996, 2000) and used by Gordon et al. (2001) and Matthews & Wood (2001). However, the clumpiness in the interstellar medium is difficult to relate to observations. For example the optical properties of grains in clouds and fundamental parameters such as cloud size and density contrast are poorly known. Because of such uncertainties in the parameterisation of clumpy media, our model does not explicitly incorporate clumps, though our radiative transfer code can treat any arbitrary distribution of emitters and absorbers. However, our model considers implicitly the clumpy distribution in form of the spectral template used for the star-forming complexes. Thus, our model does not include clumps that are not associated with stellar sources, the so called "quiescent clouds", but does include clouds associated with star-forming regions. As discussed by Popescu et al. (2000), the quiescent clumps must be optically thick to the diffuse UV/optical radiation field in the disk to have an impact on the predicted submm emission. They would then radiate at predominantly longer wavelengths than the diffuse disk emission in the FIR/submm spectral range. One can speculate that such optically thick "quiescent clouds" could be physically identified with partially or wholly collapsed clouds that, for lack of a trigger, have not (yet) begun to form stars. However, due to the lack of intrinsic sources, a very substantial mass of dust would have to be associated with the quiescent clouds to account for a substantial change in the submm SED of galaxies. On the other hand, there may be also dark clouds associated with star-forming complexes. In the Milky Way H II regions around newly born massive stars are commonly seen in juxtaposition to parent molecular clouds (e. g., M17). This is thought to be a consequence of the fragmentation of the clouds due to mechanical energy input from the winds of the massive stars. Thus, warm dust emission from cloud surfaces directly illuminated by massive stars can be seen along a fraction of the lines of sight, together with cold submm dust emission from the interior of the associated optically thick cloud fragments. These dense clumps (with or without associated sources) may contribute to the submm emission, and thus supplement the contribution of compact H II regions. Our template for localised sources was successful in reproducing the FIR-submm surface brightness distribution for the prototypical galaxy NGC 891. However, modelling of more galaxies may be needed to check whether our template is a good representation for localised sources, or whether it needs to be supplemented with a submm cold component of dark clouds. More detailed FIR-submm observations of star-forming complexes may also help in improving our H II-region templates.

Different dust populations can produce different types of extinction curves and change the colour of the FIR SED. Our model used the Laor & Draine (1993) silicate/graphite mix, compatible with a Milky Way extinction law. Our choice was based on the agreement between our model extinction curve and the empirical extinction law derived for the studied galaxies independently by Xilouris et al. (1997, 1998, 1999) via the radiative transfer modelling of the optical images. In the case that different galaxies would require different extinction laws, we would adopt a different dust model (chemical composition, grain size distribution).

Finally the metallicity was fixed in our model to be the solar metallicity. Again, this choice was confirmed by the agreement in the extinction curves. A different metallicity may also affect the redistribution of the UV photons with UV wavelengths, as given by the stellar population synthesis models.

In the SED model presented here we did not consider any interactions or effects of inflows/outflows in galaxies. Observationally there is however increased evidence for Hα extraplanar emission in galaxies (e. g. Hoopes et al. 1999, Howk & Savage 2000), of X-ray (e. g. Bregman & Houck 1997) and radio halos (e. g. Allen et al. 1978), of galactic (starburst-driven) winds (e. g. Heckman et al. 2000) or for enhanced FIR emission in interacting/merging systems (e. g. Sanders et al. 1998). A discussion of the possible contamination of the FIR disk emission from an extraplanar component has been qualitatively discussed in Popescu et al. (2000b). Moreover, the effect of the cluster environment can also affect the morphology of both stellar and dust distribution, or can produce even more dramatic effects, for example by sweeping the gas and dust material and producing tail-like companions with collisionally heated grains visible in the FIR. As noted by Popescu et al. (2000a), the FIR flux density of the transient IR emission from the dust trail is predicted to rival that of the photon-heated dust in the galactic disk, and, despite the difference in heating mechanism, have similar colours (with a spectral peak in the 100–200 μm range). This, combined with removal of dust from the disk, and hence reduced internal extinction, will create a discrete system with brighter apparent blue magnitudes and a boosted spatially integrated IR flux density. If seen in a distant cluster, where the intracluster IR component could not be resolved from the disk component, this could create the illusion of a galaxy with an enhanced star-formation activity, even though the star-formation in the galaxy may actually be somewhat suppressed by the gas removal in reality.

Our model has been applied only to edge-on galaxies, where the scale heights of the old stellar population and old dust disk can be directly determined. However our model will also be applicable to face-on systems where the scale heights cannot be so directly determined, making use of UV data as an additional constraint. Although our model requires resolved optical/NIR images to constrain the old stellar population and associated dust, it relies on geometry-sensitive colour information in the FIR/submm to constrain the spatial distributions of young stars and associated dust. The model will therefore be applicable to studies of cosmologically distant "normal" galaxies, which, though detectable, will be unresolved with forthcoming generations of space-borne FIR observatories. It is to be expected that the optical-FIR-submm SEDs of these objects will differ systematically from their local universe counterparts, not only due to the presence of younger stellar populations, but also, for example, because of evolution of stellar disk thicknesses and changes in the dust abundance and composition.

SED modelling from UV to submm has an important application to interpretation of the cosmological FIR background radiation in terms of constituent galaxies. This particularly applies to the possible contribution from normal galaxies in the early universe, since these are too faint to be observed directly. The development of SED models incorporating a self consistent physical theory for the evolution of these systems may ultimately be needed for this application.

Acknowledgements

C. Popescu would like to express her gratitude to the organisers of the JENAM 2001 and AG meeting, and especially to Dr. Magdalena Stavinschi, for their invitation to give a Highlight Talk at the meeting. We are grateful to all collaborators in this project, N. Kylafis, A. Misiriotis, J. Fischera and B. F. Madore. We would also like to acknowledge E. M. Xilouris and J. Gallagher for interesting and constructive discussions.

References

Allen, R. J., Baldwin, J. E., Sancisi, R. 1978, A&A 62, 397

Alton, P. B., Bianchi, S., Rand, R. J., et al. 1998, ApJ 507, L125

Alton, P. B., Rand, R. J., Xilouris, E. M., et al. 2000, A&A 356, 795

Bekki, K. & Shioya, Y. 2000, ApJ 542, 201

Bianchi, S., Davies, J. I., & Alton, P. B. 2000, A&A 359, 65

Bregman, J. N. & Houck, J. C. 1997, ApJ 485, 159

Bruzual, A. G. & Charlot, S. 2001, in preparation

Dale, D. A., Helou, G., Contursi, A., Silbermann, N. A., & Kolhatkar, S. 2001, ApJ 549, 215

Devriendt, J. E. G., Guiderdoni, B., & Sadat, R. 1999, A&A 350, 381

Désert, F. X., Boulanger, F., & Puget, J. L. 1990, A&A 237, 215

Draine, B. T. & Lee, H. M. 1984, ApJ 285, 89

Draine, B. T. & Anderson, N. 1985, ApJ 292, 494

Dwek, E. 1986, ApJ 302, 363

Efstathiou, A., Rowan-Robinson, M., & Siebenmorgen, R. 2000, MNRAS 313, 375

Ferrara, A., Bianchi, S., Cimatti, A., & Giovanardi, C. 1999, ApJS 123, 437

Gordon, K. D., Misselt, K. A., Witt, A. N., & Clayton, G. C. 2001, ApJ 551, 269

Guhathakurta, P. & Draine, B. T. 1989, ApJ 345, 230

Guélin, M., Zylka, R., Mezger, P. G., et al. 1993, A&A 279, L37

Heckman, T. M., Lehnert, M. D., Strickland, D. K., & Armus, L. 2000, ApJS 129, 493

Hoopes, C. G., Walterbos, R. A. M., & Rand R. J. 1999, ApJ 522, 669

Howk, J. C. & Savage, B. D. 2000, AJ 119, 644

Kennicutt, R. C. Jr. 1998, ApJ 498, 541

Kuchinski, L. E., Terndrup, D. M., Gordon, K. D., Witt, A. N. 1998, AJ 115, 1438

Kylafis, N. D. & Bahcall, J. N. 1987, ApJ 317, 637

Laor, A. & Draine B. T. 1993, ApJ 402, 441

Mathis, J. S., Rumple, W., & Nordsieck, K. H. 1977, ApJ 217, 425

Matthews, L. D. & Wood, K. 2001, ApJ 548, 150

Misiriotis, A., Kylafis, N. D., Papamastorakis, J., & Xilouris, E. M. 2000, A&A 353, 117

Misiriotis, A., Popescu, C. C., Tuffs R. J., & Kylafis, D. 2001, A&A 372, 775

Ohta, K. & Kodaira, K. 1995, PASJ 47, 17

Pohlen, M., Dettmar, R.-J., Lutticke, R. 2000, A&A 357, L1

Popescu, C. C., Tuffs, R. J., Fischera, J., & Völk, H. J. 2000a, A&A 354, 480

Popescu, C. C., Misiriotis, A., Kylafis, N. D., Tuffs, R. J., & Fischera, J. 2000b, A&A 362, 138

Popescu, C. C., Madore, B. F., Tuffs, R. J., & Kylafis, N. D. 2001, AAS 198, 76.01

Popescu, C. C., Tuffs, R. J., Völk, H. J., Pierini, D., & Madore, B. F. 2002, ApJ (in press)

Sanders, D. B., Soifer, B. T., Elias, J. H., et al. 1998, ApJ 325, 74

Silva, L., Granato, G. L., Bressan, A., & Danese, L. 1998, ApJ 509, 103

Swaters, R. A., Sancisi, R., & van der Hulst, J. M. 1997, ApJ 491, 140

Tuffs, R. J., Popescu, C. C., Pierini, D., Völk, H. J., Hippelein, H., et al. 2002, ApJS (in press)

Weingartner, J. C. & Draine, B. T. 2001, ApJ 548, 296

Witt, A. N. & Gordon, K. G. 1996, ApJ 463, 681

Witt, A. N. & Gordon, K. G. 2000, ApJ 528, 799

Xilouris, E. M., Kylafis, N. D., Papamastorakis, J., Paleologou, E. V., & Haerendel, G. 1997, A&A 325, 135

Xilouris, E. M., Alton, P. B., Davies, J. I., et al. 1998, A&A 331, 894

Xilouris, E. M., Byun, Y. I., Kylafis, N. D., Paleologou, E. V., Papamastorakis, J. 1999, A&A 344, 868

Xu, C. & Buat, V. 1995, A&A 293, L65

Xu, C. & Helou, G. 1996, ApJ 456, 163

Young, J. S., Xie, S., Kenney, J. D. P., & Rice, W. L. 1989, ApJS 70, 699

ASTRONOMISCHE GESELLSCHAFT: Reviews in Modern Astronomy **15**, 259–277 (2002)

Nature of the Cosmic Infrared Background and Cosmic Star Formation History: Are Galaxies Shy?

David Elbaz

CEA Saclay/DAPNIA/Service d'Astrophysique
F-91191 Gif-sur-Yvette Cedex, France
delbaz@cea.fr

Abstract

New evidence have emerged during the very last years that intense star formation takes place in galaxies which do not appear as such when observed in the optical. A large fraction of the UV light radiated from young and massive stars must have been absorbed by dust and reemitted in the infrared to explain the strong cosmic background detected with the COBE satellite in the far infrared to sub-millimeter range. Nearly simultaneously deep images of the sky in the mid infrared, far infrared and sub-millimeter with the ISOCAM, ISOPHOT and SCUBA instruments have unveiled a strong redshift evolution of faint galaxies consistent with this strong cosmic infrared background. Because of its better spatial resolution and because it deals with relatively nearby galaxies (z < 1.3), the mid infrared camera ISOCAM onboard ISO revealed the largest fraction of galaxies responsible for the cosmic infrared background, i.e. for the bulk of the light radiated through star formation since galaxies formed. In this paper, we present these evidence which imply that galaxies were "shy" when they formed the bulk of their present-day stars, i.e. that they were hiding themselves behind a veil of dust which had the effect of making them appear red to us.

1 Introduction

Until 1996, there was little evidence that most galaxies were "shy", i.e. that they would hide their stars behind a veil of dust and turn red when forming stars, radiating the bulk of their luminosity in the infrared (IR) at a given epoch of their history. Ten years before, IRAS had unveiled a population of luminous IR galaxies exhibiting such a "shy" behavior, the so-called LIGs and ULIGs (with $12 \geq \log_{10}\left(L_{\mathrm{IR}}/L_{\odot}\right) \geq 11$ and $\log_{10}\left(L_{\mathrm{IR}}/L_{\odot}\right) \geq 12$ respectively, where $L_{\mathrm{IR}} = L[8\text{–}1000\,\mu\mathrm{m}]$), which are responsible for the shape of the bolometric luminosity function of local galaxies above $\sim 10^{11}\ L_\odot$ (Sanders & Mirabel 1996). But integrated over the whole local luminosity function, LIGs and ULIGs only produce $\sim 2\%$ of the total integrated luminosity and overall only $\sim 30\%$ of the bolometric luminosity of local galaxies is radiated in the IR above $\lambda \sim 5\,\mu$m. The discovery of an extragalactic background in the IR at least as large as the UV-optical-near IR one, the so-called cosmic infrared background (CIRB), with the COBE satellite (Puget et al. 1996, see references below) implied that shyness must have been more common among galaxies in the past than it is today. This was confirmed with the detection of an excess of faint mid IR (MIR) galaxies by ISOCAM onboard ISO (Elbaz et al. 1999), as well as in the far IR (FIR) with ISOPHOT onboard ISO (Dole et al. 2001) and in the sub-millimeter with SCUBA at the JCMT (Smail et al. 2001 and references therein). This excess is relative to expectations based on galaxies in the local universe. It implies that galaxies were more luminous in the IR regime and/or more numerous in the past (Chary & Elbaz 2001, Franceschini et al. 2001).

2   The cosmic infrared background (CIRB)

The extragalactic background light (EBL) is a measurement of the sum of the light produced by all extragalactic sources over cosmic time. When it is integrated over the full spectral range, the so-called cosmic background is a fossil record of the overall activity of all galaxies from their birth until now. It can be considered as the global energetic budget available for any model aiming at simulating the birth and fate of galaxies during the Hubble time. However the physical origin of this light will remain unknown until we have pinpointed the individual sources responsible for it.

The cosmic infrared background (CIRB) is the EBL integrated over all wavelengths within $\lambda = 5$ to $1000\,\mu$m. It was recently detected and measured thanks to the cosmic background explorer (COBE) instruments FIRAS (Far Infrared Absolute Spectrometer) and DIRBE (Diffuse Infrared Background Experiment) (Puget et al. 1996, Fixsen et al. 1998, Lagache et al. 1999, 2000, Hauser et al. 1998, Dwek et al. 1998, Finkbeiner et al. 2000) from $100\,\mu$m to 1 mm. It peaks around $\lambda_{max} \simeq 140\,\mu$m and was found to represent at least half and maybe two thirds of the overall cosmic background (see review of Dwek & Arendt 1998). Hence the CIRB reflects the bulk of the star formation that took place over the history of the universe. By resolving it into individual galaxies, we would therefore unveil the times and places where most stars seen in the local universe were formed. Two physical processes are considered for its origin: nucleosynthesis, i.e. stellar radiation in star forming galaxies, and accretion around a black hole, i.e. active galactic nuclei. In both cases, the light is not directly coming from its physical source but is reprocessed by dust, i.e. absorbed and re-radiated by the "warm" dust. Both processes are probably related (see Genzel et al. 1998), but energetic considerations, based on the presence of massive black holes and on the amount of heavy elements in local galaxies, suggest that star formation should by far dominate in the CIRB over AGN activity (Madau & Pozzetti 2000, Franceschini et al. 2001). However, until the individual galaxies responsible for the CIRB are found and studied in detail, this result will remain theoretical.

The spectral energy distribution (SED) in the IR of local galaxies peaks above $\sim 60\,\mu$m and typically around $80 \pm 20\,\mu$m (see Sanders & Mirabel 1996). As a result, the distant galaxies responsible for the peak of the CIRB detected by COBE around $\lambda_{\mathrm{max}} \sim 140\,\mu$m should be located below $z \sim 1.3$ and present a redshift distribution peaked around $z \sim 0.8$, if their SEDs do not strongly differ from those of local galaxies.

3 Galaxies in the mid to far infrared

Above a bolometric luminosity of a few $10^{10}\,L_\odot$, the SED of a galaxy is dominated by dust radiation with respect to direct stellar emission. Stellar light is absorbed by dust and re-radiated in the IR regime above $\lambda \sim 5\,\mu$m. The full SEDs, from the UV to the radio, of the starburst galaxy, M82, and of the ULIG, Arp 220, are plotted in the Fig. 1a. For comparison, we show on Fig. 1b a series of synthetic spectra built by Chary & Elbaz (2001) in order to fit the mid to far IR SEDs of local galaxies. The continuous evolution from the low to high luminosity end results from the correlations between the mid and far IR luminosities of local star forming galaxies (see Elbaz et al. 2002a and Chary & Elbaz 2001).

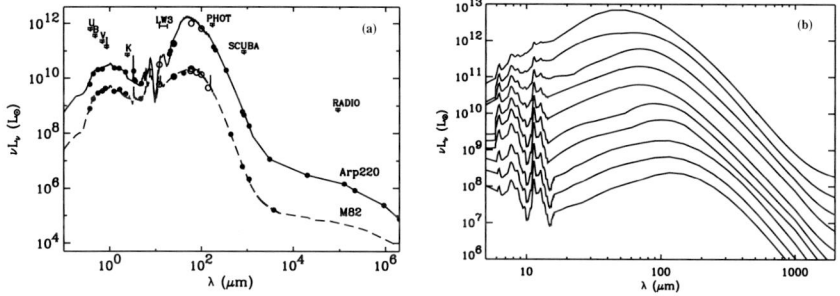

Figure 1: Spectral energy distribution of galaxies in the infrared. **a)** SEDs of M82 (dashed line) and Arp 220 (plain line) combining photometric (filled and open dots) and spectroscopic data (see Elbaz et al. 2002a). Open dots are used when they overlap with spectroscopic data. **b)** 10 SEDs with $\log_{10}(L_{\mathrm{IR}}/L_\odot) = 8.5$ to 13, with a step of 0.5, from the library of 105 SEDs of Chary & Elbaz (2001).

The MIR (5–40 μm) and FIR (40–1000 μm) emission of a galaxy combine three major components:

• Broad emission features and their associated underlying continuum, which dominate in the $\lambda \sim 5$–10 μm domain, and up to $\sim 20\,\mu$m for galaxies with moderate star formation. These bands located at 6.2, 7.7, 8.6, 11.3

and 12.7 μm, are usually quoted as PAHs (polycyclic aromatic hydrocarbons, Léger & Puget 1984, Puget & Léger 1989, Allamandola et al. 1989), although their exact nature remains uncertain, e. g. Jones & d'Hendecourt (2000).

• The "warm" dust continuum produced by very small dust grains (VSGs, smaller than 0.01 μm, Andriesse 1978, Désert, Boulanger & Puget 1990) heated at temperatures larger than 200 K without reaching thermal equilibrium. This component dominates the SED between 10–20 to 40 μm for luminous galaxies.

• The "cold" dust continuum produced by big dust grains ("big grains", of size > 0.01 μm) in the thermal equilibrium at cool temperatures (below 60–70 K typically). This component is responsible for the bulk of the FIR light where the SED peaks.

4   Deep surveys in the mid IR, far IR, and sub-millimeter

4.1   The contribution of IRAS

One of the major results from the IRAS satellite was to demonstrate that the bolometric luminosity of galaxies more luminous than $L_{bol} \sim 10^{11} \, L_\odot$ was underestimated by more than one order of magnitude before the IR luminosity was accounted for (Soifer et al. 1987). These luminous IR galaxies are not typical of the local galaxy population, since local galaxies radiate only about 30 % of their bolometric luminosity in the dust regime from 8 to 1000 μm and only $\sim 2\%$ of the local bolometric luminosity density is due to luminous IR galaxies. If we convert the IR luminosity into a star formation rate (SFR) using the formula of Kennicutt (1998),

$$\text{SFR}(\text{M}_\odot/\text{yr}) = 1.71 \times 10^{-10} \, L_{\text{IR}}[8-1000 \, \mu\text{m}](L_\odot) \tag{1}$$

we find that luminous IR galaxies ($L_{bol} \sim L_{\text{IR}} \geq 10^{11} \, L_\odot$) form stars at a rate larger than $\sim 20 \, M_\odot \, \text{yr}^{-1}$. Eq. (1) assumes continuous bursts lasting 10–100 Myr, solar abundance and a Salpeter IMF (Kennicutt 1998).

IRAS surveyed about 95 % of the sky down to a completeness limit of ~ 0.5 mJy at 60 μm (~ 1.5 Jy at 100 μm) and the differential counts are well fitted by an Euclidean slope, $dN/dS_\nu \sim S_\nu^{-2.5}$ (Soifer et al. 1987). The fluxes of all galaxies detected down to this sensitivity limit by IRAS add up to a 60 μm IGL of ~ 0.15 nW m^{-2} sr^{-1}. This is less than 1 % of the value of the CIRB measured by COBE at $\lambda_{max} \sim 140 \, \mu$m, $IGL_{140} = (25 \pm 7)$ nW m^{-2} sr^{-1}.

4.2   The contribution of ISOPHOT

Deeper FIR extragalactic surveys were performed with ISOPHOT (Lemke et al. 1996) onboard ISO at 170 μm. At this depth, the differential counts are no more fitted by an Euclidean slope, expected in the case of no evolution of galaxies with redshift, but instead a strong excess of faint sources was found. At the confusion limit of ISOPHOT of about 120 mJy, less than 10 % of the value of the CIRB at 170 μm is resolved into individual galaxies (Dole et al. 2001). A preliminary follow-up of the ISOPHOT galaxies suggests that a large fraction of these galaxies are located at low redshift.

4.3   The contribution of SCUBA

In the sub-millimeter range, the SCUBA bolometer array on the James Clerk Maxwell Telescope (JCMT) has produced deep images at 450 and 850 μm. Number counts at both wavelengths present a steep slope compatible with a strong evolution as compared to the local universe (see Smail et al. 2001 and references therein).

At 450 μm, a depth of 10 mJy is reached and the combined fluxes of all SCUBA galaxies produce about 15 % of the CIRB measured by COBE-FIRAS at this wavelength (Smail et al. 2001).

At 850 μm, SCUBA is confusion limited at $\sim$ 2 mJy (Hughes et al. 1998, Barger, Cowie & Sanders 1999, Eales et al. 2000, Smail et al. 2001), because of its large point spread function (PSF) of 15$''$ full width half maximum (FWHM). About 20 % of the value of the CIRB measured by COBE-FIRAS at 850 μm (the 850 μm EBL) is resolved into galaxies at this depth. However, using gravitational lensing this limit can be lowered to $\sim$ 1 mJy, where $\sim$ 60 % of the CIRB is resolved (Smail et al. 2001, Blain et al. 1999).

The 850 μm EBL measured by COBE-FIRAS ($\sim$ 0.5 $\pm$ 0.2 nW m^{-2} sr^{-1}) is 50 times lower than the peak value of the CIRB at $\lambda_{max} \sim$ 140 μm. In order to compute the contribution of SCUBA galaxies to the peak of the CIRB it is therefore necessary to determine their redshift distribution and SED. Until now, very few redshifts have been obtained due to the large PSF of SCUBA and optical faintness of these galaxies. Hence the fraction of the CIRB resolved into individual galaxies by SCUBA remains highly uncertain.

4.4   The contribution of ISOCAM

The 15 μm images of the ISOCAM camera (Cesarsky et al. 1996) onboard the Infrared Space Observatory (ISO, Kessler et al. 1996) have revealed sources one thousand times fainter than with IRAS, e. g. the Marano FIRBACK Deep Survey (Fig. 2).

The combination of a series of deep extragalactic surveys have revealed the presence of an excess of faint sources by one order of magnitude in comparison with predictions assuming no evolution of the 15 μm luminosity function with redshift (Elbaz et al. 1999, see Fig. 3). The presence of broad emission features in the MIR spectrum of galaxies alone cannot explain the shape of the number counts and a strong evolution of either the whole luminosity function (Xu 2000, Chary & Elbaz 2001) or preferentially of a sub-population of starburst galaxies evolving both in luminosity and density (Franceschini et al. 2001,

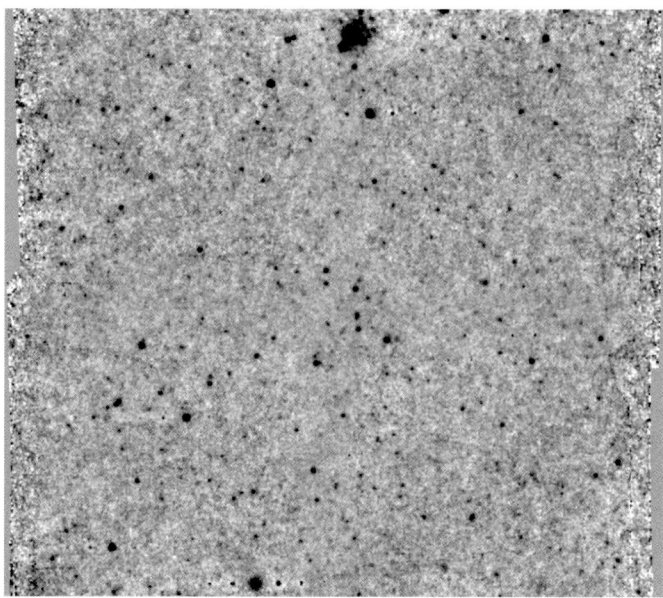

Figure 2: ISOCAM 15 μm image of the Marano FIRBACK Deep Survey used to produced number counts down to 0.4 mJy (Fig. 3).

Chary & Elbaz 2001, Xu et al. 2001) is required in order to fit the ISOCAM 15 μm counts.

Above the Earth's atmosphere, the 15 μm light is strongly dominated by the zodiacal emission from interplanetary dust and it has not yet been possible to make a direct measurement of the 15 μm background, or EBL. Individual galaxies contribute to this background and a lower limit to the 15 μm EBL can be obtained by adding up the fluxes of all ISOCAM galaxies detected per unit area down to a given flux limit. The resulting value is called the 15 μm integrated galaxy light (IGL). Below a 15 μm flux density of $\sim$ 3 mJy, about 600 galaxies were used to produce the counts in Fig. 3a. The 15 μm IGL does not converge above a sensitivity limit of $S_{15} \sim 50\,\mu$Jy, but the flattening of the curve below $S_{15} \sim 0.4$ mJy suggests that most of the 15 μm EBL should arise from the galaxies already unveiled by ISOCAM.

Above the completeness limit of $S_{15} \sim 50\ \mu$Jy, we computed an IGL of:

$$IGL_{15}(S_{15} \geq 50\,\mu\text{Jy}) = 2.4 \pm 0.5 \text{ nW m}^{-2} \text{ sr}^{-1} \tag{2}$$

where the error bar corresponds to the 68 % confidence level (i.e. 1-σ). The error bar combines the uncertainty on the measurements for each individual survey with cosmic variance. Each flux bin is covered by two to three independent surveys.

This result is consistent with the upper limit on the 15 μm EBL estimated by Stanev & Franceschini (1998) of:

$$EBL^{max}(15\,\mu\text{m}) \sim 5 \text{ nW m}^{-2} \text{ sr}^{-1} \tag{3}$$

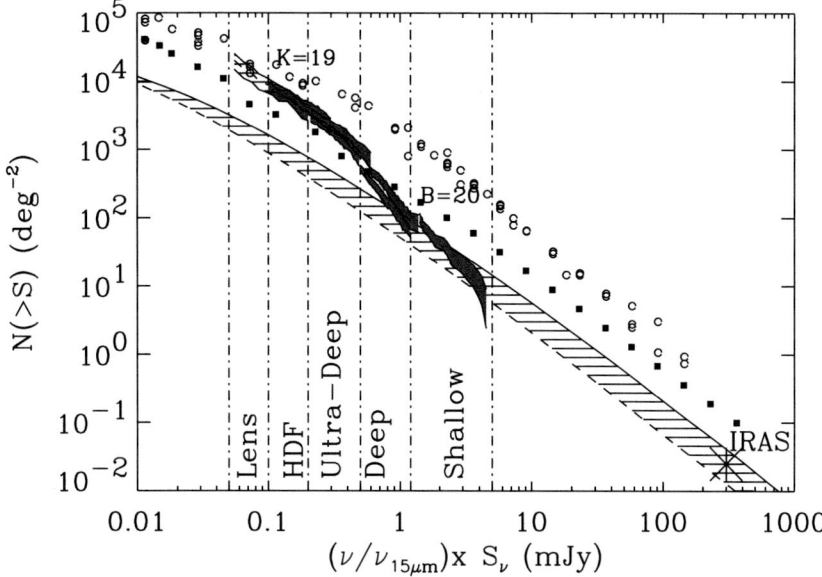

Figure 3: Integral counts, i. e. the number of galaxies, N, detected at $15\,\mu m$ above the flux S (mJy), with 68 % confidence contours (from Elbaz et al. 1999). K counts (Gardner et al. 1993) and B counts (Metcalfe et al. 1995), multiplied by the ratio ν/ν_{15} to represent the relative energy densities at high fluxes, are overplotted with open circles and filled squares, respectively. The hatched area materializes the range of possible expectations from models assuming no evolution and normalized to the IRAS $12\,\mu m$ local luminosity function (LLF). The upper limit was calculated on the basis of the LLF of Rush, Malkan & Spinoglio (1993), as in Xu et al. (1998) and shifted from 12 to $15\,\mu m$ with the template SED of M82; the lower limit uses the LLF of Fang et al. (1998) and the template SED of M51.

This upper limit was computed from the 1997 γ-ray outburst of the blazar Mkn 501 ($z = 0.034$) as a result of the opacity of MIR photons to γ-ray photons, which annihilate with them through e^+e^- pair production. It was recently confirmed by Renault et al. (2001), who found an upper limit of 4.7 nW m^{-2} sr^{-1} from 5 to $15\,\mu m$.

The models of Franceschini et al. (2001) and Chary & Elbaz (2001), which reproduce the number counts from ISOCAM at $15\,\mu m$, from ISOPHOT at 90 and $170\,\mu m$ and from SCUBA at $850\,\mu m$, as well as the shape of the CIRB from 100 to $1000\,\mu m$, consistently predict a $15\,\mu m$ EBL of:

$$EBL^{models}(15\,\mu m) \sim 3.3 \text{ nW m}^{-2}\text{ sr}^{-1}. \tag{4}$$

If this prediction from the models is correct then about 73 ± 15 % of the $15\,\mu m$ EBL is resolved into individual galaxies by the ISOCAM surveys.

5 Relative sensitivities of existing surveys to dusty starbursts

The deepest ISOCAM surveys reach a completeness limit of $\sim 100\,\mu$Jy at $15\,\mu$m (without lensing, but this limit goes down to ~ 30–$50\,\mu$Jy including lensing, see Altieri et al. 1999). For a given galaxy SED and redshift, this flux density can be converted into a bolometric IR luminosity ($L_{IR}[8$–$1000\,\mu$m$]$). There is no a priori knowledge of the exact SED of the relatively distant ($z \sim 1$) ISOCAM galaxies but the fact that local galaxies exhibit a correlation between their mid and far IR luminosities (Elbaz et al. 2002, Chary & Elbaz 2001) implies that this correlation should be valid up to a given redshift at least. Indeed we will see in Section 7 that the MIR and radio luminosities of these galaxies predict a consistent bolometric IR luminosity, L_{IR}. Any galaxy more luminous than this L_{IR} will be detected in the survey. In order to compare the sensitivity to distant luminous IR galaxies of ISOCAM, ISOPHOT and SCUBA, we have calculated L_{IR} corresponding to their sensitivity limits for a given redshift between $z = 0$ and 3.

On one hand, we used the SED of M82 (Fig. 1a) for galaxies of all luminosities. It turns out that although M82's IR luminosity is small ($L_{IR} = 4 \times 10^{10}\,L_\odot$) as compared to the ISOCAM galaxies (LIGs and ULIGs mainly), the shape of its IR SED is typical of local LIGs.

On the other hand, we used a library of template SEDs from Chary & Elbaz (2001, cf. Fig. 1b). In the following, we will call it the "multi-template" technique. This library of 105 SEDs was constructed under the constraint that it reproduces the correlations observed between the 6.75, 12, 15, 60, 100 and $850\,\mu$m luminosities of local galaxies. The SEDs were interpolated from a sample of observed galaxies SEDs, including MIR spectra obtained with the ISOCAM Circular Variable Filter. They cover the luminosity range $\log_{10}(L_{IR}/L_\odot) = 8.5$ to 13.5.

The radio continuum ($\lambda = 21$ cm, $\nu = 1.4$ GHz) is also a tracer of L_{IR} because of the tight correlation between both luminosities in the local universe (see Yun, Reddy & Condon 2001 and references therein). If this correlation remains valid in the distant universe then we can translate the sensitivity limit of the deepest radio surveys into a minimum L_{IR} accessible at a given redshift assuming a radio SED. The correlation between the radio and FIR luminosities is usually described by the "q" parameter (Condon et al. 1991):

$$q = \log_{10}\left(\frac{L_{FIR}(W)}{3.75 \times 10^{12}(Hz)} \times \frac{1}{L_{1.4GHz}(W\ Hz^{-1})}\right) \qquad (5)$$

where $L_{1.4GHz}$ is the monochromatic luminosity at 1.4 GHz and L_{FIR} is the FIR luminosity between 40 and $120\,\mu$m, as defined by Helou et al. (1988),

$$L_{FIR} = 1.26 \times 10^{-14}\ (2.58\ S_{60} + S_{100}) \times 4\pi d(m)^2 \qquad (6)$$

where S_{60} and S_{100} are the IRAS flux densities at 60 and $100\,\mu$m in Jy. The relationship between $L_{FIR}[40$–$120\,\mu$m$]$ and $L_{IR}[8$–$1000\,\mu$m$]$ was computed with the sample of 300 galaxies of the Bright Galaxy Sample (BGS, Soifer et al. 1987) detected in the four IRAS bands (12, 25, 60 and 100 μm):

$$L_{IR} = (1.91 \pm 0.17) \times L_{FIR} \qquad (7)$$

L_{IR} is defined as (Sanders & Mirabel 1996):

$$L_{IR} = [13.48\ S_{12} + 5.16\ S_{25} + 2.58\ S_{60} + S_{100}] \\ \times 1.8 \times 10^{-14} \times 4\pi d(m)^2 \qquad (8)$$

where S_{12} and S_{25} are the IRAS flux densities at 12 and 25 μm in Jy. Yun, Reddy & Condon (2001) measured a value of $q = 2.34 \pm 0.01$ from a flux limited ($S_{60} \geq 2$ Jy) sample of 1809 IRAS galaxies.

The deepest radio surveys presently available were performed in the Hubble Deep Field North (HDFN) by Richards (2000) and Garrett et al. (2000) with the VLA and WSRT respectively. Both surveys reach the same sensitivity limit of 40 μJy at 1.4 GHz (5-σ). In order to convert this flux density into a $L_{1.4GHz}$ for a given redshift, we used a power index of $\alpha = 0.8 \pm 0.15$ ($S_\nu \propto \nu^{-\alpha}$), as suggested by Yun, Reddy & Condon (2001) for starburst galaxies.

Fig. 4 presents the sensitivity of the deepest MIR, FIR, sub-millimeter and radio surveys in the form of L_{IR} (or SFR) as a function of redshift. The sensitivity limits used for ISOCAM, ISOPHOT and SCUBA are taken from the deepest existing surveys at these wavelengths, as previously described. Fig. 4a (with M82) and Fig. 4b ("multi-template" technique) both clearly show that the faintest IR galaxies are best detected at 15 μm up to $z \sim 1.3$. The right axis of the plots shows the corresponding minimum SFR that a galaxy must harbor in order to be detected at a given redshift (abscissa) in the surveys. Up to $z \sim 1$, ISOCAM (plain line) detects all LIGs while SCUBA (dot-dashed line) detects only ULIGs. The difference in sensitivity between ISOCAM and the deepest radio surveys is about a factor 2. ISOPHOT (dotted line) is only sensitive to ULIGs above $z \sim 0.5$. In Fig. 4b, where we used the library of template SEDs from Chary & Elbaz (2001), the plain line corresponding to ISOCAM rises faster above $z \sim 1$ than in Fig. 4a, where we used M82's SED. This different behavior comes from the fact that at these redshifts ISOCAM measures fluxes at about 7 μm in the rest-frame of the galaxy and the ratio $L_{IR}/L_{7\,\mu m}$ increases with L_{IR}.

Above $z \sim 2$, the sub-millimeter becomes the most efficient technique, although only galaxies more luminous than a few 10^{12} $L_\odot$ can be detected above a sensitivity limit of ~ 2 mJy at 850 μm. Hence the unlensed ISO-CAM and SCUBA surveys are not sampling the same redshift and luminosity ranges. This statement is confirmed observationally in the HDFN itself where none of the ISOCAM sources are detected by SCUBA (Hughes et al. 1998) and over a larger scale in the Canada France Redshift Survey 14 (CFRS-14, Eales et al. 2000). In this latter field of ~ 50 arcmin^2, only the two brightest 15 μm ISOCAM sources are detected at 850 μm, among a sample of 50 ISO-CAM (Flores et al. 1999) and 19 SCUBA sources. This confirms that both

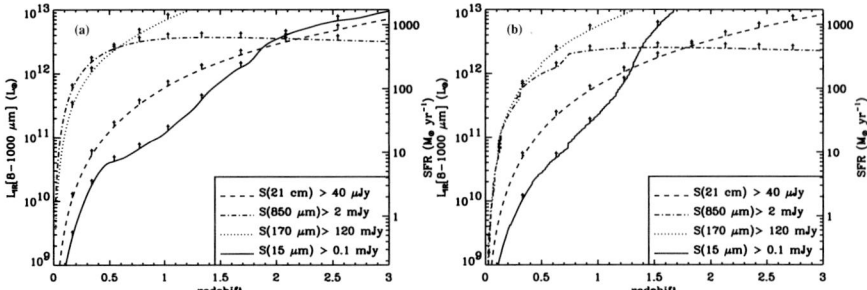

Figure 4: IR luminosity (left axis) and SFR (right axis) corresponding to the sensitivity limits of ISOCAM (15 μm, plain line), ISOPHOT (170 μm, dotted line), SCUBA (850 μm, dot-dashed line) and VLA/WSRT (21 cm, dashed line) as a function of redshift. K-correction from: **a)** the template SED of M82 (Fig. 1a), normalized to the appropriate IR luminosity, **b)** the library of template SEDs from Chary & Elbaz (2001), plotted in the Fig. 1b.

instruments detect different sets of objects. More common objects between ISOCAM and SCUBA are expected if they are gravitationally lensed, since the sensitivity limits of both instruments are then decreased by a factor of about t wo. Indeed in the clusters A 370 and A 2390, three of the four lensed SCUBA galaxies (Smail et al. 2001) are also detected with ISOCAM (Altieri et al. 1999, Metcalfe 2000).

6 Nature of the ISOCAM Galaxies

The spatial resolution (4 arcsec PSF FWHM) of ISOCAM provided the possibility to identify rather easily optical counterparts to these galaxies and to determine their redshift. As a first step, their redshift distribution was

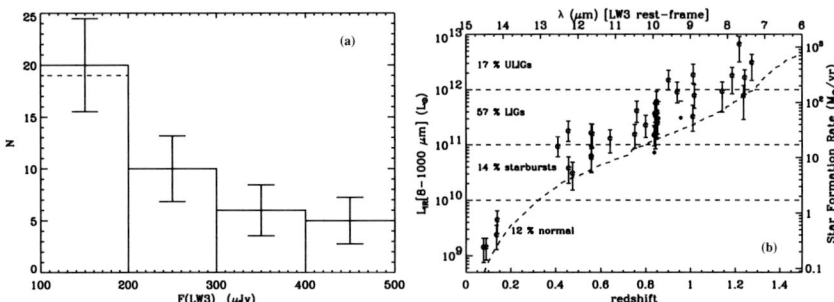

Figure 5: **a)** Histogram of the Redshift distribution of the 40 galaxies with spectroscopic redshift detected above 0.1 mJy at 15 μm with ISOCAM in the HDFN (98 % complete in redshift, poissonian error bars). **b)** L_{IR} [8–1000 μm] and SFR versus redshift and $\lambda_{rest-frame}$ for the HDFN galaxies detected above a 15 μm completeness limit of 0.1 mJy (dashed line). Filled dots: 5 AGNs (left axis only).

inferred from a sub-sample of galaxies in the HDFN (see Fig. 5a; Aussel et al. 1999) and their luminosities and star formation rates are presented in the Fig. 5b. About 75 % of the galaxies brighter than about 0.1 mJy at 15 μm, and responsible for the steep slope of the number counts, belong to the class of LIGs ($\sim$ 55 %) and ULIGs ($\sim$ 20 %). Their redshifts spread over the $z = $ 0.5–1.3 range with a median around $\bar{z} = 0.7$–0.8.

The fraction of IR light produced by active nuclei was computed from the cross-correlation of ISOCAM with the deepest X-ray surveys from the Chandra and XMM-Newton observatories in the HDFN (41 MIR galaxies) and Lockman Hole (103 MIR galaxies) respectively. Less than 20 % of the ISOCAM galaxies appear to be dominated by an AGN at 15 μm ((12 ± 5) % in the HDFN and (13 ± 4) % in the Lockman Hole, see Fadda et al. 2002).

7   The radio-MIR correlation

The radio continuum is also a tracer of the bolometric IR luminosity of star forming galaxies as a result of the tight correlation between IR and 1.4 GHz radio continuum in local galaxies (see Sect. 5). The origin of this correlation remains unclear, but it is generally assumed that massive stars are both responsible for the UV photons that heat dust before it radiates in the IR and for the synchrotron acceleration of electrons, producing the radio continuum, when they explode as supernovae.

In Fig. 6a, we have plotted the 1.4 GHz and 15 μm luminosities of the 109 local galaxies (small filled dots) observed with ISOCAM and the NRAO VLA Sky Survey (NVSS, Condon et al. 1998). As expected both luminosities are correlated with each other since both the 1.4 GHz and 15 μm luminosities are correlated with L_{IR}.

Half of the star forming HDFN-ISOCAM galaxies with spectroscopic redshift (35 galaxies, after excluding AGNs) present flux densities larger than 40 μJy at 1.4 GHz (5-σ) in the VLA and WSRT catalogs from Richards (2000) and Garrett et al. (2000) respectively. Their rest-frame 15 μm luminosities were computed using the SEDs of Chary & Elbaz (2001). The rest-frame 1.4 GHz luminosities were computed assuming a power-law as in Sect. 5: $S_\nu \propto \nu^{-\alpha}$, where $\alpha = 0.8 \pm 0.15$ as suggested for star forming galaxies in Yun, Reddy & Condon (2001). The rest-frame 15 μm and 1.4 GHz luminosities of these 17 HDFN-ISOCAM galaxies are plotted as filled dots with error bars in the Fig. 6a. The error bars on $\nu L_\nu[1.4 \text{ GHz}]$ were computed from the quadratic sum of the error bars on the measurement of the radio flux densities plus the error bar on α. Nine galaxies common in both catalogs from Richards (2000) and Garrett et al. (2000) present up to 40 % differences in their 1.4 GHz radio flux densities. For these galaxies, we used the mean value and included the difference between both measurements in the error bars. Seven galaxies from the CFRS-14 field (Flores et al. 1999) were also included (open dots).

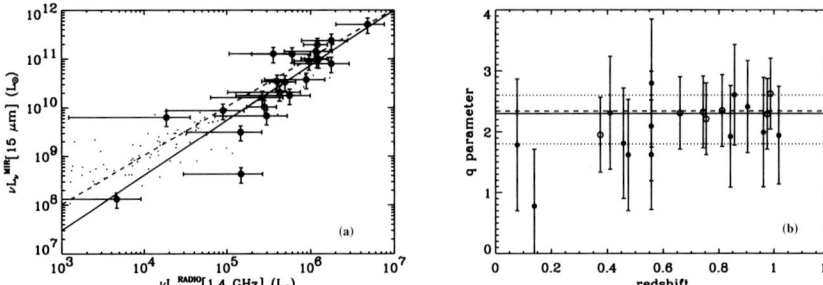

Figure 6: **a)** 15 μm versus radio continuum (1.4 GHz) luminosities. Small filled dots: sample of 109 local galaxies from ISOCAM and NVSS. Filled dots with error bars: 17 HDFN galaxies in common between ISOCAM and VLA or WSRT (see text). Open dots with error bars: 7 CFRS-14 galaxies in common between ISO-CAM and VLA (Flores et al. 1999). **b)** "q" parameter as a function of redshift for the HDFN and CFRS-14 galaxies. Plain line: median value of "q" ($2.3^{+0.3}_{-0.5}$). Dashed line: measured value from local galaxies ($q = 2.34 \pm 0.01$; Yun, Reddy & Condon 2001). Dotted line: 1-σ error bar on "q" for the 24 galaxies.

The MIR and radio luminosities of this sample of 24 distant dusty galaxies ($\bar{z} \simeq 0.8$) are strongly correlated with each other as in the local sample. The plain line in the Fig. 6a is a power-law fit to this correlation. For comparison, we have plotted in dashed line the correlation that one would expect from the correlation of $L_{15\,\mu m}$ and $L_{1.4\,GHz}$ with L_{IR}. The slope of the correlation observed for distant galaxies is marginally steeper than this dashed line. It is also slightly steeper than the correlation found for local galaxies (small filled dots), but the difference is too marginal with respect to the number of galaxies at high luminosities for detailed interpretation.

In Fig. 6b, the "q" parameter for the 24 ISOCAM-HDFN and CFRS-14 galaxies is plotted as a function of redshift. The median value of "q" for the 24 galaxies is: $\bar{q} = 2.3^{+0.3}_{-0.5}$ (plain and dotted lines in Fig. 6b), in perfect agreement with the local value of $\bar{q} = 2.34 \pm 0.01$ (Yun, Reddy & Condon 2001, dashed line in Fig. 6b). This study shows that the IR luminosities estimated from the ISOCAM 15 μm flux densities using the MIR-FIR correlations are perfectly consistent with those estimated from the radio. Although it is not clear whether the radio versus IR correlation also applyies up to $z \sim 1$, this result independently validates our estimate of the bolometric IR luminosity of the ISOCAM galaxies.

8   Evolution of the infrared volume emissivity with redshift

We have seen that the excess of faint MIR sources in number counts was due to the presence of distant ($z \sim 1$) luminous IR galaxies. As a consequence, the amount of star formation per comoving volume hidden by dust must have rapidly decreased from $z \sim 1$ to 0. Indeed, the ISOCAM-HDFN galaxies with $0.6 \leq z \leq 1.3$ and $L_{IR} \geq 10^{11} \, L_\odot$ (LIGs and ULIGs; AGNs excluded) produce a $15 \, \mu m$ luminosity density of $\mathcal{L}_{15} \, (\bar{z} \sim 1) \simeq (60 \pm 12) \times 10^6 \, L_\odot \, Mpc^{-3}$, while in the local universe luminous IR galaxies only make $\mathcal{L}_{15}(z \sim 0) \simeq (11 \pm 3) \times 10^5 \, L_\odot \, Mpc^{-3}$ (computed from the local $15 \, \mu m$ luminosity function of Xu et al. 1998; see Table 1). The comoving luminosity density produced by luminous IR galaxies at $15 \, \mu m$ was therefore about 55 times larger at $z \sim 1$ than in the local universe. The $15 \, \mu m$ local luminosity density that we have computed is consistent with the one measured at $12 \, \mu m$ from IRAS, $\mathcal{L}_{12}^{z \sim 0} \sim (1.7 \pm 0.5) \times 10^6 \, L_\odot \, Mpc^{-3}$, converted to $15 \, \mu m$ $(\nu L_\nu [15 \, \mu m] = 0.042 \times (\nu L_\nu [12 \, \mu m])^{1.12}$, Elbaz et al. 2002a).

| | $15 \, \mu m$ | | $8\text{--}1000 \, \mu m$ | | $0.44 \, \mu m$ |
	All	LIGs	All	LIGs	All
$z \sim 0$	7.5 ± 1.6	1.1 ± 0.3	120 ± 30	8.8 ± 1.6	82 ± 2
$z \sim 1$	–	61 ± 12	–	632 ± 291	240 ± 100

Table 1: Comoving luminosity density in units of $10^6 \, L_\odot \, Mpc^{-3}$ in the local universe and at $z \sim 1$ ($0.6 \leq z \leq 1.3$). Columns "All": contribution of galaxies of all luminosities are considered. Columns "LIGs": only galaxies with $L_{IR} \geq 10^{11} \, L_\odot$. AGNs were excluded for the computation of the IR values. The B-band ($0.44 \, \mu m$) values are from Lilly et al. (1996) at $z \sim 0$ and Connolly et al. (1997) at $z \sim 1$ (mean of the 0.5–1 and 1–1.5 redshift bins).

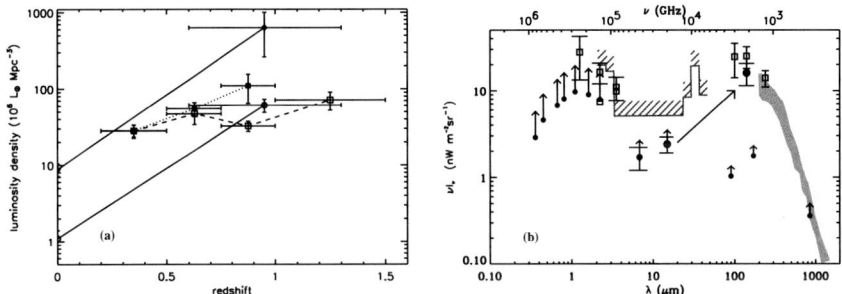

Figure 7: **a)** FIR (filled circles), MIR-15 μm (open circles) and UV-2800 luminosity density (in $L_\odot \, Mpc^{-3}$) as a function of redshift. UV-2800: open squares from Cowie, Songaila & Barger (1999); filled squares from Lilly et al. (1996). **b)** Integrated Galaxy Light (IGL, filled dots) and Extragalactic Background Light (EBL, open squares, grey area) from the UV to sub-millimeter. EBL measurements from COBE: 200–1500 μm EBL from COBE-FIRAS (grey area, Lagache et al. 1999), 1.25, 2.2, 3.5, 100, 140 μm EBL from COBE-DIRBE (open squares). IGL in the U, B, V, I, J, H, K bands from Madau & Pozzetti (2000). The upper end of the arrows indicate the revised values suggested by Bernstein et al. (2002, factor two higher). Our estimate of the 15 μm IGL (2.4 ± 0.5 nW m^{-2} sr^{-1}) is marked with a surrounded star. 6.75 μm (ISOCAM-LW2 filter) IGL from Altieri et al. (1999, filled dot). Hatched upper limit from Mkn 501 (Stanev & Franceschini 1998). The arrow from 15 to 140 μm indicates our computation of the 140 μm IGL due to ISOCAM galaxies.

If we now consider the bolometric IR luminosity (from 8 to 1000 μm) of ISOCAM galaxies, we find that the comoving density of IR luminosity radiated by dusty starbursts was about (70 ± 35) times larger at $\bar{z} \sim 1$ than today (computed from the LLF of Soifer et al. 1987). Since the IR luminosity is directly proportional to the extincted star formation rate of a galaxy, this means that the comoving density of star formation taking place in luminous IR galaxies was about (70 ± 35) times larger at $z \sim 1$ than today. In case of a pure density evolution proportional to $(1+z)^n$, this would translate into a value of $n \simeq 6$. For comparison, the B-band $(0.44\,\mu$m) luminosity density was only about three times larger at $z \sim 1$ than today. However, we want to emphasize that we are only considering here the galaxies detected by ISOCAM, not the full luminosity function.

The redshift evolution of the comoving density of IR luminosity is compared to the UV (2800 Å) one in Fig. 7a. Both wavelengths exhibit similar luminosity densities at low redshift but the IR rises faster than the UV and reaches a larger value at $z \sim 1$, implying that a much larger fraction of star formation was hidden by dust at $z \sim 1$ than today.

Finally, we note that the projected density of galaxies detected in the B-band in the HDFN (529 galaxies/arcmin^2 with $B_{AB} \leq 29$, Pozzetti et al. 1998) is 330 times greater than the projected density of ISOCAM galaxies (1.6 sources/arcmin^2, with $S_{15} \geq 0.1$ mJy or $AB(15\,\mu$m$) \leq 18.9$). However the ISOCAM galaxies produce a 15 μm IGL which is only twice lower than the B-band IGL ($IGL_B \sim 4.57^{+0.73}_{-0.47}$ nW m^{-2} sr^{-1}, Madau & Pozzetti 2000). This confirms that the very luminous galaxies detected in the MIR radiate mostly in the IR.

9 ISOCAM Mid-InfraRed Detection of HR 10: A Distant Clone of Arp 220 at $z = 1.44$

HR 10 (or ERO J 164502+4626.4, Dey et al. 1999), is the first and presently only Extremely Red Object (ERO, usually defined as galaxies with $I - K > 4$) known to be associated with the class of ULIGs. It was detected by Hu & Ridgway (1994, hereafter HR) together with another ERO (HR 14 or ERO J 164457+4626.0) in the field of the QSO PC 1643+4631A ($z = 3.79$). HR initially suggested that both galaxies with extreme colors ($I - K > 6$) could be distant ellipticals lying at $z \sim 2$–3. More generally, deep near IR (NIR) surveys indicate that EROs present the same clustering properties (Daddi et al. 2000, McCarthy et al. 2001) and surface brightness distribution (Moriondo, Cimatti & Daddi 2000) as elliptical galaxies. But dusty starbursts being potential progenitors of local ellipticals, they may also show similar clustering properties and some local examples, as NGC 7252 (Hibbard et al. 1994) or Arp 220 (Scoville et al. 2000), already show a de Vaucouleurs luminosity profile typical of ellipticals. High resolution NIR imagery and spectroscopy with the Keck telescopes (Graham & Dey 1996) revealed that HR 10 was a moderately distant ($z = 1.44$) galaxy with an asymmetric morphology and Hα

in emission. Another evidence against HR 10 being an early-type galaxy is its strong sub-millimeter luminosity as measured with SCUBA at the JCMT (Cimatti et al. 1998, Dey et al. 1999). The detection in HR 10 of a large CO luminosity, hence molecular hydrogen mass, was recently reported by Andreani et al. (2000) and presented as evidence favoring a star formation origin for the bulk of the IR luminosity rather than an AGN. However, a large gas mass may not only feed star formation but also gas accretion onto an AGN (see Papadopoulos et al. 2001).

Another test for the presence of an AGN in a dusty galaxy is the warm over cold dust, i. e. mid IR (MIR, 3–40 μm) over far IR (FIR, 40–300 μm), luminosity ratio as well as the shape of the MIR spectrum. In Elbaz et al. (2002b), we present the first detection of HR 10 in two MIR bands of ISOCAM onboard ISO corresponding to the rest-frame 3.3–6.1 and 4.9–7.4 μm wavelength ranges (LW3 and LW10 filters respectively). HR 10 is the first ERO spectroscopically identified to be associated with an ULIG detected in the radio, MIR and sub-millimeter. The rest-frame SED of HR 10 is amazingly similar to the one of Arp 220, scaled by a factor 3.8 ± 1.3. The corresponding 8–1000 μm luminosity ($L_{IR} \sim 7 \times 10^{12}~h_{70}^{-2}~L_\odot$) translates into a star formation rate of about 1200 $h_{70}^{-2}~M_\odot$ yr^{-1} if HR 10 is mostly powered by star formation.

Most authors favor the starburst hypothesis as a dominant source of energy in Arp 220: its MIR spectrum up to 40 μm shows no evidence for high ionization lines expected for AGNs (Sturm et al. 1996), its "IR excess" ($L_{IR}/L[\mathrm{Ly}\alpha] \sim 24$, Anantharamaiah et al. 2000) is much lower than for AGNs (~ 45–65) and typical of starburst galaxies (~ 12–45, Genzel et al. 1998), its radio emission is produced by several compact sources (Smith et al. 1998), the ratio of aromatic features over MIR continuum (Genzel et al. 1998) and the slope of the MIR continuum (Laurent et al. 2000) are typical of starbursts. More recently, Haas et al. (2001) suggested that the MIR luminosity of Arp 220 could be underestimated because of dust extinction in the MIR and that after dereddening, its FIR over MIR luminosity ratio would be closer to the one for AGNs. But the dereddening factor varies by a factor five depending on the dust geometry assumed. Finally, the flat 2–10 keV hard X-ray spectrum of Arp 220 implies that in order to be mostly powered by an AGN, it would need to be Compton thick with a column density larger than 10^{25} cm^{-2} (Iwasawa et al. 2001).

Even if the presence of a combination of an AGN and a starburst is still an option for both HR 10 and Arp 220, most studies favor a dominant contribution from star formation to their IR luminosities, implying that HR 10 presents the largest SFR known at present.

10   On the Shyness of Galaxies

The MIR-FIR correlations observed in the local universe (Elbaz et al. 2002a) can be used to compute the contribution of the luminous IR galaxies unveiled by ISOCAM below $z \sim 1.5$ to the CIRB once the redshift distribution is determined. The $15\,\mu m$ IGL computed from ISOCAM number counts $(IGL_{15}(S_{15} \geq 50\,\mu Jy) = 2.4 \pm 0.5$ nW m^{-2} sr^{-1}, Sect. 4.4) was converted into a $140\,\mu m$ IGL in Elbaz et al. (2002a) using the redshift distribution of the spectroscopically complete sample of ISOCAM galaxies in the HDFN. The resulting value of (16 ± 5) nW m^{-2} sr^{-1} represents a large fraction of the extragalactic background light measured with COBE at $140\,\mu m$, EBL_{140} $= (25 \pm 7)$ nW m^{-2} sr^{-1}. Hence luminous IR galaxies below $z \sim 1.5$ are responsible for the bulk of the CIRB. Since the CIRB contains most photons radiated by galaxies over the history of the universe, this means that luminous IR galaxies represent a common phase for galaxies. Chary & Elbaz (2001) have studied the range of possible parameters for the evolution of galaxies in luminosity and density over the history of the universe, that would fit number counts from ISOCAM, ISOPHOT and SCUBA as well as the CIRB and the redshift distribution of ISOCAM galaxies. The major result of this study is that although a level of degeneracy remains in the choice of the parameters ruling the evolution of galaxies, existing observations set a strong constraint on the relatively recent ($z < 1.5$) evolution of the number and luminosity density of luminous IR galaxies.

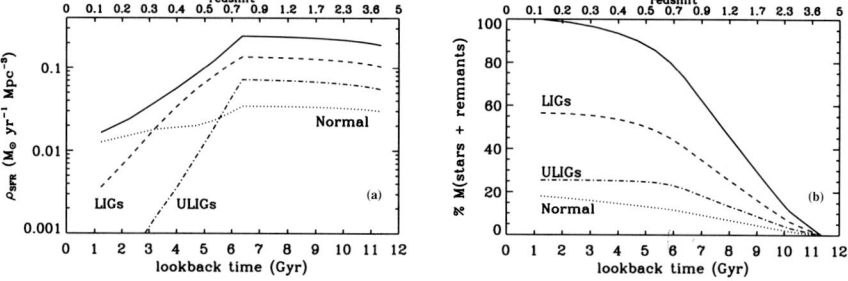

Figure 8: **a)** Cosmic density of star formation as a function of lookback time and redshift from Chary & Elbaz (2001). **b)** Fraction of present-day stars (+ remnants) formed as a function of lookback time and redshift for the scenario of Fig. 8a. Stellar lifetimes (for $Z_\odot$): Bressan et al. (1993). Remnant masses: Prantzos & Silk (1998).

The best fit is obtained for the cosmic history of star formation shown in the Fig. 8a, where the relative roles of ULIGs, LIGs and "normal" galaxies are differentiated. Fig. 8a implies that we are living at an epoch when "normal" galaxies ($L_{bol} < 10^{11} L_\odot$) contribute dominantly to the global star formation activity in the local universe, whereas above $z \sim 0.3$ the reverse was true: the bulk of the cosmic density of star formation was due to luminous IR galaxies. Hence galaxies in general must have experienced a period of shyness, such as local LIGs and ULIGs, when they formed the bulk of their present-day stars. Fig. 8b represents the fraction of present-day stars plus remnants formed as a function of lookback time or redshift for a given IMF (here from Gould et al.

1996). The total mass is comparable to the local density of baryons in the local universe ($5 \pm 3 \times 10^8$ $M_\odot$ Mpc^{-3}, Fukugita et al. 1998). The error bar on the computed stellar mass is as large $\sim$ 50 % (including the conversion from MIR to FIR and FIR to SFR), but this result suggests that the bulk of present-day stars formed at a time when their host galaxies experienced such a phase of shyness 5 to 10 Gyr ago, i.e. between $z = 0.5$ and 2 for an age of the universe of 12.6 Gyr in our cosmology (H$_\circ$ = 75 km s^{-1} Mpc^{-1}, $\Omega_{\text{matter}} = 0.3$, $\Omega_\Lambda = 0.7$). The shyness of galaxies seems to be the result of galaxy encounters since all ISOCAM galaxies in the HDFN are either merging or members of small groups of galaxies. The fact that the CIRB peaks around $\lambda \sim 140\,\mu$m was already an indication that it must originate from this redshift range since galaxies SEDs peak above $\lambda \sim 60\,\mu$m.

References

Allamandola, L.J., Tielens, A.G.G.M., Barker, J.R. 1989, ApJS 71, 733

Altieri, B., Metcalfe, L., Kneib, J.P., et al. 1999, A&A 343L, 65

Anantharamaiah, K.R., Viallefond, F., Mohan, N.R., Goss, W.M., Zhao, J.H. 2000, ApJ 537, 613

Andreani, P., Cimatti, A., Loinard, L., Röttgering, H. 2000, A&A 354, L1

Andriesse, C.D., 1978, A&A 66, 169

Aussel, H., Cesarsky, C.J., Elbaz, D., Starck, J.L. 1999, A&A 342, 313

Barger, A.J., Cowie, L.L., Sanders, D.B. 1999, ApJ 518, L5

Bernstein, R.A., Freedman, W. L., Madore, B.F. 2002, ApJ 571, 85

Blain, A.W., Smail, I., Ivison, R.J., Kneib, J.-P. 1999, MNRAS 302, 632

Bressan, A. et al. 1993, A&AS, 100, 647

Cesarsky, C.J., Abergel, A., Agnèse, P., et al. 1996, A&A 315, L32

Chary, R.R., Elbaz, D. 2001, ApJ 556, 562

Cimatti, A., Andreani, P., Röttgering, H., Tilanus, R. 1998, Nature 392, 895

Condon, J.J., Anderson, M.L., Helou, G. 1991, ApJ 376, 95

Condon, J.J., Cotton, W.D., Greisen, E.W., et al. 1998, AJ 115, 1693

Connolly, A.J., Szalay, A.S., Dickinson, M.E., et al. 1997, ApJ 486, L11

Cowie, L.L., Songaila, A., Barger, A.J. 1999, ApJ 118, 603

Daddi, E., Cimatti, A., Renzini, A. 2000, A&A 362, L45

Dey, A., Graham, J.R., Ivison, R.J., et al. 1999, ApJ 519, 610

Désert, F.-X., Boulanger, F., Puget, J.-L. 1990, A&A 237, 215

Dole, H., Gispert, R., Lagache, G., et al. 2001, A&A 372, 364

Dwek, E., Arendt, R.G., Hauser, M., et al. 1998, ApJ 508, 106

Dwek, E., Arendt, R.G. 1998, ApJ 508, L9

Eales, S., Lilly, S., Webb, T., et al. 2000, AJ 120, 2244

Elbaz, D., Cesarsky, C.J., Fadda, D., et al. 1999, A&A 351, L37

Elbaz, D., Cesarsky, C.J., Chanial, P. et al. 2002a, A&A 384, 848

Elbaz, D., Flores, H., Chanial, P. et al. 2002b, A&A 381, L1

Fadda, D., Flores, H., Hasinger, G., et al. 2002, A&A 383, 838

Fang F., Shupe D., Xu C., Hacking P., 1998, ApJ 500, 693

Finkbeiner, D.P., Davis, M., Schlegel, D.J. 2000, ApJ 524, 867

Fixsen, D. J., Dwek, E., Mather, J.C., Bennett, C.L., Shafer, R.A. 1998, ApJ 508, 123

Flores, H., Hammer, F., Thuan, T.X., et al. 1999, ApJ 517, 148

Franceschini, A., Aussel, H., Cesarsky, C.J., Elbaz, D., Fadda, D. 2001, A&A 378, 1

Fukugita, M., Hogan, C.J. & Peebles, P.J.E. 1998, ApJ, 503, 518

Gardner J.P., Cowie L.L., Wainscoat R.J., 1993, ApJ 415, L9

Garrett, M.A., de Bruyn, A.G., Giroletti, M., et al. 2000, A&A 361, L41

Genzel, R., Lutz, D., Sturm, E., et al. 1998, ApJ 498, 579

Gispert, R., Lagache, G., Puget, J.L. 2000, A&A 360, 1

Gould, A., Bahcall, J.N., Flynn, C. 1996, ApJ 465, 759

Graham, J., Dey, A. 1996, ApJ 471, 720

Haas, M., Klaas, U., Müller, S.A.H., Chini, R., Coulson, I. 2001, A&A 367, L9

Hauser, M.G., Arendt, R.G., Kelsall, T., et al. 1998, ApJ 508, 25

Helou, G., Khan, I.R., Malek, R., Boehmer, L. 1988, ApJS 68, 151

Hu, E.M., Ridgway, S.E. 1994, AJ 107, 1303 (HR)

Hughes, D. H., Serjeant, S., Dunlop, J., et al. 1998, Nature 394, 241

Iwasawa, K., Matt, G., Guainazzi, M., Fabian, A.C. 2001, MNRAS 326, 894

Jones, A.P., d'Hendecourt, L. 2000, A&A 355, 1191

Kennicutt, R.C.Jr 1998, ARA&A 36, 189

Kessler, M., Steinz, J., Anderegg, M., et al. 1996, A&A 315, L27

Lagache, G., Abergel, A., Boulanger, F., Desert, F.-X., Puget, J.-L. 1999, A&A 344, 322

Lagache, G., Haffner, L.M., Reynolds, R.J., Tufte, S.L. 2000, A&A 354, L247

Laurent, O., Mirabel, I.F., Charmandaris, V., et al. 2000, A&A 359, L887

Léger, A., Puget, J.L., 1984, A&A 137, L5

Lemke, D., Klaas, U., Abolins, J., et al. 1996, A&A 315, L64

Lilly, S.J., Lefévre, O., Hammer, F., Crampton, D. 1996, ApJ 460, L1

Madau, P., Pozzetti, L. 2000, MNRAS 312, L9

McCarthy, P.J., Carlberg, R.G., Chen, H.W., et al. 2001, ApJ 560, L131

Metcalfe, L. 2000, In: The Extragalactic Infrared Background and its Cosmological Implications, IAU Symposium 204, 18

Metcalfe N., Shanks T., Fong R., et al. 1995, MNRAS 273, 257

Moriondo, G., Cimatti, A., Daddi, E. 2000, A&A 364, 26

Papadopoulos, P., Ivison, R., Carilli, C., Lewis, G. 2001, Nature 409, 58

Pozzetti, L., Madau, P., Zamorani, G., Ferguson, H.C., Bruzual, G.A. 1998, MNRAS 298, 1133

Prantzos, N., Silk, J. 1998, ApJ 507, 229

Puget, J.-L., Léger, A. 1989, ARA&A 27, 37

Puget, J.L., Abergel, A., Bernard, J.P., et al. 1996, A&A 308, L5

Richards, E.A. 2000, ApJ 533, 611

Rush, B., Malkan, M.A., Spinoglio, L. 1993, ApJS 89, 1

Sanders, D.B., Mirabel, I.F. 1996, ARA&A 34, 749

Scoville, N.Z., Evans, A.S., Thompson, R., et al. 2000, AJ 119, 991

Smail, I., et al. 2001, MNRAS 331, 495

Soifer, B.T., Sanders, D.B., Madore, B.F., et al. 1987, ApJ 320, 238

Stanev, T., Franceschini, A. 1998, ApJ 494, L159

Sturm, E., Lutz, D., Genzel, R., et al. 1996, A&A 315, L133

Xu C., Hacking P., Fang F. et al. 1998, ApJ 508, 576

Xu, C. 2000, ApJ 541, 134

Xu, C., Lonsdale, C.J., Shupe, D.L., O'Linger, J., Masci, F. 2001, ApJ 562, 179

Yun, M.S., Reddy, N.A., Condon, J.J. 2001, ApJ 554, 803

Index of Contributors

General Table of Contents

Volume 3 (1990): Accretion and Winds

284

286

288

General Index of Contributors